KT-437-420

Programmable Controllers

237   979

In memory of Arthur Parr, 1913–1992.

Man is still the most extraordinary computer of all.

John F. Kennedy
21 May 1963

Programmable Controllers
An engineer's guide

Third edition
E.A. Parr, MSc, CEng, MIEE, MInstMC

NORWICH CITY COLLEGE

Stock No.	237 979		
Class	629.895 PAR		
Cat.	A2	Proc	3WL

Newnes

AMSTERDAM BOSTON HEIDELBERG LONDON NEW YORK OXFORD
PARIS SAN DIEGO SAN FRANCISCO SINGAPORE SYDNEY TOKYO

Newnes
An imprint of Elsevier
Linacre House, Jordan Hill, Oxford OX2 8DP
200 Wheeler Road, Burlington, MA 01803
A division of Reed Educational and Professional Publishing Ltd

℞ A member of the Reed Elsevier plc group

First published 1993
Second edition 1999
Third edition 2003

Copyright © E.A. Parr 1993, 1999, 2003. All rights reserved.

The right of E.A. Parr to be identified as the author of this work
has been asserted in accordance with the Copyright, Designs and
Patents Act 1988

No part of this publication
may be reproduced in any material form (including
photocopying or storing in any medium by electronic
means and whether or not transiently or incidentally
to some other use of this publication) without the
written permission of the copyright holder except
in accordance with the provisions of the Copyright,
Designs and Patents Act 1988 or under the terms of a
licence issued by the Copyright Licensing Agency Ltd,
90 Tottenham Court Road, London, England W1T 4LP.
Applications for the copyright holder's written permission
to reproduce any part of this publication should be addressed
to the publishers

British Library Cataloguing in Publication Data
A catalogue record for this book is available
from the British Library

ISBN 0 7506 5757 X

For information on all Newnes publications visit our website at:
newnespress.com

Typeset by Integra Software Services Pvt. Ltd, Pondicherry, India
www.integra-india.com
Printed and bound in Great Britain by Biddles Ltd *www.biddles.co.uk*

Contents

Preface

All industrial processes need some form of control system if they are to run safely and economically. In recent years a specialist control computer, called a programmable controller, has evolved and revolutionized control engineering by combining computing power and immense flexibility at a reasonable price.

This book is concerned with the application and use of programmable controllers. It is not an instructional book in programming, and is certainly not a comparative guide to the various makes of machine on the market.

To some extent, choosing a programmable controller is rather like choosing a word processor. You ask people for their views, try a few simple examples in a shop, and buy the cheapest that you think meets your requirements. Only after several months do you really know the system. From then on, all other word processors seem awkward.

Programmable controllers are similar. Unless there are good reasons for a particular choice (ready experience in the engineering or maintenance staff, equipment being supplied by an outside contractor and similar considerations), there are good and bad points with all (the really bad machines left the market years ago).

At the Sheerness Steel Company where I work, the plant control is based on about sixty programmable controllers consisting of Allen Bradley PLC 2s and 5s, GEC (now CEGELEC) GEM-80s, ASEA (now ABB) Masters and Siemens SIMATIC S5s, with small machines primarily from Mitsubishi. These controllers are somewhat like the trees at Galleons Lap in Winnie the Pooh; there never seems to be the same number on two successive days, even if you tie a piece of string around each one!

As with most plants, the background to this distribution of controllers is largely historical chance (the original Mitsubishi came on a small turn-key plant from an outside contractor, for example), but the ready access to these machines is the reason for their prominence in this book.

Even within this range of PLC families, the coverage in this book is not complete. The PLCs have been chosen to cover the application points I wish to make, not as a complete survey of a manufacturer's range.

In 'previous lives' I have worked with PLCs from AEG, GE, Landys and Gyr, Modicon, Telemecanique, Texas Instruments and many other companies. To these manufacturers I offer my sincere apologies for not giving them more coverage, but to do so would have made a tedious book and masked the application points I have tried to make. I could happily use any of these machines, and there is not a major difference in style or philosophy between them (the manufacturers would no doubt disagree!).

The guideline is therefore choose a machine that suits *you*, and do not change manufacturers for purely economic reasons. Knowledge, consistency of spares and a good relationship with a manufacturer are very valuable.

A book like this requires much assistance, and I would like to thank Peter Bark and Dave Wilson of ABB, Adrian Bishop, Bob Hunt, Julian Fielding, John Hanscombe, Hugh Pickard, Jennie Holmes and Hennie Jacobs of Allen Bradley, Peter Backenist, David Slingsby and Stuart Webb of GEC/CEGELEC, Peter Houldsworth, Paul Judge, Allan Norbury, Dickon Purvis, Paul Brett and Allan Roworth of Siemens, and Craig Rousell who all assisted with information on their machines, commented constructively on my thoughts and provided material and photographs.

My fellow engineers at Sheerness Steel also deserve some praise for tolerating my PLC systems (and who will no doubt compare my written aims with our actual achievements!).

A book takes some time to write, and my family deserve considerable thanks for their patience.

Andrew Parr
Minster on Sea
eaparr2002@yahoo.co.uk

Note for second edition

This revision incorporates additional material covering recent developments, and reflects the increasing importance of health and safety legislation.

Notes for third edition

This edition includes a new chapter giving example ladder rungs for common industrial problems. Screen shots of Windows based programming software have been included to show how programs are entered. Health and Safety issues, particularly the introduction of IEC 61508, have been updated.

1 Computers and industrial control

1.1 Introduction

Very few industrial plants can be left to run themselves, and most need some form of control system to ensure safe and economical operation. Figure 1.1 is thus a representation of a typical installation, consisting of a plant connected to a control system. This acts to translate the commands of the human operator into the required actions, and to display the plant status back to the operator.

At the simplest level, the plant could be an electric motor driving a cooling fan. Here the control system would be an electrical starter with protection against motor overload and cable faults. The operator controls would be start/stop pushbuttons and the plant status displays simply running/stopped and fault lamps.

At the other extreme, the plant could be a vast petrochemical installation. Here the control system would be complex and a mixture of technologies. The link to the human operators will be equally varied, with commands being given and information displayed via many devices.

In most cases the operator will be part of the control system. If an alarm light comes on saying 'Low oil level' the operator will be expected to add more oil.

1.2 Types of control strategies

It is very easy to be confused and overwhelmed by the size and complexity of large industrial processes. Most, if not all, can be simplified by considering them to be composed of many small sub-processes. These sub-processes can generally be considered to fall into three distinct areas.

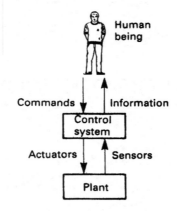

Figure 1.1 *A simple view of a control system*

1.2.1 Monitoring subsystems

These display the process state to the operator and draw attention to abnormal or fault conditions which need attention. The plant condition is measured by suitable sensors.

Digital sensors measure conditions with distinct states. Typical examples are running/stopped, forward/off/reverse, fault/healthy, idle/low/medium/high, high level/normal/low level. Analog sensors measure conditions which have a continuous range such as temperature, pressure, flow or liquid level.

The results of these measurements are displayed to the operator via indicators (for digital signals) or by meters and bargraphs for analog signals.

The signals can also be checked for alarm conditions. An overtravel limit switch or an automatic trip of an overloaded motor are typical digital alarm conditions. A high temperature or a low liquid level could be typical analog alarm conditions. The operator could be informed of these via warning lamps and an audible alarm.

A monitoring system often keeps records of the consumption of energy and materials for accountancy purposes, and produces an event/alarm log for historical maintenance analysis. A pump, for example, may require maintenance after 5000 hours of operation.

1.2.2 Sequencing subsystems

Many processes follow a predefined sequence. To start the gas burner of Figure 1.2, for example, the sequence could be:

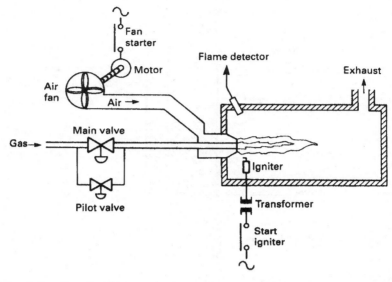

Figure 1.2 *Gas-fired burner, a sequence control system*

(a)　Start button pressed; if sensors are showing sensible states (no air flow and no flame) then sequence starts.

(b)　Energize air fan starter. If starter operates (checked by contact on starter) and air flow is established (checked by flow switch) then

(c)　Wait two minutes (for air to clear out any unburnt gas) and then

(d)　Open gas pilot valve and operate igniter. Wait two seconds and then stop igniter and

(e)　If flame present (checked by flame failure sensor) open main gas valve.

(f)　Sequence complete. Burner running. Stays on until stop button pressed, or air flow stops, or flame failure.

The above sequence works solely on digital signals, but sequences can also use analog signals. In the batch process of Figure 1.3 analog sensors are used to measure weight and temperature to give the sequence:

1　Open valve V1 until 250 kg of product A have been added.
2　Start mixer blade.
3　Open valve V2 until 310 kg of product B have been added.
4　Wait 120 s (for complete mixing).
5　Heat to 80 °C and maintain at 80 °C for 10 min.
6　Heater off. Allow to cool to 30 °C.
7　Stop mixer blade.
8　Open drain valve V3 until weight less than 50 kg.

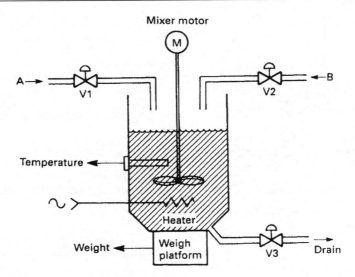

Figure 1.3   *A batch process*

1.2.3 Closed loop control subsystems

In many analog systems, a variable such as temperature, flow or pressure is required to be kept automatically at some preset value or made to follow some other signal. In step 5 of the batch sequence above, for example, the temperature is required to be kept constant to 80 °C within quite narrow margins for 10 minutes.

Such systems can be represented by the block diagram of Figure 1.4. Here a particular characteristic of the plant (e.g. temperature) denoted by PV (for process variable) is required to be kept at a preset value SP (for setpoint). PV is measured by a suitable sensor and compared with the SP to give an error signal

$$\text{error} = \text{SP} - \text{PV} \tag{1.1}$$

If, for example, we are dealing with a temperature controller with a setpoint of 80 °C and an actual temperature of 78 °C, the error is 2 °C.

This error signal is applied to a control algorithm. There are many possible control algorithms, and this topic is discussed in detail in Chapter 4, but a simple example for a heating control could be 'If the error is negative turn the heat off, if the error is positive turn the heat on.'

The output from the control algorithm is passed to an actuator which affects the plant. For a temperature control, the actuator could be a heater, and for a flow control the actuator could be a flow control valve.

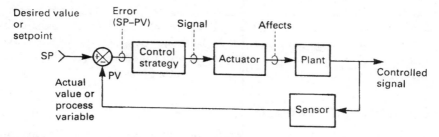

Figure 1.4   *A closed loop control system*

The control algorithm will adjust the actuator until there is zero error, i.e. the process variable and the setpoint have the same value.

In Figure 1.4, the value of PV is fed back to be compared with the setpoint, leading to the term 'feedback control'. It will also be noticed that the block diagram forms a loop, so the term 'closed loop control' is also used.

Because the correction process is continuous, the value of the controlled PV can be made to track a changing SP. The air/gas ratio for a burner can thus be maintained despite changes in the burner firing rate.

1.2.4   *Control devices*

The three types of control strategy outlined above can be achieved in many ways. Monitoring/alarm systems can often be achieved by connecting plant sensors to displays, indicators and alarm annunciators. Sometimes the alarm system will require some form of logic. For example, you only give a low hydraulic pressure alarm if the pumps are running, so a time delay is needed after the pump starts to allow the pressure to build up. After this time, a low pressure causes the pump to stop (in case the low pressure has been caused by a leak).

Sequencing systems can be built from relays combined with timers, uniselectors and similar electromechanical devices. Digital logic (usually based on TTL or CMOS integrated circuits) can be used for larger systems (although changes to printed circuit boards are more difficult to implement than changes to relay wiring). Many machine tool applications are built around logic blocks: rail-mounted units containing logic gates, storage elements, timers and counters which are linked by terminals on the front of the blocks to give the required operation. As with a relay system, commissioning changes are relatively easy to implement.

Closed loop control can be achieved by controllers built around DC amplifiers such as the ubiquitous 741. The 'three-term controller' (described further in Chapter 4) is a commercially available device that performs the function of Figure 1.4. In the chemical (and particularly the petrochemical) industries, the presence of potentially explosive atmospheres has led to the use of pneumatic controllers, with the signals in Figure 1.4 being represented by pneumatic pressures.

1.3   Enter the computer

A computer is a device that performs predetermined operations on input data to produce new output data, and as such can be represented by Figure 1.5(a). For a computer used for payroll calculations the input data would be employees' names, salary grades and hours worked. These data would be operated on according to instructions written to include current tax and pension rules to produce output data in the form of wage slips (or, today, more likely direct transfers to bank accounts).

Early computer systems were based on commercial functions: payroll, accountancy, banking and similar activities. The operations tended to be batch processes, a daily update of stores stock, for example.

The block diagram of Figure 1.5(a) has a close relationship with the control block of Figure 1.1, which could be redrawn, with a computer providing the control block, as in Figure 1.5(b). Note that the operator's actions (e.g. start process 3) are not instructions, they are part of the input data. The instructions will define what action is to be taken as the input data (from both the plant and the operator) change. The output data are control actions to the plant and status displays to the operator.

Early computers were large, expensive and slow. Speed is not that important for batch-based commercial data processing (commercial

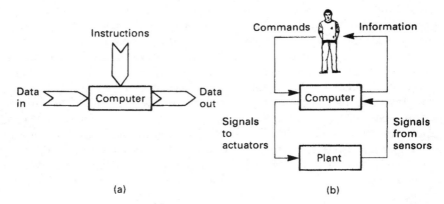

(a)   (b)

Figure 1.5   *The computer in industrial control: (a) a simple overview of a computer; (b) the computer as part of a control system*

programmers will probably disagree!) but is of the highest priority in industrial control, which has to be performed in 'real time'. Many emergency and alarm conditions require action to be taken in fractions of a second.

Commercial (with the word 'commercial' used to mean 'designed for use in commerce') computers were also based on receiving data from punched cards and keyboards and sending output data to printers. An industrial process requires possibly hundreds of devices to be read in real time and signals sent to devices such as valves, motors, meters and so on.

There was also an environmental problem. Commercial computers are designed to exist in an almost surgical atmosphere; dust-free and an ambient temperature that can only be allowed to vary by a few degrees. Such conditions can be almost impossible to achieve close to a manufacturing process.

The first industrial computer application was probably a monitoring system installed in an oil refinery in Port Arthur, USA in 1959. The reliability and mean time between failure of computers at this time meant that little actual control was performed by the computer, and its role approximated to the earlier Section 1.2.1.

1.3.1 Computer architectures

It is not essential to have intimate knowledge of how a computer works before it can be used effectively, but an appreciation of the parts of a computer is useful for appreciating how a computer can be used for industrial control.

Figure 1.5(a) can be expanded to give the more detailed layout of Figure 1.6. This block diagram (which represents the whole computing range from the smallest home computer to the largest commercial mainframe) has six portions:

1   An input unit where data from the outside world are brought into the computer for processing.
2   A store, or memory, which will be used to store the instructions the computer will follow and data for the computer to operate on. These data can be information input from outside or intermediate results calculated by the machine itself. The store is organized into a number of boxes, each of which can hold one number and is identified by an address as shown in Figure 1.7. Computers work internally in binary (see the Appendix for a description of binary, hexadecimal (hex) and other number systems) and the store does not distinguish between the meanings that could be attached to the data stored in it. For example, in an 8-bit computer (which works

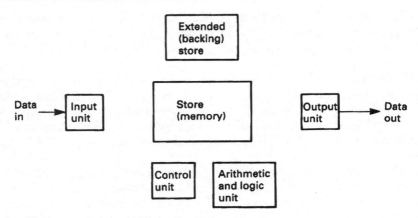

Figure 1.6 *The component parts of a computer*

with numbers 8 bits long in its store) the number 01100001 can be interpreted as:

(a) The decimal number 97.
(b) The hex number 61 (see Appendix).
(c) The letter 'a' (see Chapter 6).
(d) The state of eight digital signals such as limit switches.
(e) An instruction to the computer. If the machine was the old Z80 microprocessor, hex 61 moves a number between two internal stores.

A typical desktop computer will use 16-bit numbers (called a 16-bit word) and have over a million store locations. The industrial computers we will be mainly discussing have far smaller storage, 32 000 to 64 000

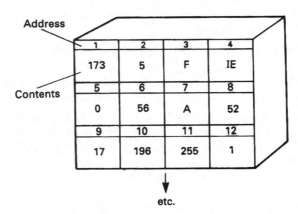

Figure 1.7 *A simple view of a computer's store*

store locations being typical for larger control machines, but even smaller machines with just 1000 store locations are common.

3 Data from the store can be accessed very quickly, but commercial computers often need vast amounts of storage to hold details such as bank accounts or names and addresses. This type of data is not required particularly quickly and is held in external storage. This is usually magnetic disks or tapes and is called secondary or backing storage. Such stores are not widely used on the types of computer we will be discussing.

4 An output unit where data from the computer are sent to the outside world.

5 An arithmetic and logical unit (called an ALU) which performs operations on the data held in the store according to the instructions the machine is following.

6 A control unit which links together the operations of the other five units. Often the ALU and the control unit are known, together, as the central processor unit or CPU. A microprocessor is a CPU in a single integrated circuit.

The instructions the computer follows are held in the store and, with a few exceptions which we will consider shortly, are simply followed in sequential order as in Figure 1.8(a).

The control unit contains a counter called an instruction register (or IR) which says at which address in the store the next instruction is to be

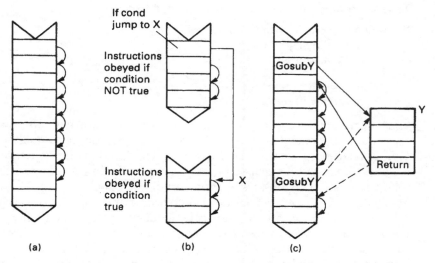

Figure 1.8 *Program flow in a computer: (a) simple sequential flow; (b) conditional jump; (c) subroutine call*

found. Sometimes the name program counter (and the abbreviation PC) is used.

When each instruction is obeyed, the control unit reads the store location whose address is held in the IR. The number held in this store location tells the control unit what instruction is to be performed.

Instructions nearly always require operations to be performed on data in the store (e.g. add two numbers) so the control unit will bring data from the store to the ALU and perform the required function.

When the instruction has been executed, the control unit will increment the IR so it holds the address of the next instruction.

There are surprisingly few types of instruction. The ones available on most microprocessors are variations on:

1 Move data from one place to another (e.g. input data to a store location, or move data from a store location to the ALU).
2 ALU operations on two data items, one in the ALU and one in a specified store location. Operations available are usually add, subtract, and logical operations such as AND, OR.
3 Jumps. In Figure 1.8(a) we implied that the computer followed a simple sequential list of instructions. This is usually true, but there are occasions where simple tests are needed. These usually have the form

IF (some condition) THEN
     Perform some instructions
ELSE
     Perform some other instructions

To test a temperature, for example, we could write

IF Temperature is less than 75 °C THEN
     Turn healthy light on
     Turn fault light off
ELSE
     Turn healthy light off
     Turn fault light on

Such operations use conditional jumps. These place a new address into the IR dependent on the last result in the accumulator. Conditional jumps can be specified to occur for outcomes such as result positive, result negative or result zero, and allow a program to follow two alternative routes as shown in Figure 1.8(b).
4 Subroutines. Many operations are required time and time again within the same program. In an industrial control system using flows measured by orifice plates, a square root function will be required many times (flow is proportional to the square root of the pressure drop across the orifice plate). Rather than write the same instruction several times (which is wasteful of effort and storage space) a subroutine

instruction allows different parts of the program to temporarily transfer operations to a specified subroutine, returning to the instruction after the subroutine call as shown in Figure 1.8(c).

1.3.2 Machine code and assembly language programming

The series of instructions that we need (called a 'program') has to be written and loaded into the computer. At the most basic level, called machine code programming, the instructions are written into the machine in the raw numerical form used by the machine. This is difficult to do, prone to error, and almost impossible to modify afterwards.

The sequence of numbers

16 00 58 21 00 00 06 08 29 17 D2 0E 40 19 05 C2 08 40 C9

genuinely are the instructions for a multiplication subroutine starting at address 4000 for a Z80 microprocessor, but even an experienced Z80 programmer would need reference books (and a fair amount of time) to work out what is going on with just these 19 numbers.

Assembly language programming uses mnemonics instead of the raw code, allowing the programmer to write instructions that can be relatively easily followed. For example, with

LOAD Temperature
SUB 75
JUMP POSITIVE to Fault_Handler

it is fairly easy to work out what is happening.

A (separate) computer program called an assembler converts the programmer's mnemonic-based program (called the source) into an equivalent machine code program (called the object) which can then be run.

Writing programs in assembly language is still labour-intensive, however, as there is one assembly language instruction for each machine code instruction.

1.3.3 High level languages

Assembly language programming is still relatively difficult to write, so ways of writing computer programs in a style more akin to English were developed. This is achieved with so-called 'high level languages' of which the best known are probably Pascal, FORTRAN and the ubiquitous BASIC (and there are many, many languages: RPG, FORTH, LISP, CORAL and C to name but a few, each with its own attractions).

In a high level language, the programmer writes instructions in something near to English. The Pascal program below, for example, gives a printout of a requested multiplication table.

```
program multtable (input, output);
var number, count : integer
begin
readln ('Which table do you want', number);
for count = 1 to 10 do
writeln (count, 'times', number, 'is', count*number);
end. (of program)
```

Even though the reader may not know Pascal, the operation of the program is clear (if asked to change the table from a ten times table to a twenty times table, for example, it is obvious which line would need to be changed).

A high level language source program can be made to run in two distinct ways. A compiler is a program which converts the entire high level source program to a machine code object program offline. The resultant object program can then be run independently of the source program or the compiler.

With an interpreter, the source program and the interpreter both exist in the machine when the program is being run. The interpreter scans each line of source code, converting them to equivalent machine code instructions as they are obeyed. There is no object program with an interpreter.

A compiled program runs much faster than an interpreted program (typically five to ten times as fast because of the extra work that the interpreter has to do) and the compiled object program will be much smaller than the equivalent source code program for an interpreter. Compilers are, however, much less easy to use, a typical sequence being:

1   A text editor is loaded into the computer.
2   The source program is typed in or loaded from disk (for modification).
3   The resultant source file is saved to disk.
4   The compiler is loaded from disk and run.
5   The source file is loaded from disk.
6   Compilation starts (this can take several minutes). If any errors are found go back to step 1.
7   An object program is produced which can be saved to disk and/or run. If any runtime errors are found, go back to step 1.

An interpreted language is much easier to use, and for many applications the loss of speed is not significant. BASIC is usually an interpreted language; Pascal, C and Fortran are usually compiled. Figure 1.9 summarizes the operation of compiled and interpreted high level languages.

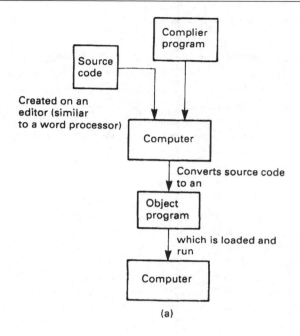

(a)

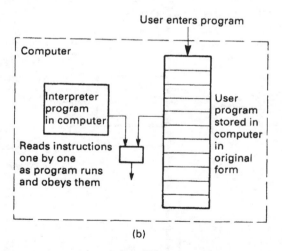

(b)

Figure 1.9 *Compiled and interpreted high level languages: (a) compiled program (e.g. Pascal, C); (b) interpreter (e.g. most BASICs)*

1.3.4 Application programs

Increasingly, as computers become more widespread, many programs have been written which allow the user to define the tasks to be performed without worrying unduly about how the computer achieves them. These are known as application programs and are typified by spreadsheets such as Lotus 123 and Excel and databases such as Approach and Access. In these the user is defining complex mathematical or database operations without 'programming' the computer in a conventional sense.

1.3.5 Requirements for industrial control

Industrial control has rather different requirements than other applications. It is worth examining these in some detail.

A conventional computer, shown schematically in Figure 1.10(a), takes data usually from a keyboard and outputs data to a VDU screen or printer. The data being manipulated will generally be characters or numbers (e.g. item names and quantities held in a stores stock list).

The control computer of Figure 1.10(b) is very different. Its inputs come from a vast number of devices. Although some of these are numeric (flows, temperature, pressures and similar analog signals) most will be single-bit, on/off, digital signals.

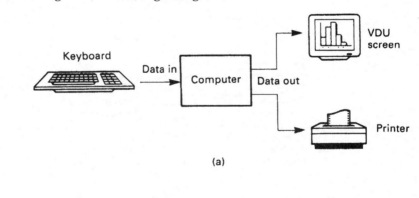

(a)

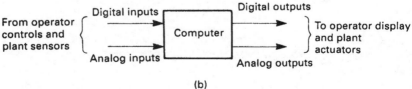

(b)

Figure 1.10 *The difference between commercial and industrial computers: (a) commercial computer; (b) industrial control computer*

There will also be a similarly large amount of digital and analog output signals. A very small control system may have connections to about 20 input and output signals; figures of over 200 connections are quite common on medium-sized systems. The keyboard, VDU and printer may exist, but they are not necessary, and their functions will probably be different to those on a normal desktop or mainframe computer.

Although it is possible to connect this quantity of signals to a conventional machine, it requires non-standard connections and external boxes. Similarly, although programming for a large amount of input and output signals can be done in Pascal, BASIC or C, the languages are being used for a purpose for which they were not really designed, and the result can be very ungainly.

In Figure 1.11(a), for example, we have a simple motor starter. This could be connected as a computer-driven circuit as in Figure 1.11(b). The two inputs are identified by addresses 1 and 2, with the output (the relay starter) being given the address 10.

If we assume that a program function bitread (N) exists which gives the state (on/off) of address N, and a procedure bitwrite (M,var) which sends the state of program variable var to address M, we could give the actions of Figure 1.11 by

```
repeat
  start:=bitread(1);
  stop:=bitread(2);
```

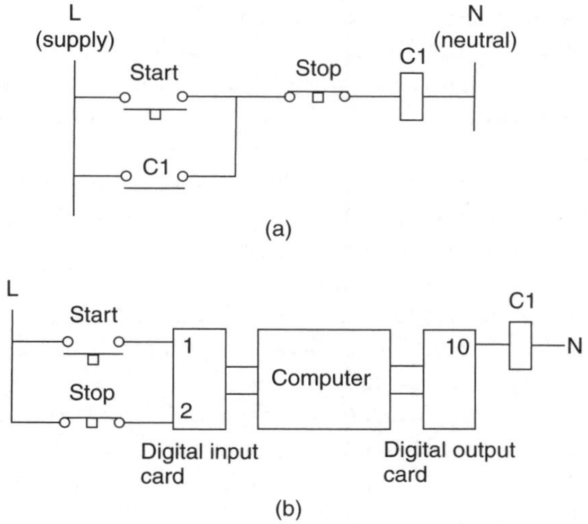

(a)

(b)

Figure 1.11 *Comparison of hardwire and computer-based schemes: (a) hardwire motor starter circuit; (b) computer-based motor starter*

```
run: = ((start) or (run)) & stop;
    bitwrite (10,run);
until hellfreezesover
```

where start, stop and run are 1-bit variables. The program is not very clear, however, and we have just three connections.

An industrial control program rarely stays the same for the whole of its life. There are always modifications to cover changes in the operations of the plant. These changes will be made by plant maintenance staff, and must be made with minimal (preferably no) interruptions to the plant production. Adding a second stop button and a second start button to Figure 1.11 would not be a simple task.

In general, computer control is done in real time, i.e. the computer has to respond to random events as they occur. An operator expects a motor to start (and more important to stop!) within a fraction of a second of the button being pressed. Although commercial computing needs fast computers, it is unlikely that the difference between one and two second computation time for a spreadsheet would be noticed by the user. Such a difference would be unacceptable for industrial control.

Time itself is often part of the control strategy (e.g. start air fan, wait 10 s for air purge, open pilot gas valve, wait 0.5 s, start ignition spark, wait 2.5 s, if flame present open main gas valve). Such sequences are difficult to write with conventional languages.

Most control faults are caused by external items (limit switches, solenoids and similar devices) and not by failures within the central control itself. The permission to start a plant, for example, could rely on signals involving cooling water flows, lubrication pressure, or temperatures within allowable ranges. For quick fault finding the maintenance staff must be able to monitor the action of the computer program whilst it is running. If, as is quite common, there are ten interlock signals which allow a motor to start, the maintenance staff will need to be able to check these quickly in the event of a fault. With a conventional computer, this could only be achieved with yet more complex programming.

The power supply in an industrial site is shared with many antisocial loads; large motors stopping and starting, thyristor drives which put spikes and harmonic frequencies onto the mains supply. To a human these are perceived as light flicker; in a computer they can result in storage corruption or even machine failure.

An industrial computer must therefore be able to live with a 'dirty' mains supply, and should also be capable of responding sensibly following a total supply interruption. Some outputs must go back to the state they were in before the loss of supply; others will need to turn off or on until an operator takes available corrective action. The designer must have the facility to define what happens when the system powers up from cold.

The final considerations are environmental. A large mainframe computer generally sits in an air-conditioned room at a steady 20 °C with carefully controlled humidity. A desktop PC will normally live in a fairly constant environment because human beings do not work well at extremes. An industrial computer, however, will probably have to operate away from people in a normal electrical substation with temperatures as low as −10 °C after a winter shutdown, and possibly over 40 °C in the height of summer. Even worse, these temperature variations lead to a constant expansion and contraction of components which can lead to early failure if the design has not taken this factor into account.

To these temperature changes must be added dust and dirt. Very few industrial processes are clean, and the dust gets everywhere (even with IP55 cubicles, because an IP55 cubicle is only IP55 when the doors are shut and locked; IP ratings are discussed in Section 8.4.2). The dust will work itself into connectors, and if these are not of the highest quality, intermittent faults will occur which can be very difficult to find.

In most computer applications, a programming error or a machine fault can at worst be expensive and embarrassing. When a computer controlling a plant fails, or a programmer misunderstands the plant's operation, the result could be injuries or fatalities. Under the UK Health and Safety at Work Act, prosecution of the design engineers could result. It behoves everyone to take extreme care with the design.

Our requirements for an industrial control computer are very demanding, and it is worth summarizing them:

1 They should be designed to survive in an industrial environment with all that this implies for temperature, dirt and poor-quality mains supply.
2 They should be capable of dealing with bit-form digital input/output signals at the usual voltages encountered in industry (24 V DC to 240 V AC) plus analog input/output signals. The expansion of the I/O should be simple and straightforward.
3 The programming language should be understandable by maintenance staff (such as electricians) who have no computer training. Programming changes should be easy to perform in a constantly changing plant.
4 It must be possible to monitor the plant operation whilst it is running to assist fault finding. It should be appreciated that most faults will be in external equipment such as plant-mounted limit switches, actuators and sensors, and it should be possible to observe the action of these from the control computer.
5 The system should operate sufficiently fast for realtime control. In practice, 'sufficiently fast' means a response time of around 0.1 s, but this can vary depending on the application and the controller used.

6   The user should be protected from computer jargon.
7   Safety must be a prime consideration.

1.3.6 The programmable controller

In the late 1960s the American motor car manufacturer General Motors was interested in the application of computers to replace the relay sequencing used in the control of its automated car plants. In 1969 it produced a specification for an industrial computer similar to that outlined at the end of Section 1.3.5.

Two independent companies, Bedford Associates (later called Modicon) and Allen Bradley, responded to General Motor's specification. Each produced a computer system similar to Figure 1.12 which bore little resemblance to the commercial minicomputers of the day.

The computer itself, called the central processor, was designed to live in an industrial environment, and was connected to the outside world via racks into which input or output cards could be plugged. In these early machines there were essentially four different types of cards:

1   DC digital input card
2   DC digital output card
3   AC digital input card
4   AC digital output card

Each card would accept 16 inputs or drive 16 outputs. A rack of eight cards could thus be connected to 128 devices. It is very important to appreciate that the card allocations were the user's choice, allowing great flexibility. In Figure 1.12(b) the user has installed one DC input card, one DC output card, three AC input cards, and two AC output cards, leaving one spare position for future expansion. This rack can thus be connected to

- 16 DC input signals
- 16 DC output signals
- 48 AC input signals
- 16 AC output signals

Not all of these, of course, need to be used.

The most radical idea, however, was a programming language based on a relay schematic diagram, with inputs (from limit switches, push-buttons, etc.) represented by relay contacts, and outputs (to solenoids, motor starters, lamps, etc.) represented by relay coils. Figure 1.13 shows a simple hydraulic cylinder which can be extended or retracted by pushbuttons. Its stroke is set by limit switches which open at the end of travel, and the solenoids can only be operated if the hydraulic pump is running. This would be controlled by the computer program of

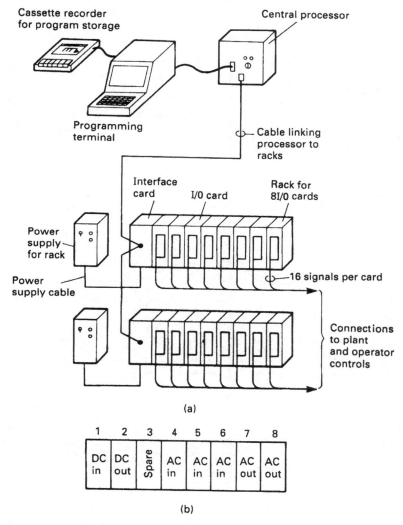

Figure 1.12 *The component parts of a PLC system: (a) an early PLC system; (b) a typical rack of cards*

Figure 1.13(b) which is identical to the relay circuit needed to control the cylinder. These programs look like the rungs on a ladder, and were consequently called 'ladder diagrams'.

The program was entered via a programming terminal with keys showing relay symbols (normally open/normally closed contacts, coils, timers, counters, parallel branches, etc.) with which a maintenance electrician would be familiar. Figure 1.14 shows the programmer

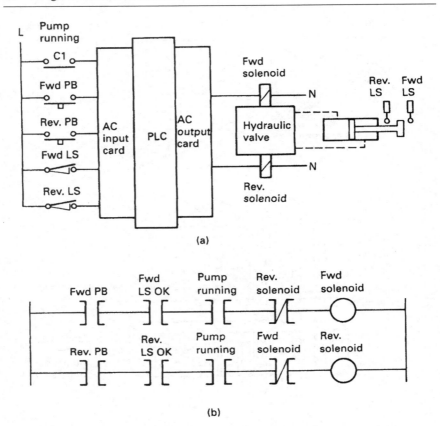

Figure 1.13   *A simple PLC application. (a) A simple hydraulic cylinder controlled by a PLC. (b) The 'ladder diagram' program used to control the cylinder. This is based on American relay symbols. –][– means that signal is present, and –]/[– means that signal is not present*

keyboard for an early PLC. The meaning of the majority of the keys should be obvious. The program, shown exactly on the screen as in Figure 1.13(b), would highlight energized contacts and coils, allowing the programming terminal to be used for simple fault finding.

The processor memory was protected by batteries to prevent corruption or loss of program during a power fail. Programs could be stored on cassette tapes which allowed different operating procedures (and hence programs) to be used for different products.

The name given to these machines was 'programmable controllers' or PCs. The name 'programmable logic controller' or PLC was also used, but this is, strictly, a registered trademark of the Allen Bradley Company. Unfortunately in more recent times the letters PC have come to be used

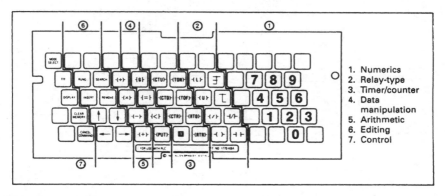

Figure 1.14 *The programming terminal keypad for an early Allen Bradley PLC (reproduced by permission of Allen Bradley)*

for personal computer, and confusingly the worlds of programmable controllers and personal computers overlap where portable and lap-top computers are now used as programming terminals. To avoid confusion, we shall use PLC for a programmable controller and PC for a personal computer. Section 2.12 gives examples of programming software on modern PCs.

1.4 Input/output connections

1.4.1 Input cards

Internally a computer usually operates at 5 V DC. The external devices (solenoids, motor starters, limit switches, etc.) operate at voltages up to 110 V AC. The mixing of these two voltages will cause severe and possibly irreparable damage to the PLC electronics. Less obvious problems can occur from electrical 'noise' introduced into the PLC from voltage spikes on signal lines, or from load currents flowing in AC neutral or DC return lines. Differences in earth potential between the PLC cubicle and outside plant can also cause problems.

The question of noise is discussed at length in Chapter 8, but there are obviously very good reasons for separating the plant supplies from the PLC supplies with some form of electrical barrier as in Figure 1.15. This ensures that the PLC cannot be adversely affected by anything happening on the plant. Even a cable fault putting 415 V AC onto a DC input would only damage the input card; the PLC itself (and the other cards in the system) would not suffer.

This is achieved by optical isolators, a light-emitting diode and photo-electric transistor linked together as in Figure 1.16(a). When current is passed through the diode D1 it emits light, causing the transistor TR1 to

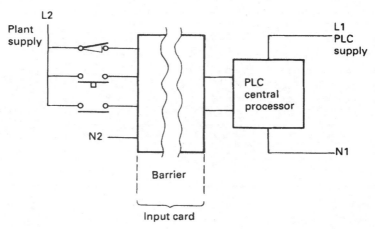

Figure 1.15   *Protection of the PLC from outside faults. The PLC supply L1/N1 is separate from the plant supply L2/N2*

switch on. Because there are no electrical connections between the diode and the transistor, very good electrical isolation (typically 1–4 kV) is achieved.

A DC input can be provided as in Figure 1.16(b). When the push-button is pressed, current will flow through D1, causing TR1 to turn on, passing the signal to the PLC internal logic. Diode D2 is a light-emitting diode used as a fault-finding aid to show when the input signal is present. Such indicators are present on almost all PLC input and output cards. The resistor R sets the voltage range of the input. DC input cards are usually available for three voltage ranges: 5 V (TTL), 12–24 V, 24–50 V.

A possible AC input circuit is shown in Figure 1.16(c). The bridge rectifier is used to convert the AC to full wave rectified DC. Resistor R_2 and capacitor C1 act as a filter (of about 50 ms time constant) to give a clean signal to the PLC logic. As before, a neon LP1 acts as an input signal indicator for fault finding, and resistor R_1 sets the voltage range.

Figure 1.17(a) shows a typical input card from the Allen Bradley range. The isolation barrier and monitoring LEDs can be clearly seen. This card handles eight inputs and could be connected to the outside world as in Figure 1.17(b).

1.4.2 Output connections

Output cards again require some form of isolation barrier to limit damage from the inevitable plant faults and also to stop electrical 'noise' corrupting the processor's operations. Interference can be more of a problem on outputs because higher currents are being controlled by

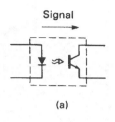

(a)

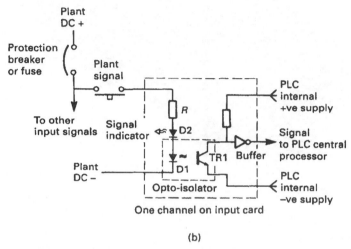

(b)

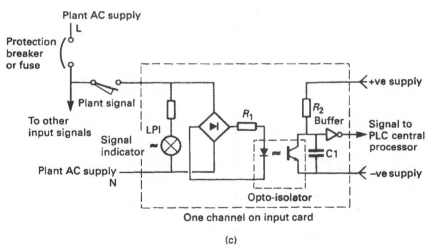

(c)

Figure 1.16 *Optical isolation of inputs: (a) an optical isolator; (b) DC input card; (c) AC input card*

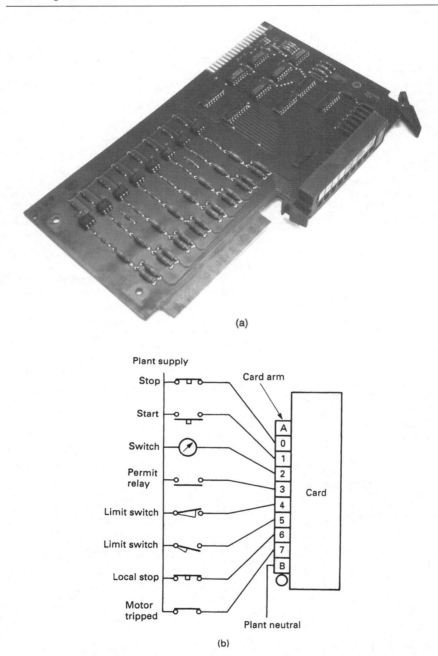

(a)

(b)

Figure 1.17   *A PLC input card: (a) Allen Bradley eight-way input card; (b) wiring of input card*

the cards and the loads themselves are often inductive (e.g. solenoid and relay coils).

There are two basic types of output card. In Figure 1.18(a), eight outputs are fed from a common supply, which originates local to the PLC cubicle (but separate from the supply to the PLC itself). This arrangement is the simplest and the cheapest to install. Each output has its own individual fuse protection on the card and a common circuit breaker. It is important to design the system so that a fault, say, on load 3 blows the fuse FS3 but does not trip the supply to the whole card, shutting down every output. This topic, called 'discrimination', is discussed further in Chapter 8.

A PLC frequently has to drive outputs which have their own individual supplies. A typical example is a motor control centre (MCC) where each starter has a separate internal 110-V supply derived from the 415-V bars. The card arrangement of Figure 1.18(a) could not be used here without separate interposing relays (driven by the PLC with contacts into the MCC circuit).

An isolated output card, shown in Figure 1.18(b), has individual outputs and protection and acts purely as a switch. This can be connected directly with any outside circuit. The disadvantage is that the card is more complicated (two connections per output) and safety becomes more involved. An eight-way isolated output card, for example, could have voltage on its terminals from eight different locations.

Contacts have been shown on the outputs in Figure 1.18. Relay outputs can be used (and do give the required isolation) but are not particularly common. A relay is an electromagnetic device with moving parts and hence a finite limited life. A purely electronic device will have greater reliability. Less obviously, though, a relay-driven inductive load can generate troublesome interference and lead to early contact failure.

A transistor output circuit is shown in Figure 1.19(a). Optical isolation is again used to give the necessary separation between the plant and the PLC system. Diode D1 acts as a spike suppression diode to reduce the voltage spike encountered with inductive loads. Figure 1.19(b) shows the effect. The output state can be observed on LED1. Figure 1.19(a) is a current sourcing output. If NPN transistors are used, a current sinking card can be made as in Figure 1.19(c).

AC output cards invariably use triacs, a typical circuit being shown in Figure 1.20(a). Triacs have the advantage that they turn off at zero current in the load, as shown in Figure 1.20(b), which eliminates the interference as an inductive load is turned off. If possible, all AC loads should be driven from triacs rather than relays.

Figure 1.21 is a photograph of the construction of AC and DC output cards; the isolation barrier, the state indication LEDs and the protection fuses can be clearly seen.

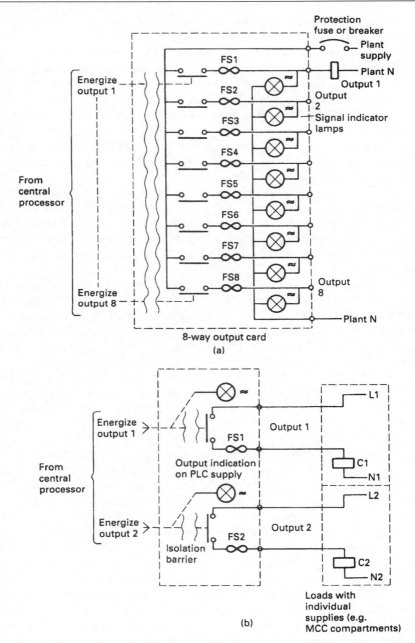

Figure 1.18   *Types of output card: (a) output card with common supply; (b) output card with separate supplies*

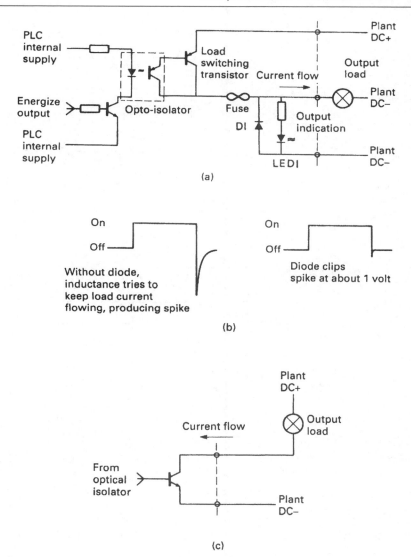

Figure 1.19 *DC output circuits: (a) DC output circuit, current sourcing; (b) effect of spike suppression diode; (c) current sinking output*

An output card will have a limit to the current it can supply, usually set by the printed circuit board tracks rather than the output devices. An individual output current will be set for each output (typically 2 A) and a total overall output (typically 6 A). Usually the total allowed for the card current is *lower* than the sum of the allowed individual outputs. It is

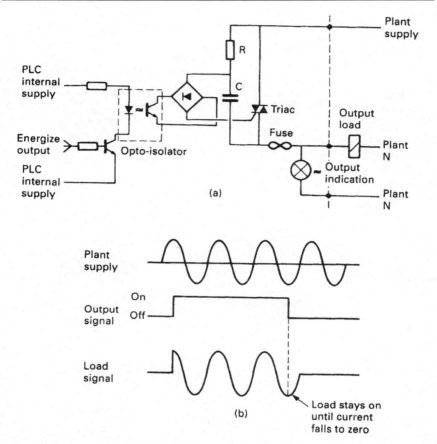

Figure 1.20 *AC output circuit: (a) AC output stage – sourcing/sinking is irrelevant on AC outputs; (b) effect of triac output*

therefore good practice to reduce the total card current by assigning outputs which cannot occur together (e.g. forward/reverse, fast/slow) to the same card.

1.4.3 Input/output identification

The PLC program must have some way of identifying inputs and outputs. In general, a signal is identified by its physical location in some form of mounting frame or rack, by the card position in this rack, and by which connection on the card the signal is wired to.

In Figure 1.22, a lamp is connected to output 5 on card 6 in rack 2. In Allen Bradley notation, this is signal

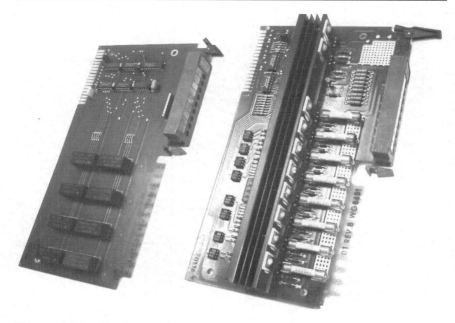

Figure 1.21 *Output cards*

O:26/05

The pushbutton is connected to input 2 on card 5 in rack 3, and (again in Allen Bradley notation) is

I:35/02

Most PLC manufacturers use a similar scheme. The topic is discussed further in Chapter 2.

1.5 Remote I/O

So far we have assumed that a PLC consists of a processor unit and a collection of I/O cards mounted in local racks. Early PLCs did tend to be arranged like this, but in a large and scattered plant with this arrangement, all signals have to be brought back to some central point in expensive multicore cables. It will also make commissioning and fault finding rather difficult, as signals can only be monitored effectively at a point possibly some distance from the device being tested.

In all bar the smallest and cheapest systems, PLC manufacturers therefore provide the ability to mount I/O racks remote from the processor, and link these racks with simple (and cheap) screened single

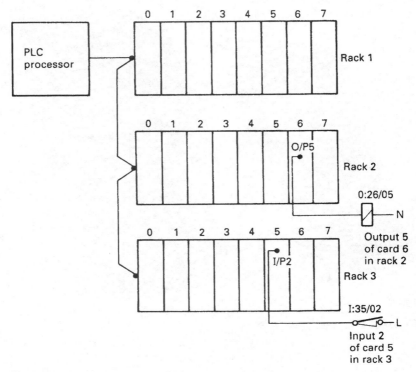

Figure 1.22 *Identification of plant signals*

pair or fibre optic cable. Racks can then be mounted up to several kilometres away from the processor.

There are many benefits from this. It obviously reduces cable costs as racks can be laid out local to the plant devices and only short multicore cable runs are needed. The long runs will only need the communication cables (which are cheap and only have a few cores to terminate at each end) and hardwire safety signals (which should not be passed over remote I/O cable, or even through a PLC for that matter, a topic discussed further in Chapter 8).

Less obviously, remote I/O allows complete units to be built, wired to a built-in rack, and tested offsite prior to delivery and installation. The pulpit in Figure 3.2 contains three remote racks, and connects to the controlling PLC mounted in a substation about 500 m away, via a remote I/O cable, plus a few power supplies and hardwire safety signals. This allowed the pulpit to be built and tested before it arrived on site. Similar ideas can be applied to any plant with I/O that needs to be connected to a PLC.

If remote I/O is used, provision should be made for a program terminal to be connected local to each rack. It negates most of the benefits if the designer can only monitor the operation from a central control room several hundred metres from the plant. Fortunately, manufacturers have recognized this and most allow programming terminals to be connected to the processor via similar screened twin cable.

We will discuss serial communication further in Chapter 5.

1.6 The advantages of PLC control

Any control system goes through four stages from conception to a working plant. A PLC system brings advantages at each stage.

The first stage is design; the required plant is studied and the control strategies decided. With conventional systems design must be complete before construction can start. With a PLC system all that is needed is a possibly vague idea of the size of the machine and the I/O requirements (how many inputs and outputs). The input and output cards are cheap at this stage, so a healthy spare capacity can be built in to allow for the inevitable omissions and future developments.

Next comes construction. With conventional schemes, every job is a 'one-off' with inevitable delays and costs. A PLC system is simply bolted together from standard parts. During this time the writing of the PLC program is started (or at least the detailed program specification is written).

The next stage is installation, a tedious and expensive business as sensors, actuators, limit switches and operator controls are cabled. A distributed PLC system (discussed in Chapter 5) using serial links and pre-built and tested desks can simplify installation and bring huge cost benefits. The majority of the PLC program is written at this stage.

Finally comes commissioning, and this is where the real advantages are found. No plant ever works first time. Human nature being what it is, there will be some oversights. Changes to conventional systems are time consuming and expensive. Provided the designer of the PLC system has built in spare memory capacity, spare I/O and a few spare cores in multicore cables, most changes can be made quickly and relatively cheaply. An added bonus is that all changes are recorded in the PLC's program and commissioning modifications do not go unrecorded, as is often the case in conventional systems.

There is an additional fifth stage, maintenance, which starts once the plant is working and is handed over to production. All plants have faults, and most tend to spend the majority of their time in some form of failure mode. A PLC system provides a very powerful tool for assisting with fault diagnosis. This topic is discussed further in Chapter 8.

A plant is also subject to many changes during its life to speed production, to ease breakdowns or because of changes in its requirements. A PLC system can be changed so easily that modifications are simple and the PLC program will automatically document the changes that have been made.

2 Programming techniques

2.1 Introduction

Chapter 1 described the evolution of the programmable controller leading to a system similar to that of Figure 1.12. This consists of a CPU linked to one or more I/O racks. These racks contain cards which are connected to the plant signals.

There are many variations on the details of Figure 1.12. Modern central processors tend to be small, live in one of the racks, and not be readily identifiable. In the smallest systems every part has been encapsulated in one unit. All, however, behave as in Figure 1.12.

In this chapter we shall consider how a PLC can be programmed. Each manufacturer, of course, has its own standards and it would be rather restrictive to deal with only one machine. This chapter is therefore written around five manufacturers' ranges:

1   The Allen Bradley PLC-5 series (Figure 2.1(a)). Allen Bradley, now owned by Rockwell, were one of the original PLC originators (and actually have the USA copyright on the name PLC). They have been responsible for much of the development of the ideas used in PLCs and have succeeded in maintaining a fair degree of upward compatibility from their earliest machine without restricting the features of the latest.

2   The Siemens Simatic S5 range (Figure 2.1(b)) which has become widely used in Europe in the early part of the 1990s.

3   The British GEM-80 (Figure 2.1(c)), originally designed by GEC through a long association with industrial computers dating back to English Electric. This part of GEC is now known as CEGELEC and is part of a French group in which Alsthom is a major shareholder.

4   The ASEA Master System (Figure 2.1(d)), now manufactured by the ABB company formed by the merger of ASEA and Brown Boveri. The Master System has features more akin to a conventional computer

(a)

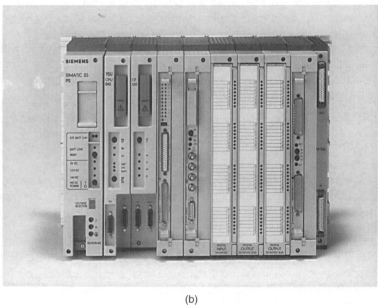

(b)

Figure 2.1   *The four medium-sized PLCs discussed: (a) the Allen Bradley PLC-5; (b) the Siemens S5-1154;*

(c)

(d)

Figure 2.1 *(continued) (c) the CEGELEC GEM-80; (d) the ABB Master.
Photographs courtesy of the manufacturers*

system and its programming language has some interesting and powerful features.

5 Many PLC systems are now very small; the author recently found it cost-effective to build a system with a PLC rather than the 12 four-pole relays that could have been conventionally used. There are many cheap small machines, and as an example of this bottom end of the market we shall consider the Japanese Mitsubishi F2-40, shown later in Figure 2.12.

Significant differences will be found in this selection (a PLC-5, for example, has three different types of timer, the Siemens 115-U has five timers, and a GEM-80 just one, which, because of its different approach, can be used in various ways). Between them most of the standards adopted by other manufacturers will be covered.

2.2 The program scan

A PLC program can be considered to behave as a permanent running loop similar to that in Figure 2.2(a). The user's instructions are obeyed sequentially, and when the last instruction has been obeyed the operation starts again at the first instruction. A PLC does not, therefore, communicate continuously with the outside world, but acts, rather, by taking 'snapshots'.

The action of Figure 2.2(a) is called a program scan, and the period of the loop is called the program scan time. This depends on the size of the PLC program and the speed of the processor, but is typically 2–5 ms per K of program. Average scan times are usually around 10–50 ms.

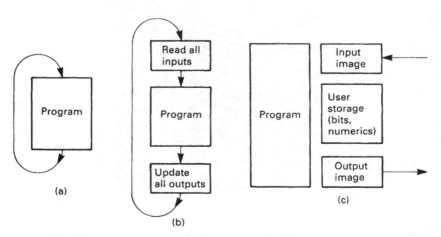

Figure 2.2 *PLC program scan and memory organization: (a) PLC operation; (b) program sequence; (c) PLC memory organization*

Figure 2.2(a) can be expanded to Figure 2.2(b). The PLC does *not* read inputs as needed (as implied by Figure 2.2(a)) as this would be wasteful of time. At the start of the scan it reads the state of *all* the connected inputs and stores their state in the PLC memory. When the PLC program accesses an input, it reads the input state as it was at the start of the current program scan.

As the PLC program is obeyed through the scan, it again does not change outputs instantly. An area of the PLCs memory corresponding to the outputs is changed by the program, then *all* the outputs are updated simultaneously at the end of the scan. The action is thus: read inputs, scan program, update outputs.

The PLC memory can be considered to consist of four areas as shown in Figure 2.2(c). The inputs are read into an input mimic area at the start of the scan, and the outputs updated from the output mimic area at the end of the scan. There will be an area of memory reserved for internal signals which are used by the program but are not connected directly to the outside world (timers, counters, storage bits, e.g. fault signals, and so on). These three areas are often referred to as the data table (Allen Bradley) or the database (ASEA/ABB).

This data area is smaller than may be at first thought. A medium-size PLC system will have around 1000 inputs and outputs. Stored as individual bits this corresponds to just over 60 storage locations in a PLC with a 16-bit word. An analog value read from the plant or written to the plant will take one word. Timers and counters take two words (one for the value, and one for the preset) and 16 internal storage bits take just one word. The majority of the store, therefore, is taken up by the fourth area, the program itself.

The program scan obviously limits the speed of signals to which a PLC can respond. In Figure 2.3(a) a PLC is being used to count a series of fast pulses, with the pulse rate slower than the scan rate. The PLC counts correctly. In Figure 2.3(b) the pulse rate is faster than the scan rate and the PLC starts to miscount and miss pulses. In the extreme case of Figure 2.3(c) whole blocks of pulses are totally ignored.

In general, any input signal that a PLC reads must be present for longer than the scan time; shorter pulses may be read if they happen to be present at the right time but this cannot be guaranteed. If pulse trains are being observed, the pulse frequency must be slower than $1/(2 \times \text{scan}$ period). A PLC with a scan period of 40 ms can, in theory, just about follow a pulse train of $1/(2 \times 0.04) = 12.5\,\text{Hz}$. In practice other factors such as filters on the input cards have a significant effect and it is always advisable to be conservative in speed estimates.

Less obviously, the PLC scan can cause a random 'skew' between inputs and outputs. In Figure 2.4 an input is to cause an 'immediate' output. In the best case of Figure 2.4(a), the input occurs just at the start

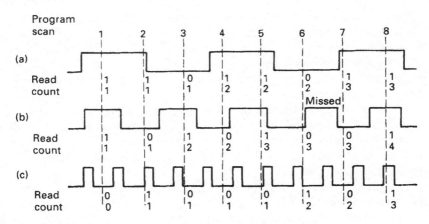

Figure 2.3 *The effect of program scan on fast pulses*

of the scan, resulting in the energization of the output one scan period later. In Figure 2.4(b) the input has arrived just after the inputs are read, and one whole scan is lost before the PLC 'sees' the input, and the rest of the second scan passes before the output is energized. The response can thus vary between one and two scan periods.

In the majority of applications this skew of a few tens of milliseconds is not important (it cannot be seen, for example, in the response of a plant

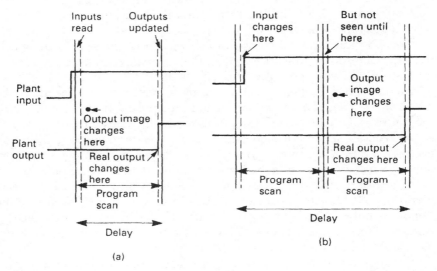

Figure 2.4 *The effect of program scan on response time: (a) best case; (b) worst case*

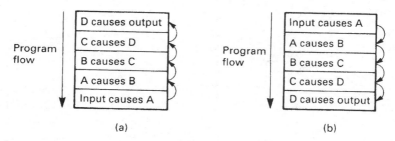

Figure 2.5 *Compounding of program scan delays: (a) logic against program flow, five scans from input to output; (b) logic with program flow, one scan from input to output*

to pushbuttons). Where fast actions are needed, however, it can be crucial. In a typical example, seen by the author, material travelling at 15 m/s was being cut by a PLC with the initiation being given by a photocell. The 30-ms scan time of the PLC resulted in a $0.03 \times 15\,000 = 450$-mm variation in cut length.

PLC manufacturers provide special cards (which are really small processors in their own right) for dealing with this type of high-speed application. We will return to these in Sections 4.7 and 4.8.

The layout of the PLC program itself can result in undesirable delays if the program logic flows *against* the PLC program scan. The PLC starts at the first instruction for each scan, and works its way through the instructions in a sequential manner to the end of the program. It then does its output update, goes to read its inputs and runs through the program again.

In Figure 2.5(a), an input causes an output, but it goes through five steps first (it could be stepping a counter or seeing if some other required conditions are present). The program logic, however, is flowing against the scan. On the first scan the input causes event A. On the next scan event A causes event B and so on until after five scans event D causes the output to energize. If the program had been arranged as in Figure 2.5(b) the whole sequence would have occurred in one single scan.

The failings of Figure 2.5(a) are self-evident, but the effect can often occur when the layout of the program is not carefully planned. The effect can also be used deliberately to produce very short (one-scan) pulses, a topic discussed in Section 6.4.

The effect of scan times can become even more complex when remote serially scanned I/O racks are present. These are generally read by an I/O scanner as in Figure 2.6, but the I/O scanner is not usually synchronized to the program scan. In this case with, say, a program

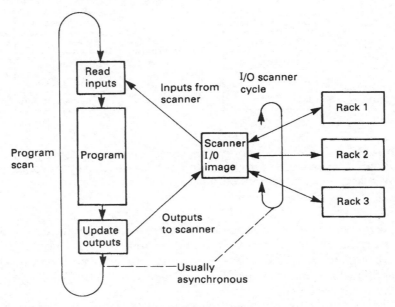

Figure 2.6   *The effect of program and I/O scan cycles*

scan of 30 ms and a remote I/O scan of 50 ms, the fastest response could be 30 ms, but the slowest response (with an input just missing the I/O scan and the I/O scan just missing the program scan) could be 130 ms.

PLC manufacturers offer many facilities to reduce the effect of scan times. Typical are intelligent high-speed independent I/O cards (discussed in Sections 4.7 and 4.8) and the ability to sectionalize the program into areas with different scan rates (desk lights, for example, probably only need a 0.5–1-s response, allowing other parts of the program to operate faster). Section 3.6 describes in more detail further methods to minimize the PLC scan time.

2.3   Identification of input/output and bit addresses

2.3.1   Racks, cards and signals

The PLC program is concerned with connections to the outside plant, and the input and output devices need to be identified inside the program. Before we can examine how the program is written we will first discuss how various manufacturers treat the I/O.

Figure 1.22 showed that a medium-sized PLC system consists of several racks, each containing cards, with each card interfacing generally to 8, 16 or 32 devices. I/O addressing is usually based on this rack/card/bit idea.

2.3.2 Allen Bradley PLC-5

The Allen Bradley PLC-5 can have up to eight racks in its 5/25 version. The rack containing the processor is automatically defined as rack 0, but the designer can allocate addresses of the other racks (in the range 1–7) by set-up switches. The racks other than rack 0 connect to the processor via a remote I/O serial communications cable.

There are three different ways in which an Allen Bradley rack can be configured, but we shall discuss the simplest (and possibly the most logical) method.

Each rack contains 16 card positions which are grouped in pairs called a 'slot'. A rack thus contains eight slots, numbered 0–7. A slot can contain one 16-way input card and one 16-way output card *or* two eight-way cards usually (but not necessarily) of the same type. In Figure 2.7, for example, slot 1 contains a 16-way input card and 16-way output card, and slot 2 contains two eight-way output cards.

Reasons why eight-way cards may be preferred to 16-way output cards are discussed in Chapter 8.

The addressing for inputs is

I:Rack Slot/Bit

with Bit being two digits. Allen Bradley use octal addressing for bits, so allowable numbers are 00–07 and 10–17. The address I:27/14 is input 14 on slot 7 in rack 2. Outputs are addressed in a similar manner:

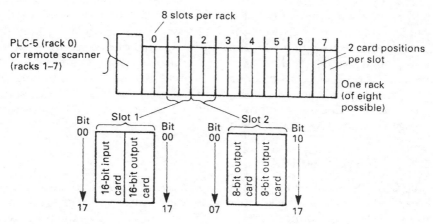

Figure 2.7 *Allen Bradley PLC-5 card layout*

O:Rack Slot/Bit

so O:35/06 is output 6 in slot 5 of rack 2. Note that if 16-way cards are used, an input and an output can have the same rack/slot/bit address, being distinguished only by the I: or the O:. With eight-way cards there can be no sharing of rack/slot/bit addressing. Figure 1.12 earlier showed the addressing of several signals.

2.3.3 Siemens SIMATIC S5

The digital I/O in Siemens PLCs is arranged into groups of 8 bits, called a byte (see Appendix). A signal is identified by its bit number (0–7) and its byte number (0–127). Inputs are denoted I<byte>. <bit> and outputs by Q<byte>. <bit>. I9.4 is thus an input with bit address 4 in byte 9, and Q63.6 is an output with bit address 6 in byte 63.

Like Allen Bradley, Siemens use card slots in one or more racks. The cards are available in 16-bit (2-byte) or 32-bit (4-byte) form. A system can be built with local racks connected via a parallel bus cable or as remote racks with a serial link. Local racks are faster and overcome some of the scan problems associated with serially connected remote racks in high-speed applications, but are, literally, local. They can be no more than a few metres from the processor.

The simplest form of addressing is fixed slot, shown in Figure 2.8(a). Four bytes are assigned sequentially to each slot; 0–3 to the first slot, 4–7 to the next slot and so on. Input 12.4 is thus input bit 4 on the first byte of the card in slot 3 of the first rack. If 16-bit (2-byte) cards are used with fixed (4-byte) addressing, the upper 2 bytes in each slot are lost.

In all bar the simplest system the user has the ability to assign byte addresses. This is known as variable slot addressing. The first byte address and the range (2 byte for 16-bit cards or 4 byte for 32-bit cards) can be set independently for each slot by switches in the adaptor module in each rack. Although *any* legitimate combination can be set up, it is recommended that a logical order is used similar to that in Figure 12.4(b).

Siemens use different notations in different countries with multilingual programming terminals. A common European standard is German, where E (for Eingang or input) is used for inputs (e.g. E4.7) and A (for Ausgang) is used for outputs (e.g. A3.5).

2.3.4 CEGELEC GEM-80

The GEM-80 again configures its I/O in terms of bits and slots within racks. The processor rack can contain eight card positions, and additional I/O can be connected into 12 position racks local to the processor via a ribbon cable (called basic I/O) or remotely via a serial link. Where

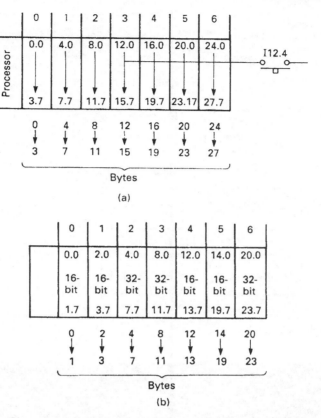

Figure 2.8 *Siemens card layout. (a) Fixed slot addressing. (b) Variable slot addressing. Switches set address and number of bytes (2 or 4) per card. Sequential addressing (as above) is not mandatory but is recommended*

a small amount of remote I/O is needed, compact 8-in/8-out units can be used rather than racks as shown in Figure 2.9. In addition to the basic I/O structure, a verification I/O highway is also available which allows the processor to check the state of the various modules.

The I/O is addressed in terms of 16-bit words, one word corresponding to one or two card positions, with the prefix A being used for inputs and B for outputs. The bit addressing runs in decimal from 0 to 15.

A3.12 is thus input bit 12 in word 3

and

B5.04 is output bit 4 in word 5

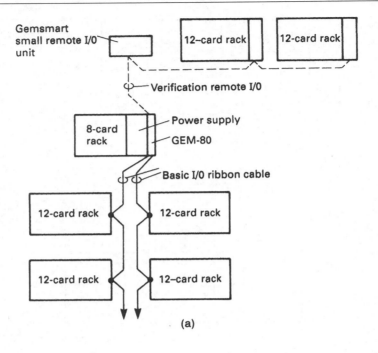

Figure 2.9   *(a) Structure of a GEM-80 system. (b) GEM smart remote I/O units*

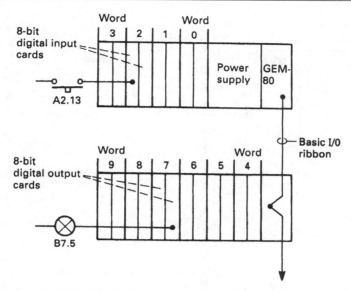

Figure 2.10 *GEM-80 addressing with 8-bit cards*

A word can only be an input or an output; duplication of word addresses is not allowed. I/O cards are available in 8-bit, 16-bit and 32-bit form, so one slot can be half a word, one word or two words according to the cards being used. Individual slot addresses are set by rotary switches on the back plane of each rack. The user has a more or less free choice in this allocation, but as usual it is best to use a logical sequential progression. Figure 2.10 shows a typical small arrangement.

2.3.5 ABB Master

The ABB (originally ASEA) Master system is a more complex system than any we have discussed so far. Its organization brings the user closer to the computer, and its language is more akin to the ideas used by programmers. If the PLCs discussed so far are taken to be represented by the home computer language BASIC, the ABB Master is analogous to Pascal or C. This comparison is actually closer than might, at first, be thought. BASIC is quick and easy to use, but can degenerate into a web of spaghetti programming if care is not taken. Pascal and C are more powerful but everything has to be declared and the language forces organization and structure on the user.

The Master system is arranged with processor cards and racks as in Figure 2.11(a). Each I/O card has two back connectors, the top connecting

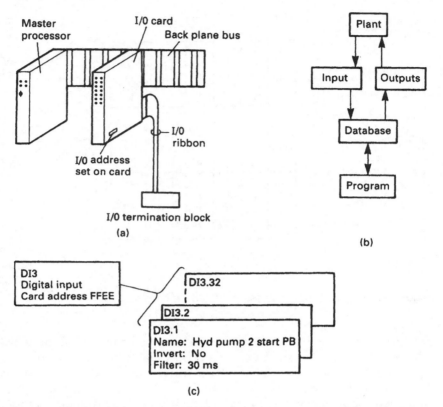

Figure 2.11 *The ABB Master system: (a) layout of ABB Master system; (b) program structure; (c) part of database (for digital input card)*

to the processor bus and the lower to a separate terminal block, one per card, which is mounted on the back plane of the cubicle.

The I/O cards are *not* identified by position in the rack, but by an address set on the card by a small plug with solder links. The I/O addressing does not, therefore, relate to card position, and a card can, in theory, be moved about (with its lower connector socket, of course) without changing its operation.

The processor memory is arranged as in Figure 2.11(b). The I/O is connected to a processor database, but unlike PLCs described earlier, the designer can specify different scan rates for different cards.

The designer also has considerable power over how the PLC program is organized. This is heavily modularized as we shall see later, and the user can specify different scan rates for different modules of the program.

Figure 2.11(c) indicates the database for one input card. There are two levels of the definition, the top level relating to details of the board itself such as address and scan rate, and the lower levels relating to details of each channel on the board such as its name and whether the signal is to be inverted. The database holds details for all the I/O which can then be referenced in the program either by its database identification (e.g. DI3.1) or by its unique name (e.g. HydPump2-StartPB).

2.3.6 Mitsubishi F2

The Mitsubishi F2 range is typical of small PLCs with I/O connection, power supply and processor all contained in one unit as in Figure 2.12. The smallest unit, the F2-40M, has 24 inputs and 16 outputs (it is characteristic of process control systems that the ratio of input to outputs is generally 3:2).

The 24 inputs are designated X400 to X427 in octal notation and the 16 outputs Y430–Y447. The apparently arbitrary numbers are directly related to the storage locations used to hold the image of the inputs and outputs. Further addresses are used in larger PLCs in the series.

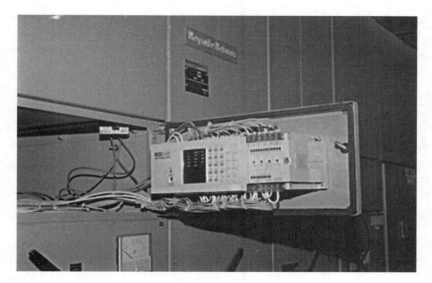

Figure 2.12 *Mitsubishi F2 programmable controller mounted (with program terminal) in the door of a motor control centre cubicle*

2.3.7   Internal bit storage

As well as inputs and outputs, the PLC will need to hold internal signals for data such as 'standby pump running', 'system healthy', 'lubrication fault' and so on. It would be very wasteful to allocate real outputs to these signals, so the PLCs all provide some form of internal bit storage. These are known variously as auxiliary relays (Mitsubishi), flags (Siemens), general workspace (GEM-80) and bit storage (Allen Bradley). The notation used within the programs varies, of course, from manufacturer to manufacturer.

Mitsubishi use Mnnn with nnn representing numbers within the predefined area M100 to M377 octal. Like most small PLCs the memory layout is fixed and cannot be defined by the user. In the other, larger, PLCs we discuss, the user can define how many storage bits are needed.

The Siemens notation is FByte.Bit (e.g. F27.06). The GEM-80 has a variety of general workspaces. The commonest is called the G table, and appears in programs as GWord.Bit (e.g. G52.14). The G table is cleared when the PLC goes from a stopped state to a run state. Storage in the R table (e.g. R12.03) retains its state with the processor halted or with power removed.

Bit storage in the PLC-5 is denoted by B3/n where n denotes the signal (e.g. B3/192). The B denotes bit storage and the 3 is mandatory and arises out of the way that the PLC-5 holds data in files. Bit storage is file 3, timers are file 4 (T4) and counters file 5 (C5) as we shall see later.

The ABB Master programming language does not really require internal storage bits, the function being provided by elements and connections within its database and the programming language.

2.4   Programming methods

2.4.1   Introduction

The programming language of a PLC will be used by engineers, technicians and maintenance electricians. It should therefore be based on techniques used in industry rather than techniques used in computer programming. In this section we shall look at the various ways in which PLCs from different manufacturers can be programmed.

It is, perhaps, worth mentioning at this point the rather interesting approach adopted by Siemens, who provide *three* different programming methods for their machines, allowing the user to choose. Even more remarkable, with a few exceptions a program written in one format can be viewed in another.

2.4.2 Ladder diagrams

Early PLCs, designed for the car industry, replaced relay control schemes. The symbols used in relay drawings, –] [– for a normally open (NO) contact, –]/[– for a normally closed (NC) contact, and –( )– for a plant output, were the basis of the language. Figure 1.14 shows the keyboard for a programmer for this type of PLC; the relationship to relay symbolism is obvious.

Suppose we have a hydraulic unit, and we wish to give a Healthy Lamp indication when

1 The pump is running (sensed by an auxiliary contact on the pump starter).
2 There is oil in the tank (sensed by a level switch which makes when the switch is covered).
3 There is oil pressure (sensed by a pressure switch which makes for adequate pressure).

With conventional relays, we would wire up a circuit as in Figure 2.13(a).

To use a PLC, we connect the input signals to an input card, and the lamp to an output card, as in Figure 2.13(b). The I/O notation used is Allen Bradley.

The program to provide the function is shown in Figure 2.13(c). The line on the left can be considered to be a supply, and the line on the right a neutral. The output is represented by a coil –( )– and is energized when there is a route from the left-hand rail. Output 0:22/01 will come on when signals 1:21/00, 1:21/01 and 1:22/02 are all present.

The program is entered from a terminal with keys representing the various relay symbols. The terminal can also be used to monitor the state of the inputs and outputs, with 'energized' inputs and outputs being shown highlighted on the screen.

In Figure 2.14, a hydraulic cylinder can be extended or retracted by operation of two pushbuttons. The notation this time is for a GEM-80. It is undesirable to allow both solenoids to be operated together; this will almost certainly result in blown fuses in the supply to the output card, so some protection is needed. The program to achieve this is shown in Figure 2.14(b).

Normally closed contacts –]/[– have been used here. Output B2.9, the extend solenoid, will be energized when the extend pushbutton is pressed, providing the retract solenoid is not energized or the retract button pressed, and the extend limit switch has not been struck.

There are two points to note in Figure 2.14. Contacts can be used off outputs as well as inputs, and contacts can be used as many times as needed in the program. Figure 2.14 also shows where the name 'ladder

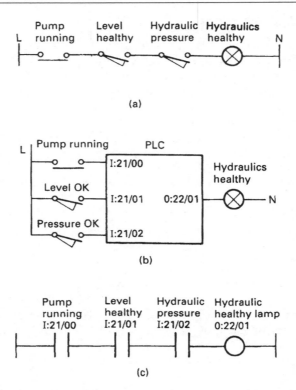

Figure 2.13 *From relay circuit to ladder diagram: (a) simple relay circuit; (b) PLC wiring; (c) ladder diagram*

program' comes from. A program in this form looks like a ladder, with each instruction statement forming a 'rung' and the power rail and neutral the supports. The term 'rung' is invariably applied to the contacts leading to one output.

Let us return to the hydraulics healthy light of Figure 2.13 and add a lamp test pushbutton (a useful feature that should be present on all panels; it not only allows lamps to be tested, but can also be used to check the PLC itself is running). To do this we add the lamp test push-button to the PLC and modify the program to Figure 2.15.

Here we have added a branch, and the output will energize if our three plant signals are all present *or* the lamp test button is pressed. The way in which the branch is programmed need not concern us here as it varies between manufacturers. Some use Start Branch and End Branch keys (the keypad of Figure 1.14 uses this method; the corresponding keys can readily be identified). Others use a Branch From/To approach. All are simple to use.

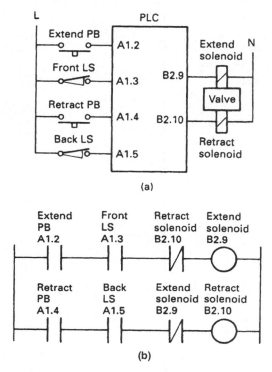

(a)

(b)

Figure 2.14 *Ladder diagram in GEM-80 notation: (a) PLC wiring; (b) ladder diagram*

A further use of a branch is shown in Figure 2.16. This is probably the commonest control circuit, a motor starter, shown using Siemens notation. The operation is simple; pressing the start pushbutton causes the output Q8.2 to energize, and the contact of the output in the branch keeps the output energized until the stop button is pressed. The program, like its relay equivalent, remembers which button was last pressed.

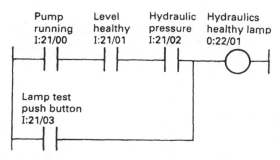

Figure 2.15 *Adding a branch*

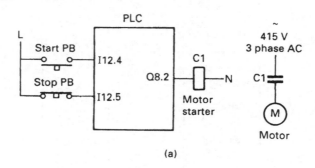

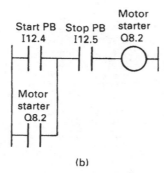

Figure 2.16 *A latching motor starter program: (a) PLC wiring; (b) ladder diagram*

There is, however, a very important point to note about the pushbutton wiring and the program. For safety, a normally closed stop button has been used giving an input signal on I12.5 when the stop button is *not* pressed. A loss of supply to the button, or a cable fault, or dirt under the contacts, will cause the signal to be lost, making the program think the stop pushbutton has been pressed, causing the motor to stop. If a normally open stop pushbutton has been used, the PLC program could easily be made to work, but a fault with the stop button or its circuit could leave the motor running, with the only way of stopping it being to turn off the PLC or the motor supply.

This topic is discussed further in Section 8.2, but note the effect on the program in Figure 2.15. The sense of the stop button input (I12.5) inside the program is the *opposite* of what would be expected in a relay circuit. The input is really acting as 'permit to run' rather than 'stop'.

2.4.3 Logic symbols

Logic gates based on TTL (transistor–transistor logic) and CMOS (complementary metal oxide semiconductor) integrated circuits are widely used in digital systems (including the boards used inside PLCs). The circuits used on these boards are represented by logic symbols, and these symbols can also be used to describe the operations of a PLC program. Logic symbols are used by Siemens and ABB; initially we will use Siemens notation.

The output from an AND gate, shown in Figure 2.17(a), is true if (and only if) all its inputs are true. The operation of the gate of Figure 2.17(a) can be represented by the table of Figure 2.17(b). In Figure 2.17(c) we have the hydraulics healthy lamp of Figure 2.13

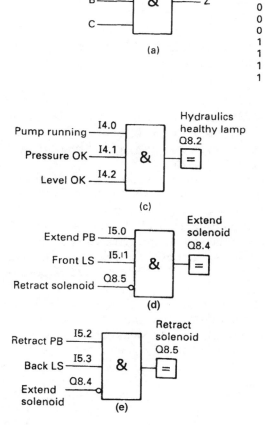

C	B	A	Z
0	0	0	0
0	0	1	0
0	1	0	0
0	1	1	0
1	0	0	0
1	0	1	0
1	1	0	0
1	1	1	1

(b)

Figure 2.17 *PLC programs using logic symbols (based on Siemens notation): (a) an AND gate; (b) truth table for AND gate; (c) hydraulic healthy lamp of Figure 2.13 in logic notation; (d) an inverted input; (e) hydraulic cylinder of Figure 2.14 in logic notation*

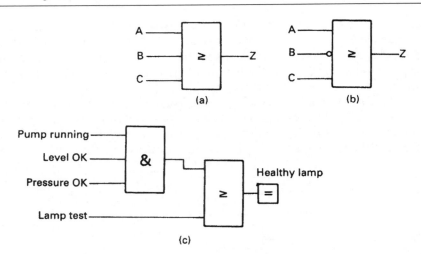

Figure 2.18 *The OR gate: (a) OR gate; (b) OR gate with inverted input; (c) lamp test added to Figure 2.17(c)*

redrawn connected to a Siemens PLC. Using logic symbols, we would program this as shown. The output block, denoted by equals (=), is energized when its input is true, so the lamp Q8.2 is energized (lit) when all the inputs to the AND gate are true.

Often a test has to be made to say a signal is *not* true. This is denoted by a small circle ○. The output of the AND gate in Figure 2.17(d) is true if (and only if) A and B are true and C is not true.

In Figure 2.14 we illustrated the control of a hydraulic cylinder with a program which prevented the extend and retract solenoids from being energized simultaneously. This is redrawn for a Siemens PLC controlled with the program of Figure 2.17(e). Note the NOT inputs on each AND gate.

The output of an OR gate, Z in Figure 2.18(a), is true if any of its inputs are true. The symbol in the gate means 'the output is true if one or more inputs are true'. The inverse of a signal can be tested, as before, with a small circle ○. The output Z of the gate in Figure 2.18(b) is true if A is true or B is false or C is true. In Figure 2.18(c) we have used an OR gate to add a lamp test to our hydraulics healthy lamp.

The circuit of Figure 2.18(c) is an AND/OR combination. The ABB Master has logic combination blocks as well as the basic gates. Figure 2.19(a) is the Master block corresponding to Figure 2.18(c) (with a Master program referring to the names in its database). Similarly, for an OR/AND combination the OR/AND block of Figure 2.19(b) can be used in a Master program.

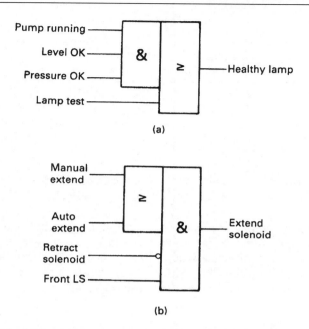

(a)

(b)

Figure 2.19 *ABB Master composite gates: (a) AND/OR gate (equivalent to Figure 2.18(c)); (b) OR/AND gate*

2.4.4 Statement list

A statement list is a set of instructions which superficially resemble assembly language instructions for a computer. Statement lists, available on the Siemens and Mitsubishi range, are the most flexible form of programming for the experienced user but are by no means as easy to follow as ladder diagrams or logic symbols.

Figure 2.20 shows a simple operation in both ladder and logic formats for a Siemens PLC. The equivalent statement list would be as shown in Table 2.1. Here A denotes AND, AN denotes AND/NOT and = sends the result to the output address Q4.11.

Figure 2.20 *Equivalent ladder and logic statements in Siemens notation*

Table 2.1

Instruction number	Operation	Address	
00	:A	I3.7	Forward pushbutton
01	:A	I3.2	Front limit OK
02	:AN	Q4.2	Reverse solenoid
03	:=	Q4.11	Forward solenoid

An OR operation is shown in Figure 2.21. The equivalent statement list is shown in Table 2.2. Here ON denotes OR/NOT and O denotes OR.

Where a set of statements can be anomalous, brackets can be used to define the operation precisely. This is similar to the use of brackets in conventional programming where the sequence $3+5/2$ can be written as $(3+5)/2=4$ or $3+(5/2)=5.5$. Although the latter is the default assumed by a program, the brackets do make the operation clear to the reader.

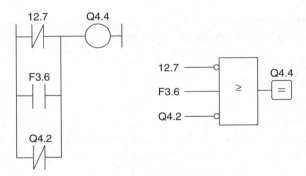

Figure 2.21 *OR gate equivalence in Siemens notation*

Table 2.2

Instruction number	Operation	Address	
00	:ON	I2.7	Local pump running auxiliary
01	:O	F3.6	Remote pump running flag
02	:ON	Q4.2	Local pump starter
03	:=	Q4.4	Pump healthy lamp

Here ON denotes OR/NOT and O denotes OR.

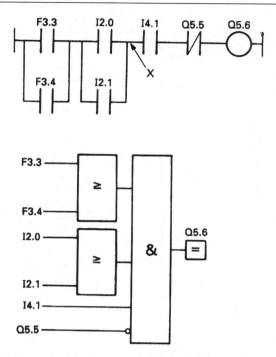

Figure 2.22 *More complex program in both notations*

Figure 2.22 shows a typical operation, as usual in both logic and ladder diagram format. The equivalent statement list is shown in Table 2.3. Computer programmers will recognize this as being similar to the operation of a stack, a topic discussed further when we consider FORTH in Section 7.3.

The Mitsubishi PLC also uses statement lists, although the manual recommends the designer to construct a ladder diagram first and then translate it into a statement list. The PLC system shown in Figure 2.23 with Mitsubishi notation becomes the statement list in Table 2.4.

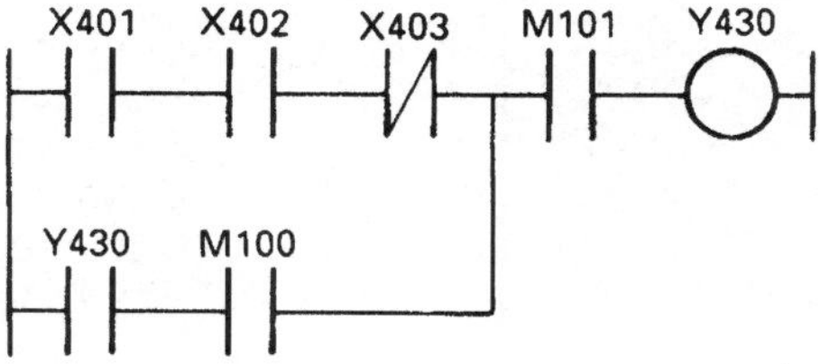

Figure 2.23 *Mitsubishi ladder program*

Table 2.3

Instruction number	Operation	Address	
00	:A(		First set of brackets
01	:A	F3.3	Manual forward
02	:O	F3.4	Automatic forward
03	:)		Result of first set of brackets
04	:A(		AND result with second set of brackets
05	:A	I2.0	Motor 1 selected
06	:O	I2.1	Motor 2 selected
07	:)		Now at point X
08	:A	I4.1	Front limit switch healthy
09	:AN	Q5.5	Reverse starter
10	:=	Q5.6	Forward starter

2.5   Bit storage

Some form of memory circuit is needed in practically every PLC program. Typical examples are catching a fleeting alarm and the motor starter of Figure 2.16, where the rung remembers which button (start or stop) has been last pressed. These are known, for obvious reasons, as storage circuits.

The commonest form is shown in ladder and logic form in Figure 2.24(a). Here output C is energized when input A is energized, and stays energized until input B is de-energized. The sense of input B is chosen for safety reasons; it acts as a 'permit to energize' signal as discussed in Sections 2.4.2 and 8.2.

Table 2.4

Line	Instruction	Comment
0	LD X401	LD starts rung or branch
1	AND X402	Xnnn are inputs
2	ANI X403	ANI is And/Not
3	LD Y430	LD starts a new branch leg
4	AN M100	Mnnn are internal storage
5	ORB	OR the two branch legs
6	AND M101	
7	OUT Y430	End of rung

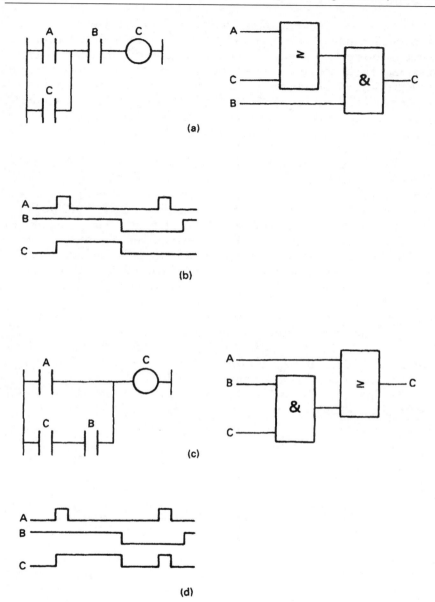

Figure 2.24 *Bit storage circuits: (a) commonest storage circuit, stop B overrides start A; (b) operation of program (a); (c) storage circuit where start A overrides stop B; (d) operation of program (c)*

The operation is summarized on Figure 2.24(b). As can be seen, input B overrides input A, the action required of a start/stop circuit. In some circuits, however, the start is required to override the stop. A typical example can be found in cars; the windscreen wipers run when they are switched on, but continue to run to the park position when they are turned off. The PLC equivalent is Figure 2.24(c), where A would be the run switch, B the park limit switch and C the wiper motor. B has again been shown energized to allow running. The operation is summarized in Figure 2.24(d).

In logic design, storage is provided by a device called a flip-flop shown in Figure 2.25(a). This has two inputs, S (for set) and R (for reset). The device remembers which input was last a 1. If both inputs occur together, the top (S) input wins. Such a circuit is called an SR flip-flop. If the device is drawn with the R input at the top, as in Figure 2.25(b), the R input will override the S input if both are present together.

The flip-flop is used in logic symbol PLC programming. A motor starter using a Siemens PLC is shown in Figure 2.26(a). Note that the RS version has been used to ensure that the stop logic overrides the run logic, and the stop signal acts as a permit to run.

Clarity is of prime importance in writing PLC programs, to help a tired engineer fault finding in the middle of the night see how a program works. The memory feature of the circuit of Figure 2.24 is not immediately clear. It helps if the latch contact is *always* the lowest in the branch (for ladder circuits) or the gate (for logic circuits).

The flip-flop symbol is often found in ladder diagrams. Figure 2.26(b) is the direct Siemens ladder equivalent of Figure 2.26(a).

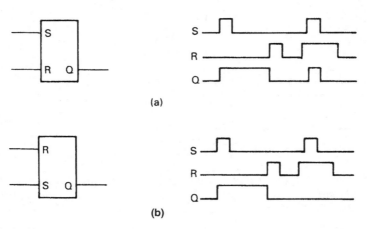

Figure 2.25 *The two types of flip-flop storage. (a) The S–R flip-flop. Set overrides reset. (b) The R–S flip-flop. Reset overrides set*

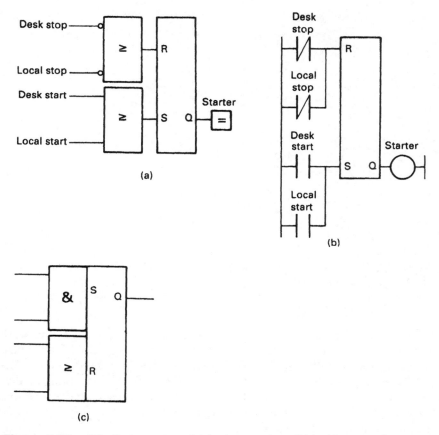

Figure 2.26 *Flip-flop storage: (a) logic notation; (b) ladder notation; (c) ABB SRAO flip-flop*

In these circuits, the preferred form of normally closed stop button has been used. Note how these appear in the program, and compare them with the earlier ladder program of Figure 2.24.

The ABB Master uses an almost identical symbol for the flip-flop, except that there are five versions. The first of these is the simple SR type shown in Figure 2.25. The other versions are based on the fact that flip-flops are invariably preceded by AND/OR combinations (Figure 2.26 is typical). The additional flip-flops are one-unit blocks consisting of a flip-flop with built-in AND/OR gates of user-defined size. Figure 2.26(c), for example, is an ABB SRAO with an AND gate on the set input and an OR gate on the reset inputs. Other units are SRAA (AND/AND), SROA and SROO.

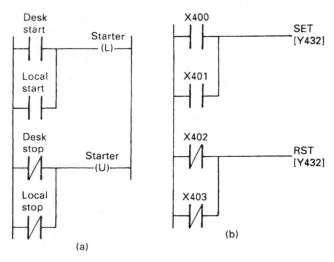

Figure 2.27 *Other forms of storage: (a) Allen Bradley latch/unlatch; (b) Mitsubishi set/reset*

In Allen Bradley ladder diagrams, program clarity can be improved by the use of latch and unlatch outputs shown in Figure 2.27(a). These work on the same bit, setting the bit when a 1 is presented to the –(L)– and resetting the bit when a 1 reaches the –(U)–. When both receive a 0, the bit holds its last state. The Mitsubishi F2 uses a similar idea, but calls them S and R outputs as in Figure 2.27(b). This would be coded into the statement list in Table 2.5.

With both the Allen Bradley latch/unlatch, and the Mitsubishi set/reset, the priority goes to whichever is last in the program because of the program scan. Both the examples of Figure 2.27 correctly give priority to the stop signals.

Table 2.5

Line	Instruction	Comment
0	LD X400	
1	OR X401	
2	S Y432	Set output
3	LDI X402	
4	ORI X403	
5	R Y432	Reset output

Power failure or halting of the PLC causes a problem with memories. When the PLC restarts should a memory bit hold the state it was in before the PLC halted, or should the memory be cleared? This is always a question of safety and convenience. A water pump in a pump house by a river 5 km from the main site should probably be allowed to restart itself if it was running before the power fail; an automatic stamping machine should almost certainly not restart itself.

The PLC manufacturers therefore allow the designer to choose whether a storage bit holds its state after a power fail (called retentive memory) or is cleared when the PLC is first run (called non-retentive memory).

In the Allen Bradley PLC-5, this is determined by the circuit; the simple coil of Figure 2.24 is non-retentive, the latch/unlatch of Figure 2.27 is retentive.

Other PLCs use the bit address. On a Siemens 115, flag addresses F0.0–F127.7 can be made retentive. On the Mitsubishi PLC, auxiliary relays M100–277 are non-retentive, and M300–M377 are retentive. In the GEM-80, the general bit storage G table is non-retentive, while a similar R table is retentive, so a circuit similar to that of Figure 2.24 constructed with R3.4 as the coil and retaining contact would hold its state after a power failure.

The ABB Master uses a very structured PLC language, and forces a disciplined style on the programmer. The nature of sub-elements such as memories and their behaviour when the PLC is first run is defined when the program elements are first declared.

Retentive storage can be very hazardous, as plants can unexpectedly leap into life after a power fail. The designer should take care that the design does not accidentally introduce retentive features by an inadvertent selection of bit addresses. The use of the R table in the GEM-80 is particularly praiseworthy, as an R table address is unlikely to be chosen in error.

2.6 Timers

Time is nearly always a part of a control system. Typical examples are: 'Lift parking brake, wait 0.5 seconds for brake to lift, drive to forward limit and stop drive, wait 1 second and apply parking brake', and 'Start hydraulic pump, if auxiliary contact not in within 0.7 seconds signal drive fault, if drive runs wait 2 seconds and energize loading valve, if hydraulic pressure not established within 3 seconds signal hydraulic fault and stop pump'. A PLC system must therefore include timers as part of its programming language. There are many types of timer, some of which are shown in Figure 2.28.

By far the commonest is the on delay of Figure 2.28(a). All the other timer blocks can be simulated with this block and a bit of thought. A 0 to 1 transition is delayed for a preset time T, but a 1 to 0 transition

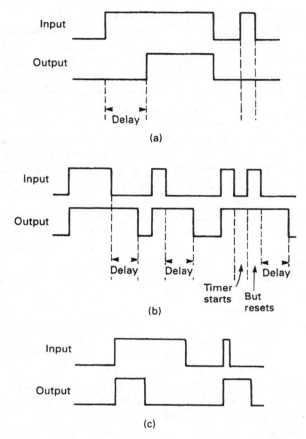

Figure 2.28   *Different forms of timer: (a) on delay; (b) off delay; (c) fixed width pulse*

is not delayed at all. An input signal shorter than T is ignored. The GEM-80 has only this type of timer, calling it a delay.

The off delay of Figure 2.28(b) passes a 0 to 1 transition instantly but delays the 1 to 0 transition. A common use of the off delay is to remove contact bounce or noise from an input signal. An off delay can be obtained from an on delay by using the inverse of the input signal and taking the inverse of the timer output signal (although the resulting program lacks some clarity).

Figure 2.28(c) is an edge-triggered pulse timer; this gives a fixed-width pulse for every 0 to 1 transition at the timer input. The PLC-5 has a one-scan pulse timer which produces a pulse lasting one (and only one) program scan. Pulses are useful for resetting counters or gating some information from one location to another. The annunciator (described in Section 6.4) is a typical example.

A timer of whatever type has some values that need to be set by the user. The first of these is the basic unit of time (i.e. what units the time is measured in). Common units are 10 ms, 100 ms, 1 s, 10 s and 100 s. The base unit does not affect the accuracy of the timer; normally the accuracy is similar to the program scan.

Next the timer duration (often called the preset) is defined. This is normally set in terms of the time base; a timer with a preset of 150 and a time base of 10 ms will last 1.5 s, for example. In small PLCs this preset is set by the programmer; in the larger PLCs the duration can be changed from within the program itself. A delay off timer used to apply a parking brake, for example, could have different preset times depending on whether the drive concerned is travelling at low speed or high speed.

When a timer is used there are several signals that may be available. Figure 2.29 shows the signals given for a PLC-5 delay on timer (called a TON) and a delay off timer (called a TOF).

- EN (for enable) is a mimic of the timer input
- TT (for timer timing) is energized whilst the time is running
- DN (for done) says the timer has finished

In larger PLCs the elapsed time (often called the accumulated time) may be accessed by the program for use elsewhere (a program may be required to record how long a certain operation takes).

PLC manufacturers differ on how a timer is programmed. Some, such as the GEM-80, treat the timer as a delay block similar to Figure 2.28(a), with the preset being stored in a VALUE block. Siemens use a similar idea, but have different types of timer. The PLC-5, however, uses the timer as a terminator for a rung, with the timer signals being available as contacts for use elsewhere.

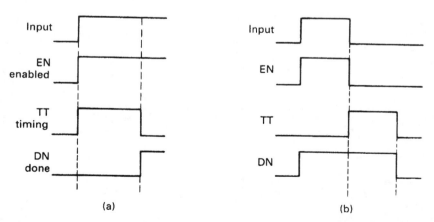

Figure 2.29 *Allen Bradley timer notations: (a) TON timer; (b) TOF timer*

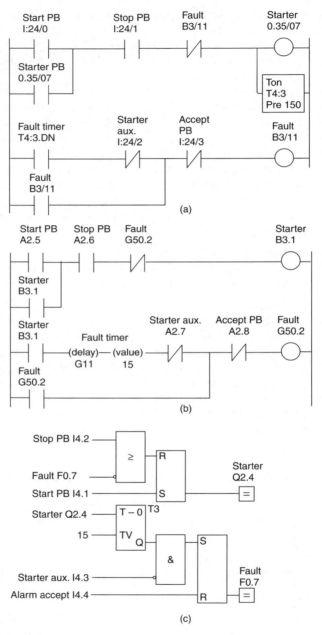

Figure 2.30 *Various timer types in the same application: (a) PLC-5 notation; (b) GEM-80 notation; (c) Siemens logic notation*

Figure 2.30 is a typical application programmed for a GEM-80, a Siemens 115U in logic symbols and a PLC-5. The program controls a motor starter which is started and stopped via pushbuttons. The motor starter has an auxiliary contact which makes when the starter is energized, effectively saying the motor is running. If the drive trips because of an overload, or because an emergency stop is pressed, or there is a supply fault, the auxiliary contact signal will be lost. The contact cannot, however, be checked until 1.5 s after the starter has been energized to allow time for the contact to pull in. Figure 2.30 checks the auxiliary contact and signals a drive fault if there is a problem. Note the difference in the ways the timer is used and the fault signal is stored.

The accumulated time in the timers discussed so far goes back to zero each time the input goes to a zero as summarized in Figure 2.31(a). This is known as a non-retentive timer. Most PLC timers are of this form. Occasionally it is useful to have a timer which holds its current value even though the input signal has gone. When the input occurs again the timer continues from where it stopped as in Figure 2.31(b). This, not surprisingly, is known as a retentive timer. A separate signal must be used to reset the timer to zero. If a retentive timer is not available on a particular PLC, the same function can be provided with a counter, a topic discussed in the next section.

A typical timer can count up to 32 767 base time units (corresponding to 16 binary bits). Some older PLCs working in BCD can only count to 999. With a 1-s time base the maximum time will be just over 546 min or about 9 h. Where longer times are needed (or times with a resolution better than 1 s) timers and counters can be used together as described in the next section.

2.7 Counters

Counting is a fundamental part of many PLC programs. The PLC may be required to count the number of items in a batch, or record the number of times some event occurs. With large motors, for example, the number of starts has to be logged. Not surprisingly, all PLCs include some form of counting element.

A counter can be represented by Figure 2.32, although not all PLCs will have the facilities we will describe. There will be two numbers associated with the counter. The first is the count itself (often called the accumulated value) which will be incremented when a $0 \rightarrow 1$ transition is applied to the count up input, or decremented when a $0 \rightarrow 1$ transition is applied to the count down input. The accumulated value (count) can be reset to zero by applying a 1 to the reset input. Like the elapsed time in a timer, the value of the count can be read and used by other parts of the program.

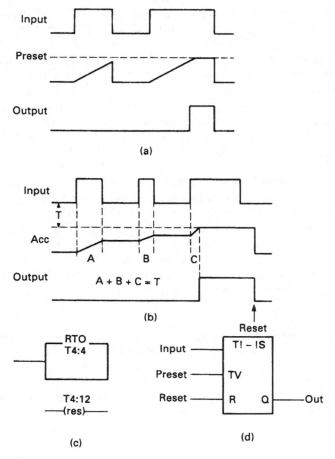

Figure 2.31   *Retentive timers: (a) non-retentive timer; (b) retentive timer; (c) Allen Bradley notation; (d) Siemens notation*

The second number is the preset, which can be considered as the target for the counter. If the count value reaches the preset value, a count complete or count done signal is given. The preset can be changed by the program; a batching sequence, for example, may require the operator to change the number of items in a batch by a keypad or VDU entry. Similarly, a signal 'zero count' is sometimes available, giving an operation which is summarized in Figure 2.32(b).

PLC manufacturers handle counters, like timers in slightly different ways. The PLC-5 and the Mitsubishi use count up (CTU), count down (CTD) and reset (RES) as rung terminators with the count done signal (e.g. C5:4.DN) available for use as a contact.

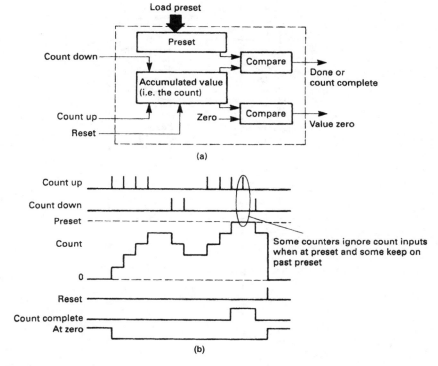

Figure 2.32 *The up/down counter: (a) counter diagram; (b) counter operation*

The Siemens S5, ABB Master and GEM-80 treat a counter as an intermediate block in a logic diagram or rung from which the required output signals can be used.

Figure 2.33 shows a simple count application performed by a PLC-5, a Siemens S5 and a GEM-80. Items passing along a conveyor are detected by a photocell and counted. When a batch is complete, the conveyor is stopped and a batch complete light is lit for the operator to remove the batch. When he does this, a restart button sets the sequence running again.

Although smaller, the GEM and Siemens programs both suffer from a small problem that is not at first apparent. If a lamp test PB is added, when pressed it will cause the conveyor to stop. In both cases this could be overcome by using an internal store saying 'count complete'. A –]/[– contact would then be used for the conveyor, and a –] [– contact for the lamp. This would add one rung to each program. Considerations such as this are known as 'software engineering', a topic we will discuss further in the next chapter.

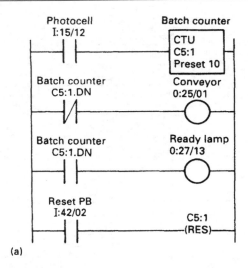

(a)

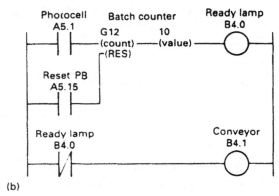

(b)

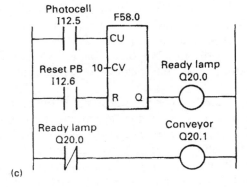

(c)

Figure 2.33 *Counter application in Allen Bradley, GEM and Siemens notations: (a) Allen Bradley; (b) GEM-80; (c) Siemens*

Like timers, most PLCs allow a counter to count up to 32 767. Where larger counts are needed, counters can be cascaded with the complete (or done) signal from the first counter being used to step the second counter and reset the first. Suppose counter 1 holds the range 0–999, and counter 2 the thousands. If counter 2 holds 23 516 and counter 1 holds 457, the total count is 23 516 457.

Figure 2.34 is a variation on the same idea used to give a very long timer. It is shown for a PLC-5, but the same idea could be used on any PLC.

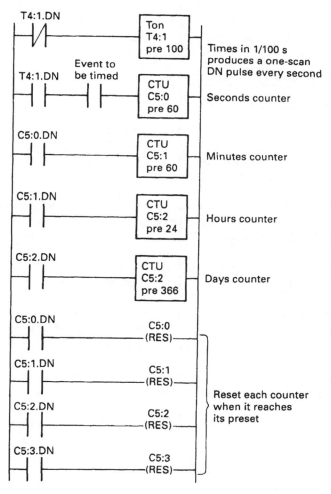

Figure 2.34 *Cascaded counters*

The first rung generates a free-running one-scan pulse with inter pulse period set by the timer. (When the timer has not timed out, the DN signal is not present and the timer is running. When it reaches the preset, the DN signal occurs, resetting and restarting the timer.)

The resulting 1-s pulse is counted by successive counters to give accumulated seconds/minutes/hours/days/years. As each counter reaches its preset it steps the next counter and resets itself. This technique is widely used to log hours run for pumps, fans and similar devices for maintenance scheduling. In this case the 'event' in the second rung will be an auxiliary contact on the motor starter.

Long-duration timers built from counters are normally retentive (i.e. they hold their value when the controlling event is not present). They can be made non-retentive by resetting the counters when the controlling event is not present, but this is rarely required.

2.8 Numerical applications

2.8.1 Numeric representations

So far we have been primarily discussing single-bit operations. Numbers are also often part of a control scheme; a PLC might need to calculate a production rate in units per hour averaged over a day, or give the amount of liquid in a storage tank. Such operations require the ability to handle numeric data.

Most PLCs work with a 16-bit word, allowing a positive number in the range 0 to +65 535 to be represented, or a signed (positive or negative) number in the range −32 768 to +32 767 in two's complement (see Appendix). In the latter case the most significant bit represents the sign, being 1 for negative numbers and 0 for positive numbers. Two's complement representation is usually (but not exclusively) used in PLC programs.

Numbers such as these are known as integers, and obviously can only represent whole numbers in the above range. Where larger whole numbers are required, two 16-bit words can be used, allowing a range of −2 147 483 648 to +2 147 483 647. This type of integer is available in the ABB Master (where it is known as a 'long integer') and the 135U and 155U in the Siemens family (where the term 'double word integer' is used).

Where decimal fractions are needed (to deal with a temperature of 45.6 °C for example) a number form similar to that found on a calculator may be used. These are known as real or floating point numbers, and generally consist of two 16-bit words which contain the mantissa (the numerical portion) and the exponent. In base ten, for example, the number 74 057 would have a mantissa of 7.4057 and an exponent of 4,

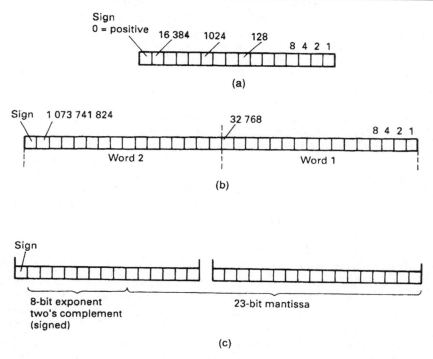

Figure 2.35 *Numerical representations: (a) standard 16-bit integer; (b) long 32-bit integer; (c) IEEE 32-bit real*

representing 10^4. PLCs, of course, work in binary and represent mantissa and exponent in two's complement form. There are inevitably variations between manufacturers, but an emerging standard is the IEEE single precision 32-bit format shown in Figure 2.35(c). This gives a range of $\pm(1.175E-38$ to $3.402E38)$. Some manufacturers trade off the range of the exponent to give greater precision in the mantissa. In the ABB Master, for example, the range is $\pm(5.4E-20$ to $9.2E18)$, which allows extra precision.

Real numbers are very useful but their limitations should be clearly understood. There are two common problems. The first occurs when large numbers and small numbers are used together. Suppose we have a system operating to base ten with four significant figures, and we wish to add 857 800 (stored as 8.578E5) and 96 (stored as 9.600E1). Because the smaller number is outside the range (four significant figures) of the larger, it will be ignored, giving the result 857 800+96=857 800.

The second problem occurs when tests for equality are made on real numbers. The conversion of decimal numbers to binary numbers can only be made to the resolution of the floating point format $(1.175E-38$

for IEEE single precision). Most home computers hold numbers in floating point form, and the effect can be demonstrated with the simple BASIC program:

```
100   A=3
110   B=6.4
120   C=9.4
130   IF(A+B)=C THEN PRINT ("3+6.4=9.4"): GOTO 150
140   PRINT ("3+6.4 does not=9.4, it=");:: PRINT C
150   PRINT ("Take care with real numbers!")
```

This simple program does not do what you might expect! If real numbers must be used for comparison, a simple equates (=) is very risky. The composites >= (greater than or equals to) and <= (less than or equal to), are safer, but it is generally better practice to use integers for tests if at all possible.

The final representation, BCD for binary coded decimal, is used for connection to outside world devices such as digital displays or thumbwheel switches. Such devices are arranged in a decimal format, with 4 binary bits per decade. This representation is wasteful, as six 'numbers' are not used per 4 bits (10 to 15 inclusive). It is, however, a convenient form to use with external wiring. Most PLCs therefore have instructions which convert BCD to the internal binary format of the PLC, and binary back to BCD. The PLC-5, for example, has FRD (from decimal) and TOD (to decimal). Figure 2.36 shows a schematic of a typical operation. Sections 9.11 and 9.12 describe how to perform Binary/BCD conversion for machines without these functions.

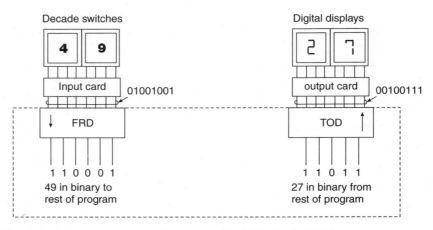

Figure 2.36   *Application of binary coded decimal (BCD)*

The types of numbers available in each PLC range vary considerably according to the model (and obviously the price). The Mitsubishi F2, for example, only allows movement, comparison and output of numerical data from counters or timers, making it essentially a bit-operation machine.

In the Siemens range, the popular 115-U uses only 16-bit integer numbers but the next model in the range, the 135-U, can handle 16-bit and 32-bit integers and floating point numbers. A similar spread of capabilities will be found amongst the Allen Bradley, GEM-80 and ABB families.

2.8.2 Data movement

Numbers are often required to be moved from one location to another; a timer preset may be required to be changed according to plant conditions, a counter value may need to be sent to an output card for indication on a digital display or the result of some calculations may be used in another part of a program.

The Allen Bradley PLC-5 uses one rung per move operation, and is possibly the simplest to explain first. Its simplicity of one rung per operation is continued in all the arithmetic functions we shall consider, but it can lead to more rungs being used for a given operation than in other machines.

Figure 2.37(a) shows the form of the rung. It starts with some binary conditions; if these are all made the output MOV (for move) is obeyed, transferring data from the source to the destination. The source and destination can be any location where numerical data can occur, for example:

Counter or timer preset	e.g. C5:17.PRE or T4:52.PRE
Counter or timer accumulated value	e.g. C5:22.ACC or T4:6.ACC
Input or output WORD data	e.g. I:23 (card 3 in rack 2, all 16 bits) 0:47 (card 7 in rack 4, all 16 bits)

Note that these data are interpreted as binary; if BCD data are needed, the FRD and TOD instructions are available (see Figure 2.36).

Internal integer storage	e.g. N7:24
Internal floating point storage	e.g. F8:32

If data are transferred between integer and floating point forms, the conversion is performed automatically. However, care must be taken when transferring floating point numbers to integers as an error can occur if

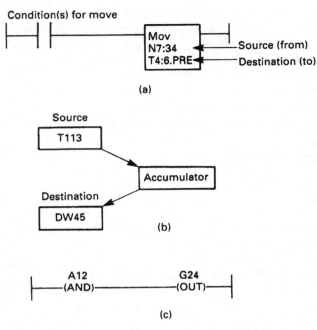

Figure 2.37   *Data movement: (a) Allen Bradley; (b) Siemens; (c) GEM-80*

the floating point number is outside the integer range. Finally, as a source only, a constant (such as 3, 17 or 4057) can be used.

The example of Figure 2.37 thus moves the number held in N7:34 to the preset of timer T4:6 when the rung conditions are met.

Siemens and GEC use a slightly different approach which leads to more compact programs and a small improvement in rounding errors at the expense of a less direct way of working. Both treat a data movement as two separate instructions via a separate accumulator (a single word storage location). Siemens use the instructions Load to move data from a source to the accumulator, and Transfer to move data from the accumulator to the destination, as in Figure 2.37(b). The data can come from (or go to) any data storage area, some of which are

IW   a 16-bit input word
QW   a 16-bit output word
T   a timer word
C   a counter word
DW   a 16-bit data storage word

Figure 2.37 would thus be programmed as

:L   T113   (timer value to accumulator)
:T   DW45   (accumulator to data word 45)

The use of the accumulator is not obvious in the GEM-80. The – <AND> – instruction puts the binary number from the specified location (again internal storage or I/O) into the rung, and the – <OUT> – instruction puts the value from the rung to the specified address. In Figure 2.37(c) the (binary) value from 16-bit input word A12 is placed into 16-bit storage word G24. BCD/binary conversion is available with – <BCDIN> – and – <BCDOUT> – instructions, the direction of the conversion being obvious.

The difference between Figure 2.37(b) and (c) and Figure 2.37(a) will become apparent when we discuss arithmetic operations in Section 2.8.4.

In the ABB Master, the points between which data are to be transferred are simply linked on the logic diagram.

2.8.3 Data comparison

Numerical values often need to be compared in PLC programs; typical examples are a batch counter saying the required number of items have been delivered, or alarm circuits indicating, say, a temperature has gone above some safety level.

These comparisons are performed by elements which have the generalized form of Figure 2.38, with two numerical inputs corresponding to the values to be compared, and a binary (on/off) output which is true if the specified condition is met.

Many comparisons are possible; most PLCs provide

A greater than B
A greater than or equal to B
A equal to B
A less than or equal to B
A less than B

where A and B are numerical data. With real (floating point) numbers the 'equal to' test should be avoided for the reasons given in the previous section. There are many other possible comparisons; a PLC-5, for example, has a Limit instruction which tests for A lying between B and C and the GEM and Siemens have a 'not equal' test.

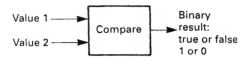

Figure 2.38 *Basics of data comparison*

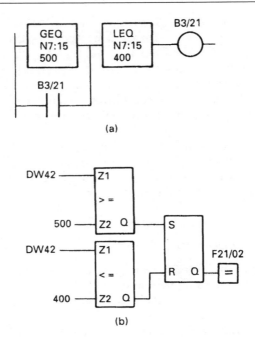

Figure 2.39 *Use of data comparison: (a) Allen Bradley; (b) Siemens (logic notation)*

Figure 2.39 shows the setting and resetting of an alarm flag B3/21 (for a PLC-5 ladder diagram) and F21/02 (for Siemens logic symbols). The alarm bit is set if temperature (read from an analog input card in format nn.n °C and held in N7:15 in the PLC-5 or DW42 in the Siemens 115-U) goes above 50.0 °C. Once set, the alarm is stored until the temperature goes below 40.0 °C.

2.8.4 Arithmetical operations

Numerical data implies the ability to do arithmetical operations, and all PLCs we are considering (apart from the simple F2) provide the ability to do at least four function mathematical operations (add, subtract, multiply and divide).

In Section 2.8.1 we discussed integer and floating point numbers. Care needs to be taken with integer operations. The range of a 16-bit two's complement number is $-32\,768$ to $+32\,767$ (see Appendix). If an arithmetical operation goes outside this range, the number will overspill, for example:

26732
8647
−30157 in 16-bit two's complement

which is not quite the expected result. The PLCs have an overspill flag which can be examined and used to flag an alarm, or set the result to, say, zero with a Move instruction. Similar precautions need to be taken with subtraction and multiplication (the latter being particularly vulnerable to giving an overspill; $200 \times 200 = 40\,000$, well over range).

It should be borne in mind that an arithmetical overspill could arise from a fault on an analog input card, a plant sensor or even the plant itself, and the fault could be otherwise undetected. There is a true story of a false missile attack warning which occurred in the USA in the 1960s when a radar system received echoes from the moon. The target distance (calculated by dividing the echo delay by the speed of light) grossly overspilled, but no check was made and the result was an apparently legitimate distance with echoes corresponding to incoming missiles. This caused the USA defences to go to a first state of alert. Fortunately human beings intervened after a few minutes.

Even greater care needs to be taken with division. A programming error or a fault condition on external plant or a PLC input card can lead to a divide by zero error. This will stop many PLCs dead in their tracks with a 'program fault'. It is therefore good practice to precede any vulnerable divide instruction with a limit check to ensure it will only be obeyed when a sensible result is obtained.

Figure 2.40 illustrates a typical example that caught me out. It was required to measure the speed of an object, and this was achieved by timing the nose between two photocells, the velocity then being $v = d/T$. All worked well until PEC2 failed some months later, returning a time

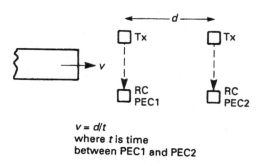

$v = d/t$
where t is time
between PEC1 and PEC2

Figure 2.40   *An example of a divide by zero error*

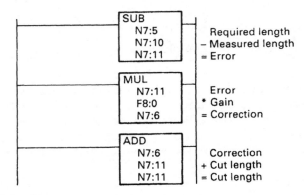

In high level language
Cut length: = Gain * (required length – measured length) + cut length

Figure 2.41   *Arithmetic in the PLC-5*

of zero and producing a divide by zero PLC fault in the early hours of the morning. The maintenance staff changed several items (including the PLC itself as a PLC fault light was on) before the fault was found. The author was duly shamefaced.

Each PLC manufacturer handles arithmetic in a slightly different way with varying degrees of ease and readability. None are as simple as a high level language such as BASIC or Pascal, and the facilities are generally limited to four function maths plus square root in all bar the most expensive machines.

A PLC-5 uses maths blocks such as ADD, SUB, MULT and DIV, giving a simple, if somewhat lengthy, program. Figure 2.41 shows how a simple calculation could be performed for a self-correcting length-cutting program. More powerful PLC-5s (such as the 5/40) have a block compute instruction which allows a mathematical expression to be evaluated in a single instruction.

The 115-U evaluates arithmetic instruction in STL (statement list) format. It will be remembered from our discussion of the accumulator that the load (L) and transfer (T) instructions use an internal accumulator. There are, in fact, two accumulators, and a load instruction moves the contents of accumulator 1 to accumulator 2 and then moves the contents of the source to accumulator 1, as shown in Figure 2.42. An arithmetic instruction (add, subtract, etc.) works on the contents of both accumulators. Figure 2.42 thus adds two numbers and transfers the result to storage.

The Siemens equivalent of Figure 2.41 would be

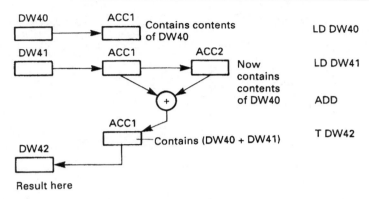

Figure 2.42 *Arithmetic in a Siemens S5 PLC*

L DW30	(required length)
L DW31	(measured length)
SUB	(leaving error in Acc 1)
L DW32	(gain)
MULT	(leaving correction)
L DW40	(the old cut length)
ADD	(add change to give new length)
T DW40	(put back to store)

The most understandable forms of representation are possibly the GEM-80 ladder and the ABB Master formats shown in Figures 2.43(a) and 2.43(b) respectively. These require little, if any, elaboration. People familiar with FORTH (discussed in Section 7.3) will realize that Figures 2.42 and 2.43 are both based on the idea of a pushdown stack.

All maths operations, particularly those involving floating point numbers, are time consuming, and it is good programming practice to only obey instructions when they are needed, and not waste time repetitively obeying them on every PLC scan.

2.9 Combinational and event-driven logic

2.9.1 Combinational logic

Any control system based on digital signals can be represented by Figure 2.44(a), with a set of outputs Z, Y, X, W, etc. whose state is determined by inputs A, B, C, D, etc. The control scheme can operate in a combination of two basic ways.

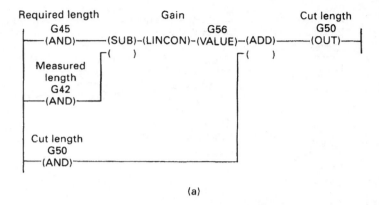

(a)

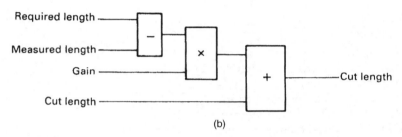

(b)

Figure 2.43 *Mathematical functions in GEM-80 and ABB Master. (a) GEM-80 Arithmetic LINCON is an arithmetic function used to avoid truncation errors with integer mathematics. (b) ABB Master. Variables are accessed by database names*

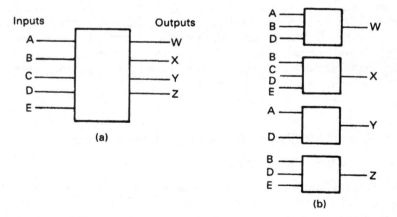

Figure 2.44 *Combinational logic: (a) top level view; (b) broken into smaller blocks*

The simplest of these is combinational logic, where the scheme can be broken down into smaller blocks as in Figure 2.44(b) with one output per block, each output state being determined *solely* by the corresponding input states. The loading valve for a hydraulic pump, for example, is to be energized when

The pump is running AND
    (Raise is selected AND top limit SW is not struck) OR
    (Lower is selected AND bottom limit SW is not struck)

The operation of this loading valve can be implemented with the simple ladder and logic program of Figure 2.45, but it is worth developing a standard way of producing a combinational logic program.

The first stage is to break the control system down into a series of small blocks, each with one output and several inputs. For each output we now draw up a so-called truth table, in which we record all the possible input states and the required output state. In Figure 2.46(a) we have an output Z controlled by four inputs ABCD. There are 16 possible

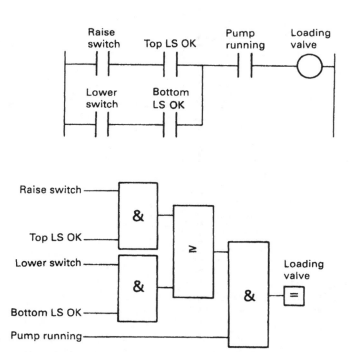

Figure 2.45 *Combinational logic in ladder diagram and logic notation*

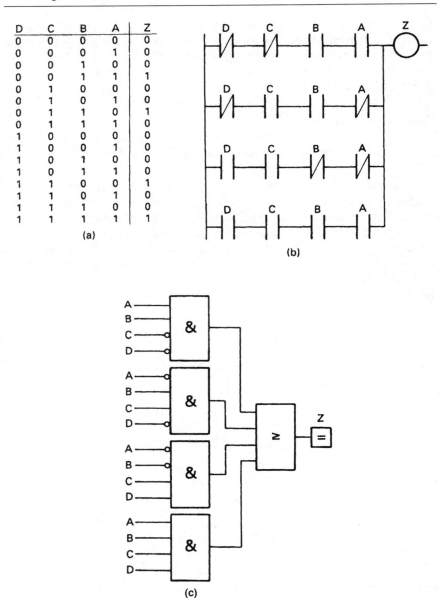

D	C	B	A	Z
0	0	0	0	0
0	0	0	1	0
0	0	1	0	0
0	0	1	1	1
0	1	0	0	0
0	1	0	1	0
0	1	1	0	1
0	1	1	1	0
1	0	0	0	0
1	0	0	1	0
1	0	1	0	0
1	0	1	1	0
1	1	0	0	1
1	1	0	1	0
1	1	1	0	0
1	1	1	1	1

(a)

(b)

(c)

Figure 2.46 *Combinational logic from a truth table: (a) truth table; (b) ladder implementation of truth table; (c) logic implementation*

input states, and Z is energized for four of these. This can be translated directly into the ladder diagram of Figure 2.46(b) or the logic circuit of Figure 2.44(c), with each rung branch or AND gate corresponding to one row in the truth table. The use of a truth table method for the

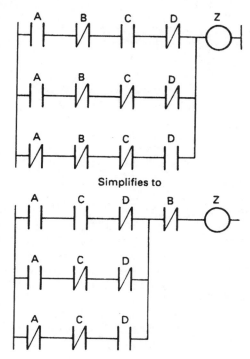

Figure 2.47   *Clarifying combinational logic*

design of combinational logic circuits leads directly to an AND/OR arrangement called, technically, a sum of products (S of P) circuit.

An inevitable question is 'Is this the simplest arrangement?' The answer is 'probably not', and techniques such as Karnaugh maps and Boolean algebra exist to give a simpler solution. When a circuit is built from logic gates or relays it can be very important to design circuits with the minimum number of gates or contacts to reduce construction costs. With a PLC program, however, the cost of additional contacts is zero, so *clarity* of operation rather than *simplicity* should be the aim. The one simplification that should always be made (again for clarity) is to pull a common contact out of the branches as shown in Figure 2.47.

Consider, for example, the motor starter desk layout of Figure 2.48(a). For cheapness, the three-position switch has been wired with just two contact blocks as in Figure 2.48(b) (bad practice, as a supply fault will cause both pumps to run). The truth table gives the ladder diagram of Figure 2.48(c), but the minimal ladder is Figure 2.48(d). The simplest ladder, however, masks the operation of the switch and would make fault finding just a little bit more difficult.

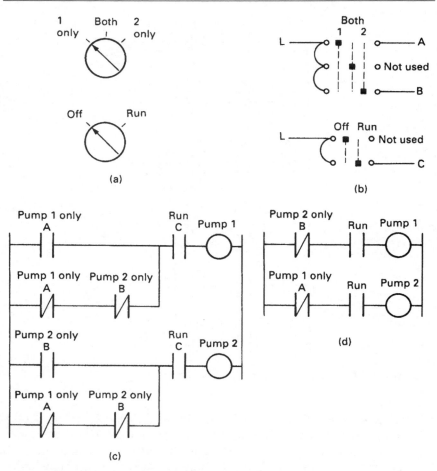

Figure 2.48   *Minimization does not aid clarity: (a) desk controls; (b) desk wiring; (c) truth-table-derived program; (d) minimal program*

2.9.2   Event-driven logic

The states of outputs in combinational logic are determined solely by the input signals. In event-driven logic (also known as a sequencer) the state of an output depends not only on the state of the inputs, but also on what was occurring previously. It is not therefore possible to draw a truth table from which the required logic can be deduced.

Consider, for example, the simple motor starter circuit of Figure 2.49(a). With neither button pressed, the motor could be running or stopped depending on what occurred last. The operation can be described by Figure 2.49(b), which is known as a state transition diagram (often shortened to state diagram).

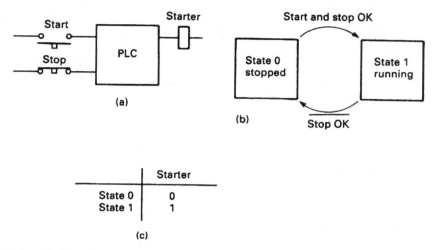

Figure 2.49 *A state transition diagram: (a) motor starter; (b) state transition diagram; (c) output table*

The square boxes are the states the system can be in (the motor can be running or stopped) and the arrows are the transitions that cause the system to change states. If the motor is running, pressing the stop button will cause the motor to stop. A bar above a signal (e.g. $\overline{StopPBOK}$) means signal not present; note the wiring of the stop PB and the signal sense. It is a useful convention to label states with numbers and transitions with letters.

State transition diagrams can be constructed from storage elements, with one less storage element than there are states, the one default state being inferred from the absence of others. It therefore requires just one storage element (e.g. latch or SR flip-flop) to implement the motor starter of Figure 2.49.

Figure 2.50 is a more complex example (based on a real lime silo). A preset weight of lime is fed into a weigh hopper ready for the next discharge, which is initiated (not surprisingly) by a discharge pushbutton. A hood then lowers (to reduce dust emissions) and the lime discharges. After the discharge, the hood retracts and the weigh hopper refills. An abort pushbutton stops a discharge, and a feed permit switch stops the feed.

There are two fault conditions; failure to get the batch weight in a given time (probably caused by material jamming in the feeder) and failure to get zero weight from the discharge (again in a given time and again probably caused by a material jam). Both of these trip the system from automatic to manual operation to allow the cause of the fault to be determined.

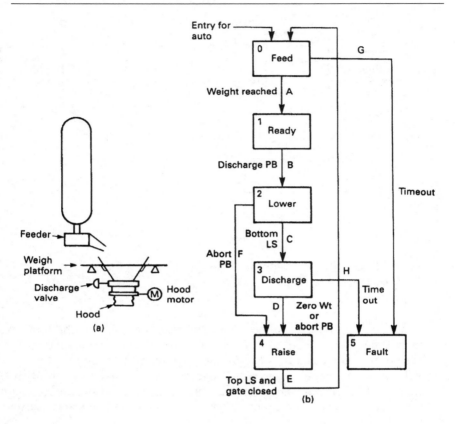

Figure 2.50 *State transition diagram for a real plant: (a) plant layout; (b) state transition diagram; (c) output table*

	State	Feed	Lower	Raise	Disch	Auto Permit	Ready Lamp	Fault Lamp
Feed	0	1	0	0	0	1	0	0
Ready	1	0	0	0	0	1	1	0
Lower	2	0	1	0	0	1	0	0
Discharge	3	0	0	0	1	1	0	0
Raise	4	0	0	1	0	1	0	0
Fault	5	0	0	0	0	0	0	1

(c)

We can now draw the state diagram of Figure 2.50(b). It is good practice, but not essential, to label the common, normal route with successive numbers (for states) and letters (for transitions). The default state is the state that the system will enter from manual, and care needs to be taken in its selection. Here feed is the sensible choice; if the hopper is already full the system will immediately pass to state 1 (Ready); if not, the hopper will be filled. The choice of any other state as default could lead to a wasted cycle through all the states with no material in the weigh hopper.

The definition of the transitions needs care because parallel routes are normally not allowed. If transition A was defined as 'Feed Complete' and transition G as 'Time Out OR Not Feed Permit', the system would work correctly, but inevitably two signals will one day occur together causing anomalous operation of the plant and great embarrassment for the programmer.

The correct definition of transition A is

Feed Complete AND Not Time Out AND Feed Permit

and for transition G

Time Out OR Not Feed Permit

This gives the fault transition priority over the normal transition. Similar considerations apply to transitions F, D and H.

We can now construct a table linking the outputs to the states. This is straightforward and is given in Figure 2.50(c).

The next stage is to translate this state diagram into a PLC program. The steps so far are common to all methods of PLC programming. We will produce the complete program in ladder format but each operation could be produced in equivalent logic format.

The program relies very much on the idea of the program scan, described in Section 2.2. By breaking down the program for our state diagram into four areas as in Figure 2.51 we can control the order in which each stage operates. The actual layout of Figure 2.51 is not critical, but it is essential for transition and states to be kept separate and not mixed. A formal layout of a state-diagram-based program is one of the fundamentals of EDDI, described in Section 8.5.7.

Automatic/manual selection comes first; this is achieved with the simple rung of Figure 2.52. Automatic mode is only allowed if there are no faults and the hood is raised.

Next come the transitions, the first three of which are shown in Figure 2.53. These are straightforward and need little comment. Note that the first contact in each rung is a state, so inputs are only examined at the correct point in a sequence.

The states themselves are given in Figure 2.54. With the exception of state 0, simple latches have been used throughout for the states and for the auto/manual selection so that after a power failure the system will resume in manual mode. Note that these are set and reset by the transitions.

Finally we have the outputs as in Figure 2.55. An output is energized during the corresponding state(s) in automatic or from a manual maintenance pushbutton in manual.

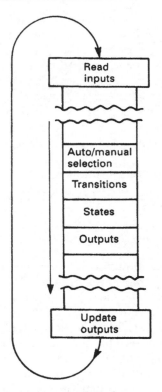

Figure 2.51   *The program scan and state transition diagrams*

We have described the basic ideas of transition, states and outputs in ladder diagram form. The method is equally easily implemented in logic notation.

The state diagram technique is very powerful, but it can lead to confusion if the basic philosophy is not understood. The often-quoted argument is that it takes more rungs or logic elements than a direct approach

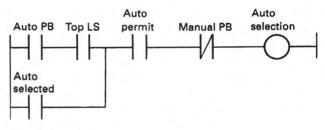

Figure 2.52   *Auto/manual selection*

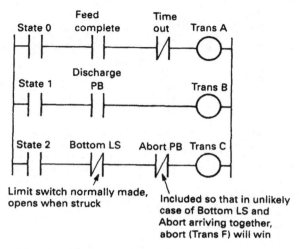

Figure 2.53 *The first three transitions*

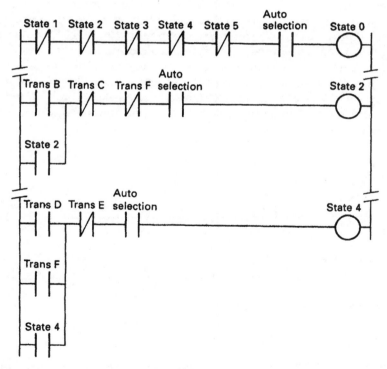

Figure 2.54 *Three of the six states*

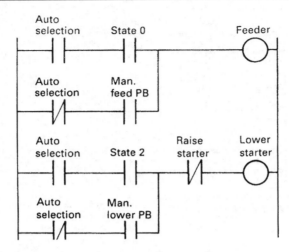

Figure 2.55   *Two of the plant outputs*

programmed around the outputs. This is true, but programming around the outputs can lead to very twisted and difficult-to-understand programs. Figure 2.56 is one rung roughly corresponding to state 2 of our state diagram. It mixes manual and automatic operation and its action is by no means clear (it is known as spaghetti programming). Problems can arise where transitions go against the program scan, like transition E in Figure 2.50(b). If care is not taken, a sequence based purely on outputs can easily end up doing two things at once, or nothing at all because of the way in which the program scan operates.

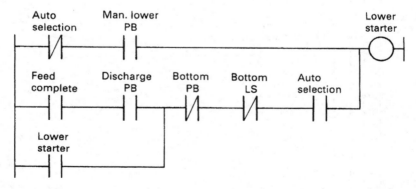

Figure 2.56   *An example of spaghetti programming approximating to state 2*

Modifications are also tricky with a direct approach, but simple with a state diagram. Suppose (as happened on the real plant) it is required to add a dust extraction fan to reduce emissions. This should work during the discharge and for about one minute after the discharge ends before the hood is raised.

The new state diagram is shown in Figure 2.57. All that is needed is a new state 6 (Post Run Fan). (In practice, the states and transitions

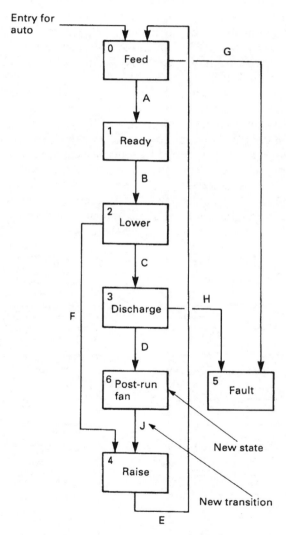

Figure 2.57 *Modifications to a state transition diagram*

should be relabelled to keep a sequential order for state numbers and transition letters, but we are showing it in the modify/test situation.) The fan is to run in state 3 or state 6. The program changes are therefore simple; a new transition J and a new state 6, modifications to state 0 (adding –]/[– for state 6) and state 4 (entry via J rather than D), adding a 1-min timer (for transition J) and a new output for the fan starter.

State diagrams are being formalized by the International Electrotechnical Commission and the British Standards Institute, and already exist with the French standard Grafset. These are basically identical to the approach outlined above, but introduce the idea of parallel routes which can be operated at the same time. Figure 2.58(a) is called a divergence; state 0 can lead to state 1 for condition s *or* to state 2 for condition t with transitions s and t mutually exclusive. This is the form of the state diagrams described so far.

Figure 2.58(b) is a simultaneous divergence, where state 0 will lead to state 1 *and* state 2 simultaneously for transition u. States 1 and 2 can now run further sequences in parallel.

Figure 2.58(c) again corresponds to the state diagrams described earlier, and is known as a convergence. The sequence can go from state 5 to state 7 if transition v is true *or* from state 6 to state 7 if transition w is true.

Figure 2.58(d) is called a simultaneous convergence (note again the double horizontal line); state 7 will be entered if the left-hand branch is in state 5 *and* the right-hand branch is in state 6 *and* transition x is true.

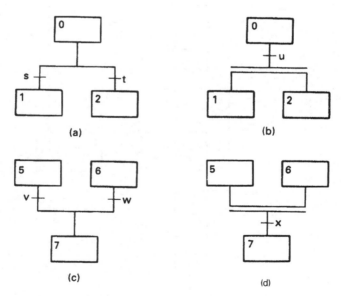

(a)   (b)   (c)   (d)

Figure 2.58 *Grafset symbols: (a) divergence; (b) simultaneous divergence; (c) convergence; (d) simultaneous convergence*

The state diagram is so powerful that most medium-sized PLCs include it in their programming language in one form or another. Telemecanique give it the name Grafcet (with a 'c'), and others use the name Sequential Function Chart (SFC, Allen Bradley) or Function Block (Siemens). We will return to these in the next chapter.

Even the simple Mitsubishi F2 supports state diagrams with its STL (Stepladder) instruction. These have the prefix S and can range from S600 to S647. They have the characteristic that when one or more are set, any others energized are automatically reset. A RET instruction ends the sequence. The state diagram of Figure 2.59(a) thus becomes the ladder diagram of Figure 2.59(b) which would be programmed for the first few instructions

```
LD    X   400
S     S   601
STL   S   601
OUT   Y   431
LD    X   401
S     S   602    etc.
```

Where there are no branches and the sequence is a simple ring (operating rather like a uniselector), a sequence can be driven by a counter which selects the required step. The counter is stepped when the transitions for the current step are met. The GEM-80 has a SEQR (sequence) instruction which acts as a 16-step uniselector.

The PLC-5 has two instructions which fulfil the same role. These are called a Sequencer Input (SQI) and Sequencer Output (SQO) and are controlled by a counter which gives the current step (or state). Each instruction has a table with one row corresponding to each step (state) number. For the SQI the table holds the inputs corresponding to the required transitions to exit each state. For the SQO the table holds the pattern of outputs to be energized in each state. The SQI output steps the counter in the SQO when the inputs corresponding to the current state occur as shown in Figure 2.60. Although the SQI and SQO give very compact programs, the fact that the controlling data are only visible in table form can, in the author's opinion, make fault finding a little cumbersome.

2.10 Micro PLCs

A recent innovation has been the introduction of very small PLCs with a limited number of inputs and outputs. These have been designed for applications such as greasing, heating and air conditioning systems where the programs are written once then installed and sold as part of

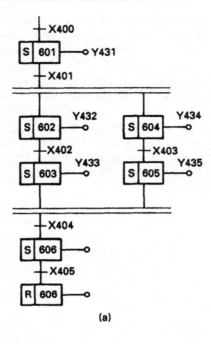

(a)

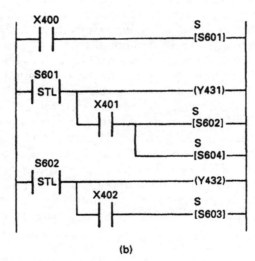

(b)

Figure 2.59   *State diagrams on the Mitsubishi F2: (a) state diagram (based on Grafset); (b) part of ladder diagram corresponding to the start of (a)*

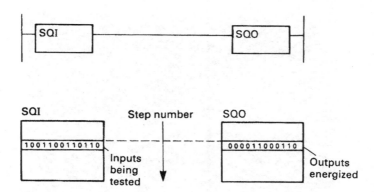

Figure 2.60 *The Allen Bradley sequencer instructions*

the final product or system. In many cases the end-user will not be aware that a PLC is controlling the system. The low price of these micro PLCs (under £100 at the time of writing) makes them very cost effective even compared to one or two relays.

The Siemens LOGO! shown in Figure 2.61 is a typical micro PLC. The unit shown has six inputs and four outputs. A range of LOGO! PLCs are available with different numbers and forms of input and outputs.

The LOGO! is programmed using function blocks, which can be considered to belong to three groups:

- *Connections (Co)*
  These cover Inputs (I1–I6), Outputs (Q1–Q4), Block outputs (B01, B02, etc.) and fixed High and Low signals.
- *General functions (GF)*
  Standard logic functions are covered here, such as AND, OR, NAND, NOR, XOR and Inverters.
- *Special functions (SF)*
  There are 11 special functions including the usual types of timer (delay on, delay off and retentive, etc.) plus set–reset flip-flops, counters and a very useful real time clock with adjustable cams. Figure 2.62 shows a clock used for an application such as air conditioning, where the cams have been set to make an output to come on between 9 a.m. and 4 p.m. on Mondays to Fridays, 10 a.m. to 1 p.m. on Saturdays and not at all on Sundays. The display in Figure 2.61 is showing a clock block with its three individually programmed cams. The outputs from several independent clocks can be further combined using the GF logic gates.

Figure 2.61   *Photograph of a LOGO! PLC. All inputs and outputs connect directly to the PLC and no separate cards or power supply are required. Photograph courtesy of Siemens*

Unusually, the program entry starts at an output. The programmer selects what the output is fed from by first selecting Co, GF or SF then the specific type. The inputs from this block are then selected and so on until the full logic sequence for the output has been built. The logic for each output is constructed in a similar manner.

Figure 2.63 shows a simple circuit for automatically opening and closing a door. Typical applications are found at supermarkets, stores and garages. The door is opened when motion detectors I1 or I2 detect movement on either side of the door. Outputs Q1 and Q2 are connected to the door open and close pneumatic solenoids. Limit switches I4 and I5 say, respectively, the door is fully open or fully closed. These remove the corresponding outputs when the door is fully open or closed, thereby saving air. Note that each output inhibits the other to prevent both solenoids being energized at once in the event of a fault.

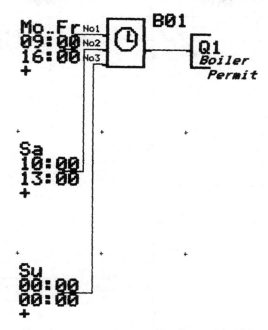

Figure 2.62 *A simple program using the clock function. The LOGO! is competitively priced with a conventional cam timer*

The 'x' connections to blocks B03 and B04 mean these block inputs are unused. The 'R' connection on the timer is a reset, and the standard OR gate has three inputs, only two of which are used in this application.

Block B03 is a delay off timer. Its output comes on immediately when any motion is detected on either side of the door, and stays on for 10 seconds when motion ceases. Note that blocks B03 and B04 are used in both output circuits. The circuit for output Q1 was built in its entirety. Output Q2's circuit was only built back to the inverter B06 whose input was then selected as Co (for connection) followed by B03 (denoting output from existing block B03).

The LOGO! can, surprisingly, be programmed easily from just the six buttons on its face. It can also be programmed offline on a normal PC. Once debugged, the program is usually stored in an EPROM plugged into the front of the PLC.

2.11 IEC 1131-3, towards a common standard

PLCs can be programmed in several different ways. In recent years the International Electrotechnical Commission (IEC) have been working

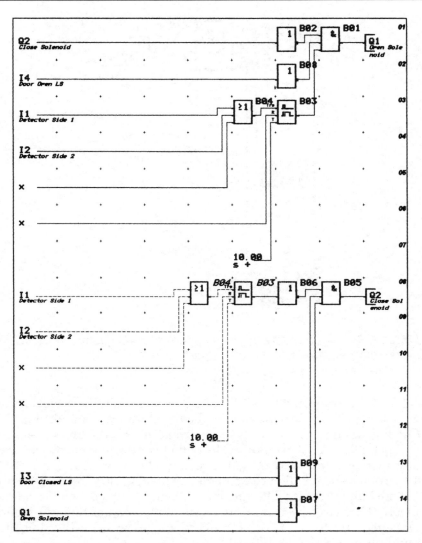

Figure 2.63 *A LOGO! program for controlling an automatic door. This program is based on a standard Siemens circuit*

towards defining standard architectures and programming methods for PLCs. The result, published in Spring 1993, is IEC 1131-3, a standard-ized approach which will help at the specification stage and assist the final user who will not have to undergo a mind-shift when moving between different machines. IEC 1131 parts 1 and 2 cover aspects of hardware design.

The earliest, and probably still the commonest, programming method described in IEC1131-3 is the Ladder Diagram (or LD in IEC1131).

Function Block Diagrams (FBDs) use logic gates (AND/OR etc.) for digital signals and numeric function blocks (arithmetic, filters, controllers, etc.) for numeric signals. FBDs are similar to PLC programs for the ABB Master and Siemens SIMATIC families. There is a slight tendency for digital programming to be done in LD, and analog programming in FBD.

Many control systems are built around State Transition Diagrams, and IEC 1131-3 calls these *Sequential Function Charts* (SFCs). The standard is based on the French Grafset standard shown earlier in Section 2.9.2 and Figure 2.58.

Finally there are two text based languages. *Structured Text* (ST) is a structured high level language with similarities to Pascal and C. *Instruction List* (IL) contains simple mnemonics such as LD, AND, ADD, etc. IL is very close to the programming method used on small PLCs described earlier in Section 2.4.4, where the user draws a program up in ladder form on paper, then enters it as a series of simple instructions.

Figure 2.64 illustrates simple examples of all these programming methods. The next four figures show how IEC1131 is used in modern programming software.

Like most programming software Siemens S7 is IEC1131 compliant and Figure 2.65 shows the same simple instructions in LAD, FBD and IL formats.

The next three examples are from Rockwell's ControlLogix software. Figure 2.66 is a more complex FBD diagram for a tank level alarm, Figure 2.67 shows structured text programming and finally an example of SFC programming is given in Figure 2.68.

A given project does not have to stick with one method; they can be intermixed. A top level, for example, could be an SFC, with the states and transitions written in ladder rungs or function blocks as appropriate.

The aim of IEC1131-3 is to break the link between a PLC program and a manufacturer's specific hardware. It should allow, in principle, a PLC program written for, say, a Siemens PLC to be transferred to an ABB PLC with little effort. In practice, what seems more likely to occur is that PLC programs will be written on personal computers in the non-specific IEC1131 format then converted by PC based application software to a PLC manufacturer's specific format. This approach will be of great benefit to manufacturers of fairly standard PLC controlled machinery whose customers, understandably, specify the type of PLC required to ensure standardization at their site. An IEC1131 conversion program allows the machinery manufacturers to write the PLC program only once, and simply convert it for each customer's PLC.

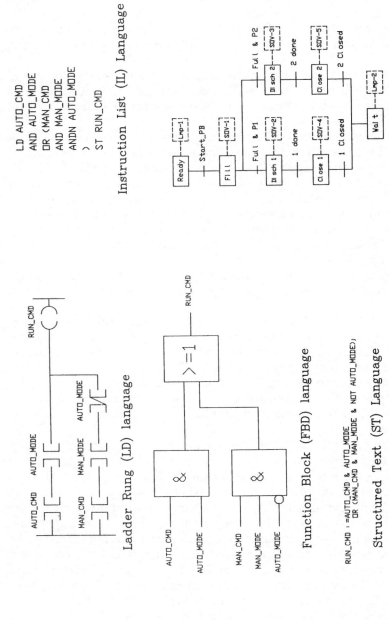

Instruction List (IL) Language

```
LD AUTO_CMD
AND AUTO_MODE
OR <MAN_CMD
AND MAN_MODE
ANDN AUTO_MODE
>
ST RUN_CMD
```

Sequential Function Chart (SFC) Language

Ladder Rung (LD) language

Function Block (FBD) language

Structured Text (ST) Language

```
RUN_CMD : =AUTO_CMD & AUTO_MODE
          OR <MAN_CMD & MAN_MODE & NOT AUTO_MODE)
```

Figure 2.64   The five programming methods defined in IEC1131

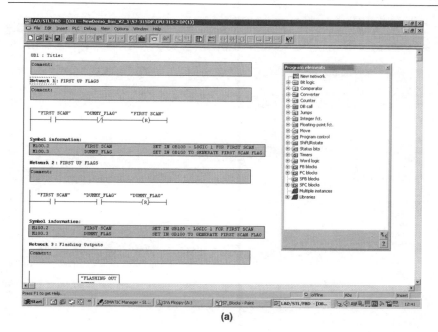

(a)

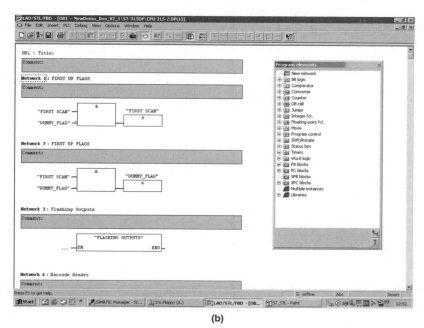

(b)

Figure 2.65   *Example of IEC1131 programming using Siemens S7 programming software: (a) Ladder (LD) programming; (b) Function Block Diagram (FBD) programming*

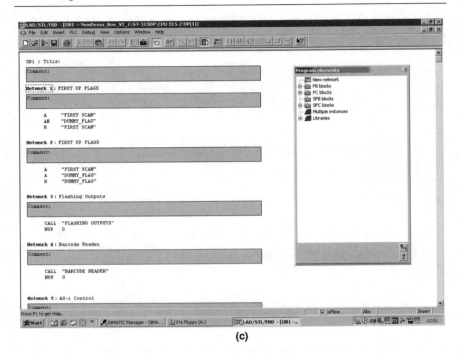

(c)

Figure 2.65   *(Continued) (c) Instruction List (IL) programming*

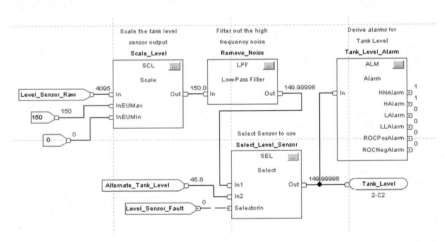

Figure 2.66   *More complex FBD programming using Rockwell ControlLogix software. This produces a tank level alarm*

It will be interesting to see the effect of IEC 1131-3. Most attempts at standardization fail for reasons of national and commercial pride. MAP, and latterly Fieldbus, have all had problems in gaining wide acceptance. A standard will be useful at the design stage, and could be accepted by

```
(* Wait for operator to select product on HMI *)

IF HMI_Select_P23_Frothy THEN
    (* Check that we have enough Raw materials *)
    IF (P23BulkTank.Level >= QuantityRequested/24.356) and Frothavailable THEN
        (* OK, select the product *)
        Product:=P_23_Frothy;
        CASE P23_Consistency OF
            1 : MixSpeed:=123;
                WaterAdd:=0;
            2 : MixSpeed:=150;
                WaterAdd:=24;
            3 : MixSpeed:=175;
                WaterAdd:=45;
        END_CASE;
    ELSE
        (* Indicate error on HMI *)
        HMI_IndicateErrorInSelection:=42;
    END_IF;
ELSIF HMI_Select_F29_Smooth THEN
    (* Check that we have enough Raw materials *)
    IF (F29BulkTank.Level >= QuantityRequested/12.554) THEN
        (* OK, select the product *)
        Product:=F_29_Smooth;
    ELSE
        (* Indicate error on HMI *)
        HMI_IndicateErrorInSelection:=43;
    END_IF;
ELSE
    (* No product selected yet *)
    Product:=0;
END_IF;
```

Figure 2.67 *Structured Text (ST) programming with Rockwell ControlLogix software*

the end user if programming terminals presented a common face regardless of the connected machine. It is to be hoped it doesn't act as a brake on design ingenuity and inhibit development.

2.12 Programming software

Originally PLC manufacturers provided dedicated program terminals which were specific to their PLCs. The Allen Bradley keypad shown earlier in Figure 1.14 is typical.

As portable computers became cheaper and more readily available, manufacturers moved to providing software which can run on a standard portable PC. Often the link to the PLC is via a simple standard RS232 point-to-point link using COM1. Where the PLC and PC communicate via a multi-station network some form of driver is required either as an internal card in a spare PCI or ISA slot or via an external device plugged into the PCMCIA port.

The programming terminal will have a hard life and care should be taken to ensure it is sufficiently rugged. The author does not like small notebook computers; they are not very robust, and, because of their small size, you end up with external power supplies and PCMCIA

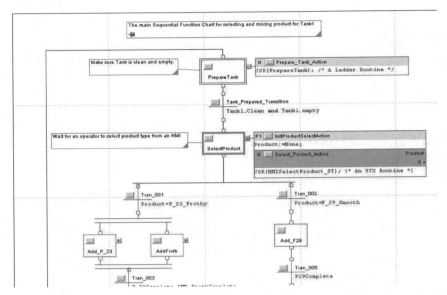

The main Sequential Function Chart for selecting and mixing product for Tank1

Make sure Tank is clean and empty.

PrepareTank

N   Prepare_Tank_Action
JSR(PrepareTank); /* A Ladder Routine */

Tank_Prepared_Transition
Tank1.Clean and Tank1.empty

Wait for an operator to select product type from an HMI

SelectProduct

P1   InitProductSelectAction
Product:=None;

N   Select_Product_Action   Product
0

JSR(HMISelectProduct_ST); (* An STX Routine *)

Tran_001
Product=P_Z3_Frothy

Tran_002
Product=F_29_Smooth

Add_P_23   AddFroth

Add_F29

Tran_003

Tran_005
F29Complete

Figure 2.68   *Sequential Function Chart (SFC) programming with Rockwell ControlLogix software*

devices hanging about outside it. The power supplies on many notebooks will also only operate on 240 V, not the 55/055 sockets found on site in industry. They, can, of course, run on batteries but only for a limited time. Batteries tend to go flat just when they are needed! Notebooks are also very easy to steal. A ruggedized industrial 'luggable' will accept internal ISA or PCI cards and run on any supply voltage. Being large and heavy they are much less attractive to thieves.

Early versions of PC based software ran under MSDOS, and many still do. DOS based software is simple, fast, robust and does not put heavy demands on the processor. It is quite feasible to run DOS programming software on a 286 machine with a 20 MByte hard drive. A typical example is the Allen Bradley DOS based AI Series software shown in Figure 2.69. This is based on a "tree" driven by the ten function keys. From the top level, for example, Edit mode is selected by key [F3] then Append by [F2] then the instruction type selected by one of the keys shown.

Nowadays all software is, of course, Windows based. Figure 2.70 shows the current RSLogix5 software. Contacts and coils can be dragged onto the rung being edited. The Tree at the left gives access to data tables, program files and processor functions. Examples of Siemens S7 software were shown earlier in the previous section. Windows based software is more intuitive and visual but, being mouse driven, can be slower to use than DOS software. Using a mouse can be very difficult on

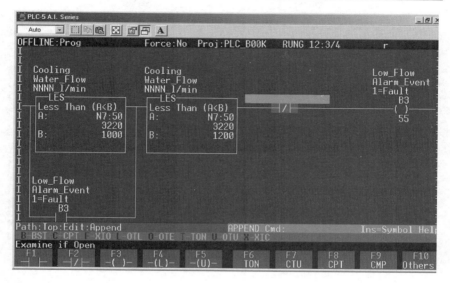

Figure 2.69 *Editing on MSDOS based software. This is driven by function keys*

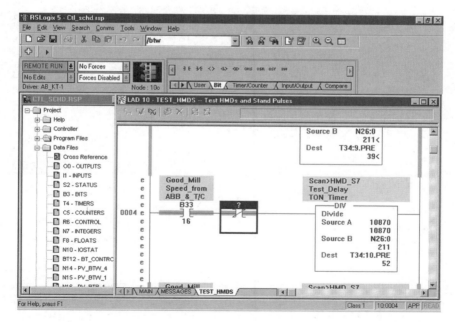

Figure 2.70 *Editing on Windows based RSLogix5 software*

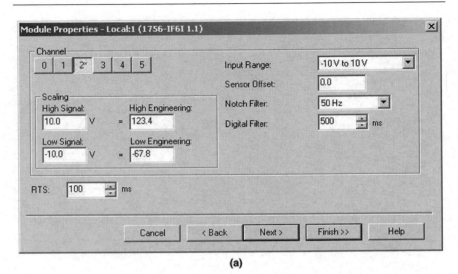

(a)

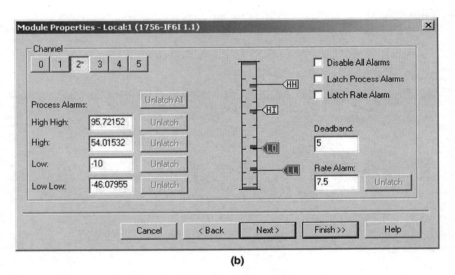

(b)

Figure 2.71 *Simple configuration of an Analog Input Card with ControlLogix programming software: (a) configuration of range to engineering units; (b) setting of alarm levels using mouse dragging*

site with the programming terminal perched precariously on top of a valve skid. Various forms of built-in mice exist on notebooks but these are all a bit awkward. Hand-held mice with a small trackerball work fairly well.

The visual nature of Windows does, though, simplify many previously complex tasks. For example, Figure 2.71 shows the two screen data entry and drag set-up for scaling of an analog input card and alarm setting on one channel of an analog input card. Previously this would have been done with several rungs of program. Modern programming software also comes with many useful diagnostic aids as described in the following section.

2.13 Programming software tools

The programming software is not just used for the obvious task of writing the program, it is also a valuable aid for fault finding. This section describes some of the typical maintenance and fault finding features found on modern programming software.

Once the controlling software has been debugged, it is almost certain that any faults will be related to the plant devices (e.g. limit switches, sensors, solenoids, contactors, etc.). All software will show the state of digital devices on the ladder diagram or function block diagram. Usually the state is shown by a colour change or high-lighting. Similarly the value of numerical signals will be shown to allow numeric signals to be monitored. Figure 2.72 is a typical on-line display, on which the state of each signal, digital and numeric, can be clearly seen. Digital signal which are 'ON' have an emphasized line either side of the contact.

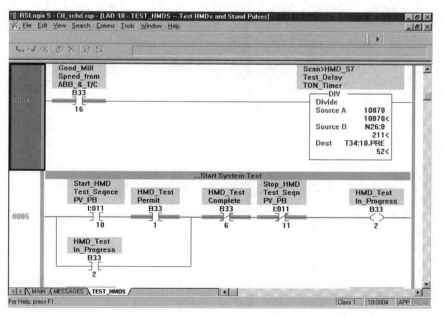

Figure 2.72 *RSLogix online monitoring*

It is also useful to monitor data in a tabular form. In Figure 2.73 display of a bit data table for a PLC5 has been selected allowing the state of several bits to be observed at once. Several windows can be open at once allowing the signal path from an input to an output to be followed without jumping around the program. Custom displays can also be built to collect signals relevant to a particular task. Figure 2.74 shows an example from the Siemens S7 programming software. The monitoring software will also allow values to be written into the program for test purposes.

Most faults on PLC controlled plant will occur with the plant devices. After these, the Input/Output cards are the most vulnerable. If there is a failure in the PLC system the software will identify the cause of the failure, usually down to the point of identifying which device has failed. Figure 2.75 shows the response of the Siemens S7 software to a deliberately induced demonstration fault.

Forces are a common fault finding aid. These allow the state of an input or output to be put into a known state from the program terminal and over-riding the true state of the plant signals. These are used in three circumstances.

The first, used during commissioning, allows the software to be tested without the input and output devices being present. A more common

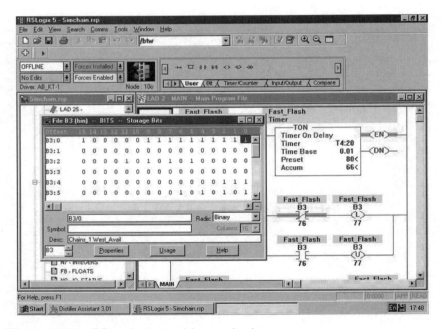

Figure 2.73   *RSLogix data table monitoring*

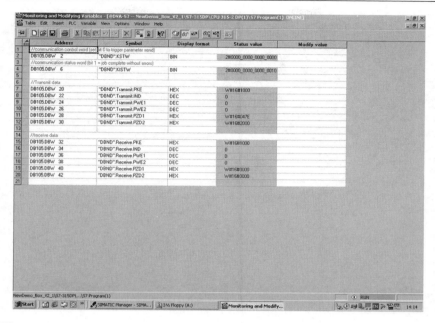

Figure 2.74 *Data monitoring with Siemens S7 programming software*

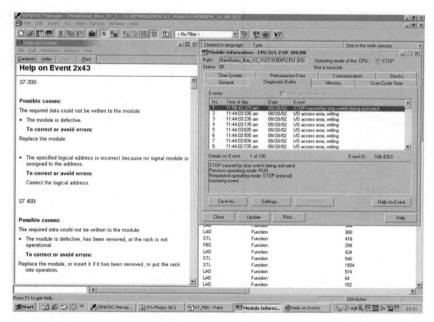

Figure 2.75 *Processor diagnostics on Siemens S7 programming software*

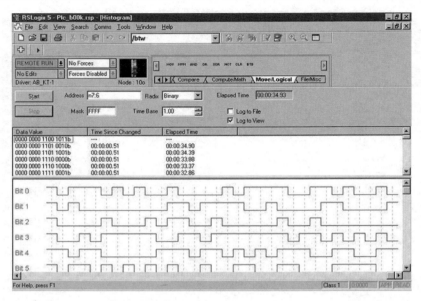

Figure 2.76   *RSLogix Histogram*

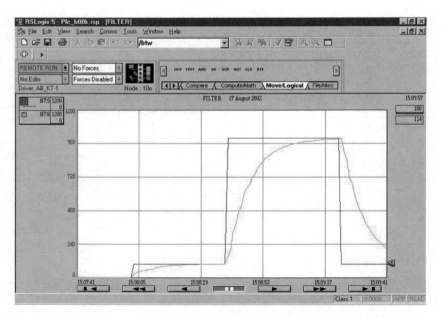

Figure 2.77   *RSLogix Trending. The screen shows the operation of the filter from Section 9.5*

use, though, is during fault finding where outputs can be turned on directly from the terminal for test purposes. For example, a hydraulic loading valve output signal could be forced on to see if the pump pressure rises to the correct value. Similarly a sequence start could be initiated by forcing a material present photocell input signal. Finally, in extreme circumstances, forces can be used to temporarily allow a plant to continue to run where some plant device has failed.

Forces are very useful but are also very dangerous. Usually the programming terminal is remote from the plant and the controlled devices cannot be seen. An applied force can often have unexpected results, particularly if the plant is in a fault condition. The possible implications of a force should always be carefully considered before the force is applied. Warning tape should be put around any plant items that may move without warning and lookouts employed if the plant cannot be seen from the programming terminal.

Temporary forces to overcome plant failures may be acceptable in the short term if there are no safety implications, but often the only person who knows about the force is the person who installed it and the plant runs without some protection for months or maybe even years. A log book of forces should be kept and the reason for the installation of

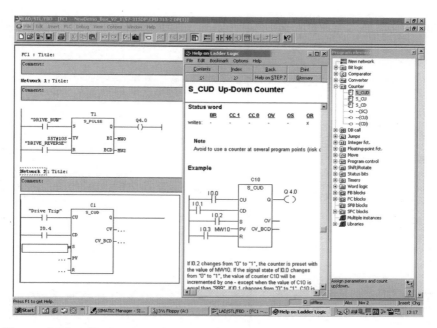

Figure 2.78   *On line help for a counter instruction in Siemens S7 programming software*

forces checked. If a particular sensor is failing regularly look at the design. Is some form of mechanical protection required for the sensor or its cabling? Is the sensor really needed at all?

Intermittent faults can be very difficult to trace. Programming software often includes trending facilities allowing signal states to be observed. Figure 2.76 shows the Histogram feature with RSLogix5. This has one window showing the value in numeric form with the time at every change. The data can be displayed in any radix, binary has been chosen here. Below, the data is shown graphically. The data can also be stored to a file and viewed later; very useful for chasing intermittent faults.

A trend feature is also provided for numeric data. Figure 2.77 shows trends for the first order filter described in detail in the later Section 9.5.

Modern PLCs are very powerful and consequently have vast manuals which describe all the features. Pull down Help is therefore included. Usually this is context sensitive, for example clicking on a counter in the Siemens S7 software then clicking on Help will bring up a window describing the counter instruction and how it is used as shown in Figure 2.78.

3 Programming style

3.1 Introduction

'...and the hydraulic system will have three hydraulic pumps plus an oil circulation pump'. So ends a typical specification for a control system. Like most specifications this simple statement leaves many unanswered questions; are all three pumps to be run, or just one, or two? If less than three, how is (are) the duty pump(s) to be selected? If less than three pumps are used, are unused pump(s) to act as standby with automatic changeover? How are pumps started, individually or all together? Does the emergency stop operate on all (if not why not?). Does the circulation pump start with the main pumps, or is it a precondition for starting the main pumps? If the circulation pump trips, should the main pump(s) trip? What protective signals are there (e.g. temperature/level)? If none, why not, and are you sure? Should these stop the pumps or merely produce alarms? Often such questions will reveal that the suppliers have thought only of what equipment is needed, and not how it is to be used.

The designer of a PLC system has to produce a program which fulfils these often poorly defined requirements. Allied with this is the need to assign I/O to plant signals and operator controls, and decide how the all-important link between human beings and the plant is to be performed.

Programmers involved with commercial software have similar problems, and have coined the term 'Software engineering' to describe how a software project goes from the user's original (and probably imprecise) ideas to a successful working system. In this chapter we will examine the factors that need to be considered in the design of a PLC control system.

3.2 Software engineering

Figure 3.1 shows the six stages that any software project must go through during its life. Although few projects are compartmentalized as neatly as this, the principles apply to all.

The first stage is analysis of the problem that is to be solved. The supplier/programmer of the PLC system must meet with the other contractors and the user to determine what controls are needed and how the control actions are to be provided. Important considerations such as operator controls need to be established at this stage. Ambiguous descriptions (such as the hydraulic pumps of Section 3.1) should be resolved.

Of all the stages, analysis is the most difficult, as the ultimate end-user and the other contractors probably have not considered the intricacy of the control strategy, and do not have the experience to decide if an item of plant is best controlled with joysticks, pushbuttons or a touchscreen VDU.

An important point which is often overlooked at this stage is the need to provide some form of manual 'maintenance' controls to test, or rescue, a fully automated plant or sequence which has failed in some obscure manner.

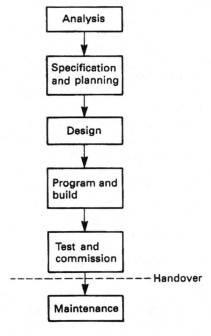

Figure 3.1 *The stages of a project*

The output from the analysis stage should be a description of how the plant works, what operator stations and controls are needed (and how these are to be implemented), what maintenance/fault-finding aids and facilities are to be included and finally (but not least) a complete list of the I/O signals with voltage/current specifications and their locations on the plant.

The difficulties (and the importance) of this first stage cannot be over-emphasized. If the ambiguities and problems are resolved at the start, the following stages are easy. Finding out at the commissioning stage that the user wanted variable speed fans and an underpressure alarm and 'thought you knew that' is not the way to ensure a smooth plant start-up. If in doubt, ask; even if you are not in doubt, still ask, and assume nothing.

At this stage, the final testing requirements should also be defined. If you do not know how you are going to test it, how will you know if the plant meets the user's requirements?

With the worst stage over, the designer should produce a description of what the control system contains, how it is going to perform and how it will be tested. This is really recording what was agreed at stage 1.

The next stage is to design the system; the cubicles, desks, and the structure of the program. This latter action, known as top-down design, is considered in the following section.

At last the programming can be done, built around the structure laid down at the design stage. No program should be constructed *ad hoc* at the keyboard; that way lies spaghetti programming. Commercial programmers estimate that this stage generally involves no more than 10% of the total effort.

With the programming completed and the plant built, testing and commissioning can start. The operation should be checked against the specifications produced at stage 2. With all bar the simplest system, it can be very time consuming to check all routes and actions given in the specifications. There is generally pressure to 'hand over' the plant when the basic operation has been tested but the ancillary, rarely used, options are untried. Too often these tests are skipped, and the first time a 'firkling fault' mode is tested is when the 'firkling fault' first occurs, possibly years after the plant has started up. Inevitably, commissioning of the control system will always be the last stage in a new plant, so the control engineer ends up carrying everyone elses' delays. It is therefore important to establish what testing *must* be carried out before a plant can start and what can be tested later, on line. On line testing, however, can be very difficult and time consuming.

Safety-related checks should never be skipped; finding out that an emergency stop sequence does not work when it is used for the first time in an emergency will ensure a visit from the Health and Safety Executive.

The final stage is usually overlooked. Once the plant is handed over, its control system must be maintained, a term used here not to mean serviced in the mechanical sense, but covering fault finding, resolving of bugs ('we never meant it to work like that') and (hopefully minor) changes arising from modifications in the way the plant operates. No plant is fixed, all change during their life in response to market or technology changes, and these modifications require changes in the control strategy.

In commercial programming it is generally thought that maintenance takes over 50% of the effort in a project's life cycle. It is therefore essential that the control strategy and program are constructed and documented so they can be changed and modified easily at a later stage, possibly by people who had no involvement with the previous five stages.

3.3 Top-down design

It is not uncommon for a PLC to contain several thousand ladder rungs or logic segments. An unstructured program of this length can be very difficult to write, and even more difficult to follow for maintenance and fault finding.

The programmer should not, therefore, write a single long program, but break it down into many small program segments. Ideally, each small segment should contain no more than ten ladder rungs or logic elements as this is about the maximum that the human mind can hold at any one time. The structure of these segments is one of the more important aspects of the design stage of Figure 3.1.

The best way of achieving a sensible split is to use a technique called top-down design. This splits a control system into areas, which are subdivided into sub-areas and so on until manageable sizes have been achieved. The idea is best shown by an example. Figure 3.2 shows a plant called a ladle furnace. This is controlled by a single PLC with about 1750 ladder diagram rungs.

The plant control can be broken down into the nine areas of Figure 3.3, each of which can be broken down further, the full structure of the power system being shown. The bottom levels can be programmed in a few rungs. Figure 3.4 shows two of the bottom level blocks and their plant and internal signals. The kWh block, for example, consists of two counters stepped by the 100-kWh pulses from a power transducer when the power is on (an internal signal from another block). Both counters are reset at the start of a treatment, and the sequence counter is reset at the start of a new sequence. The output of the block is two totals used by other blocks for operator displays and automatic control.

The resulting program structure should be recorded as part of the control system documentation, and used as the foundation of the actual

Figure 3.2 *Ladle furnace at Sheerness Steel controlled by a PLC (Courtesy of Sheerness Steel)*

programming effort. One, not immediately obvious, bonus is that a well-laid-out structure chart with signals clearly identified can be easily split amongst several programmers.

3.4 Program structure in various PLCs

In high level computer languages, programmers tend to prefer languages such as Pascal or C, which are inherently structured by their inbuilt constructs, and view 'non-structured' languages such as BASIC or FORTRAN with a certain amount of disdain. To some extent this is unfair; it is possible to write perfectly structured programs in BASIC, but the onus *is* on the programmer, and a Pascal or C program *can* easily degenerate into spaghetti without due care.

Similar observations apply to PLC programs. Many middle-range machines (and all small machines) have no built-in structure elements at all, leaving it to the programmer to decide on the layout and follow a self-imposed discipline. Figure 3.5 shows an index for a small PLC system controlling a three-unit water softener; the program structure is

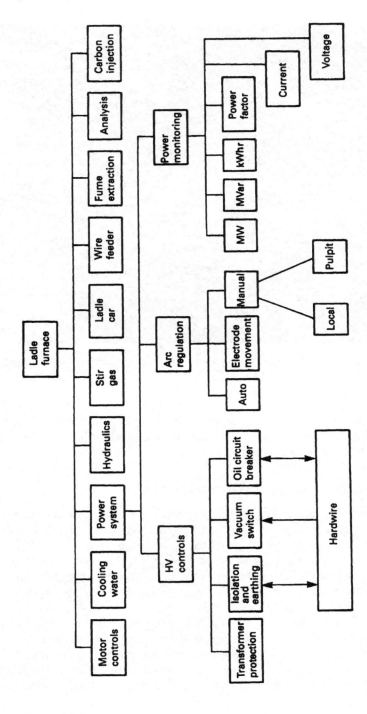

Figure 3.3 Top view of ladle furnace PLC and breakdown of one leg

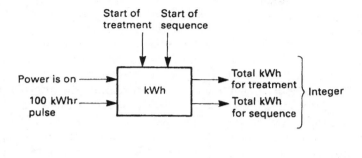

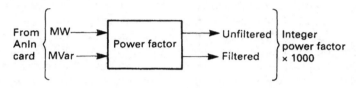

Figure 3.4 *Two blocks from Figure 3.3 with signals*

```
----- Cooling Water System...Water Softeners -----

          Program Listing 6th June 1998

    Page: 1    Rung: 1     Misc Common Control
    Page: 3    Rung: 7     Water Pumps
    Page: 6    Rung: 13    Softener 1 Common
    Page: 7    Rung: 15    ...Transitions
    Page: 10   Rung: 27    ...States Duty
    Page: 11   Rung: 28    ......Backwash
    Page: 12   Rung: 29    ......Brine Inject
    Page: 13   Rung: 30    ......Fast Rinse
    Page: 14   Rung: 31    ......Standby
    Page: 15   Rung: 32    ...Mimic Display Lamps
    Page: 17   Rung: 45    ...Valve Outputs
    Page: 19   Rung: 60    Softener 2 Common
    Page: 20   Rung: 62    ...Transitions
    Page: 23   Rung: 74    ...States Duty
    Page: 24   Rung: 75    ......Backwash
    Page: 25   Rung: 76    ......Brine Inject
    Page: 26   Rung: 77    ......Fast Rinse
    Page: 27   Rung: 78    ......Standby
    Page: 28   Rung: 79    ...Mimic Display Lamps
    Page: 30   Rung: 92    ...Valve Outputs
    Page: 28   Rung: 107   Softener 3 Common
    Page: 29   Rung: 109   ...Transitions
    Page: 32   Rung: 121   ...States Duty
    Page: 33   Rung: 122   ......Backwash
    Page: 34   Rung: 123   ......Brine Inject
    Page: 35   Rung: 124   ......Fast Rinse
    Page: 36   Rung: 125   ......Standby
    Page: 37   Rung: 126   ......Mimic Display Lamps
    Page: 39   Rung: 139   ...Valve Outputs
    Page: 41   Rung: 154   Brine Overtime Alarm
    Page: 42   Rung: 157   Brine Level Alarm
    Page: 43   Rung: 160   Sum  Valve Control
               Rung: 172   Alarm Marshalling

    Page: 46 Data Table Report
```

Figure 3.5 *A well-structured and documented PLC program split into areas of about ten rungs. This makes faults easy to find*

straightforward, and a fault on, say, Unit 2 Fast Rinse, could easily be located in the program.

Larger, and more modern, machines have built-in structure constructs. To some extent these constrain the programmer in the same way that a programmer in Pascal or C has a lot less freedom to make mistakes. These PLCs generally provide methods for breaking the program down into small understandable modules (with some larger machines such as the Allen Bradley 5/250 having the Repeat/Until, Whiledo/Endwhile, For/Next constructs for repeating the same operation on a block of data).

The most structured language is, possibly, the ABB Master, which is similar to a compiled high level language in that all variables and procedures need to be declared. The PLC program is split into one (or more) programs labelled PC1, PC2, etc. It is recommended that each deals with a different area of plant, and each program can have different scan rates.

These programs then contain control modules which can be enabled/ disabled or again run at specified time intervals. Within the control modules there are function modules and sequencer modules, the latter containing steps corresponding to the state diagrams described in Section 2.9.2. The actual logic elements are contained in the function modules or the sequencer steps. A complete program can thus be viewed as in Figure 3.6(a) with a specific example being laid out rather like an MSDOS or UNIX tree as in Figure 3.6(b). The actual structure is more flexible than this description implies; function modules can contain control modules, and a hierarchy of master/slave modules can be built, but the basic idea should be apparent.

Modules are labelled in a hierarchical manner down to the element level, so the AND gate PC1.2.3.2 is the second logic element in function block 3 of control module 2 of program PC1 (and has to be declared as such as part of building the database) (Figure 3.7). This formal nature imposes a discipline on the programmer.

Siemens use a structure consisting of organization blocks (OBs), program blocks (PBs), function block (FBs), subroutine files, which we will discuss shortly, and sequence blocks (corresponding to state diagrams). The basic building modules are the program blocks which equate to the bottom units of top-down design. When first started, the machine commences with organization block OB1 from which further PBs and FBs can be called, as in Figure 3.8. This description is again somewhat simplified.

Allen Bradley use a similar idea with the PLC-5. The programmer can break the top-level program down into smaller program modules which can be called as needed. In the PLC-5, the programs are linked by a sequence function chart (or SFC) which is again very similar to the state transition diagram described in Section 2.9.2. Each state and transition is a small ladder diagram program, a typical example being given in Figure 3.9.

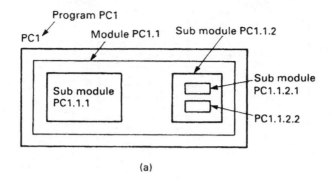

(a)

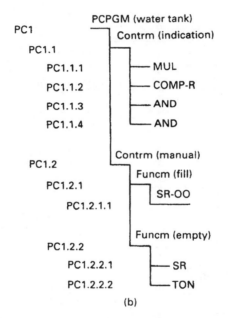

(b)

Figure 3.6 *Structure inside the ABB Master: (a) typical structure of Master program; (b) tree view of program*

Allen Bradley and Siemens both support the concept of subroutines (called function blocks (FB) by Siemens). These are small programs used to perform specific tasks which can be called repetitively by higher level programs (Figure 3.10). For example, few PLCs support trigono-metrical functions (sine, cosine, tangent) directly. It is relatively easy to calculate the sine of an angle using the expansion series

$$\sin x = x - \frac{x^3}{3!} + \frac{x^5}{5!} - \cdots \tag{3.1}$$

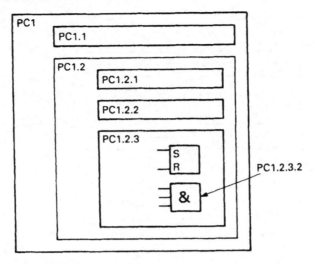

Figure 3.7   *Specifying a logic gate in the ABB Master*

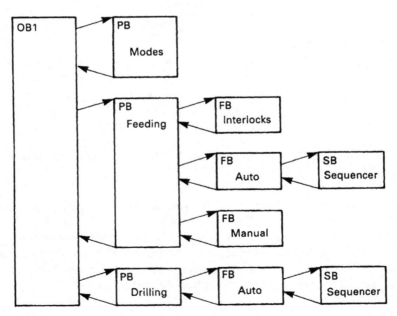

Figure 3.8   *Typical organization inside a Siemens PLC*

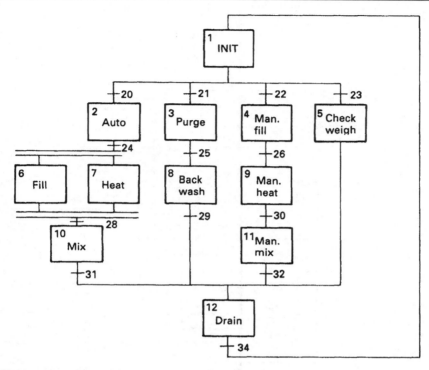

Figure 3.9 *Allen Bradley PLC-5 SFC diagram*

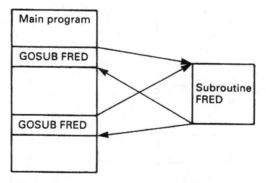

Figure 3.10 *Subroutines, available in many PLCs*

where *x* is the angle in radians. For most applications the first three terms will give sufficient accuracy. The sine of an angle theta (in degrees) could therefore be found with a sequence similar to Figure 3.11 achieved with a few ladder rungs or logic segments. Let us make this

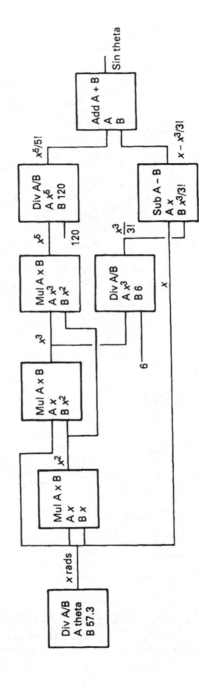

Figure 3.11 *Subroutine for sine (theta)*

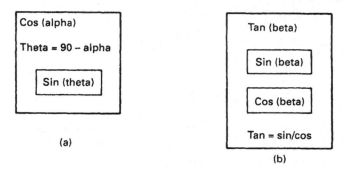

Figure 3.12 *Subroutines Cosine and Tangent (using subroutine Sine): (a) Cosine; (b) Tangent*

a subroutine called Sine (what else?) with an input angle in degrees, and returning the sine to the designated variable as Figure 3.11. The input and output variables are called parameters. Whenever we want the sine of an angle we can now call the subroutine program Sine.

We can go further, though. The cosine of an angle is given by

$$\cos (\text{theta}) = \sin (90 - \text{theta}) \tag{3.2}$$

where theta is in degrees, so we can write another subroutine to calculate the cosine of an angle as in Figure 3.12(a). Note that this only has two blocks, and calls the subroutine Sine (which does most of the work). Finally, by observing that

$$\tan (\text{theta}) = \sin (\text{theta}) / \cos (\text{theta}) \tag{3.3}$$

We can construct a Tangent subroutine as in Figure 3.12(b). This calls both the Sine and Cosine subroutines. A subroutine calling further subroutines is called 'Nesting'.

The advantage of subroutines, of course, is the saving of processor memory and the minimization of programming effort. They also make the program easier to follow, as the maintenance staff or the programmer making changes only need to examine a (possibly complex) routine at one place only.

Many PLCs allow a program to be split into executable blocks; in Allen Bradley PLC-5s, for example, this can be done with an MCR instruction (for Master Control Relay) and in GEM-80s a Start/End Block command can be used. These allow a set of rungs to be skipped (ignored) if the controlling instruction at the start of the block is not true, as shown in Figure 3.13. These instructions again serve to structure the program into small blocks, and also help to improve the scan time by ignoring rungs which are not relevant at the current time. They can,

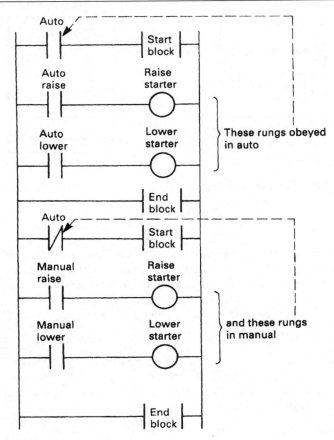

Figure 3.13 *Example of start/end block structure*

however, cause confusion to maintenance staff in the middle of the night as the split of the program is not immediately obvious.

3.5 Housekeeping and good software practice

All computer software (whether data processing, commercial or control) should:

(a) perform its function reliably
(b) behave in a predictable defined manner when the incoming data are faulty (a behaviour described as 'robust' in the jargon)
(c) be simple to understand and maintain

The first of these requirements is obvious, and most PLC software will (hopefully) do the job for which it has been designed. The other two points,

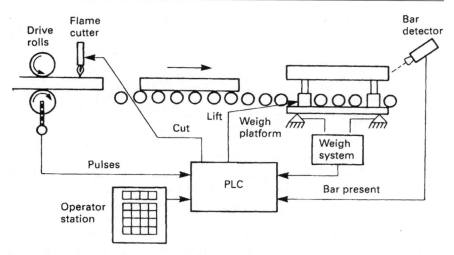

Figure 3.14   *An automatic cutting system*

however, are often overlooked, and their absence may not be apparent until the first time a problem occurs months (years?) after the plant has been commissioned and the design team disbanded.

Robust software has inbuilt protection against bad data from faulty plant sensors or miskeyed operator inputs. Figure 3.14 shows an application based on a system at the author's plant. Material is supplied to a customer's specified weight, but is cut by length, by counting pulses from the drive rolls. The operator enters a desired weight, and the PLC converts this to an equivalent length. The resulting cut material is weighed and checked against the desired weight, any error being used to correct the next cut.

This system contains many places at which bad data can occur; bad operator input giving a ridiculous weight, bad readings from the weigh system, electrical interference on the pulses from the drive rolls, to name but three. Any of these could cause problems if the resultant faulty data were handled as being correct.

It is essential to include some form of checking. The operator in Figure 3.14 can only enter weights within a specified range, and only readings from the weigh system within a window specified as a percentage of the target weight are used for the trimming function. Bad operator input or weights outside the 'window' are flagged as system alarms. Similarly a time window for cuts can be calculated, with pulse-initiated cuts only being allowed within this window and an emergency time cut being initiated (and an alarm signalled) if no pulse cut is given before the maximum time occurs.

Inevitably, robust software is more lengthy and complex; about 25% of the software for Figure 3.14 is concerned with normal operation, the remaining 75% dealing with abnormal conditions that may rarely (if ever) occur. The protection is essential, however, to give operators and production staff confidence in the system.

Programmers take a certain perverse pride in the ingenuity of their efforts and the minimization of the number of instructions used. Such tendencies should be resisted even more than in commercial programming, as the actual plant and program maintenance will be done by people who will require clarity of operation. 'Keep it simple' should be the motto; do not use complex methods and steer clear of the more obtuse instructions available in a PLC instruction set. Remember that some poor individual may have to see how it all works at three o'clock in the morning.

Figure 3.15 shows an example from my experience of how *not* to write a PLC program. This application had one PLC controlling three identical plants. The programmer started by constructing a software multiplexer, effectively a three-way rotary switch, which copied all the inputs for one selected plant to internal storage locations. There was then one program (for all three plants) which worked on the internal storage, and sent its outputs to internal storage again for sending to the outside world via a software de-multiplexer. The multiplexer and de-multiplexer were stepped per program scan, so the program dealt with plant A on scan 1, plant B on scan 2, plant C on scan 3 then back to plant A on scan 4 and so on. This was very clever and economical of memory, but impossible to understand and fault find. In normal operation the operation could not be followed, as all that could be seen on the programming terminal was a blur as the multiplexers cycled between the different plants. When a plant failure occurred, the multiplexers had to be locked into the faulty plant (shutting down the *good* plants) to allow the operation to be observed. Tricks like this should be avoided (and it is worth noting that this particular program was rewritten completely within a year).

The program should also try to reflect the plant operation. Figure 2.48 showed a common situation where either or both of two motors are to be run as selected by a desk switch. For economy, a less than ideal switch has been used. The simplest program is the single rung of Figure 2.48(d) but I would suggest that the two rungs of Figure 2.48(c) make the operation clear to the person encountering it for the first time.

Good documentation is essential for clarity. Most PLCs can be programmed offline on an MSDOS computer, and have the facility for individual signals to be documented and explanatory comments added to explain how the program works. This feature should be used fully;

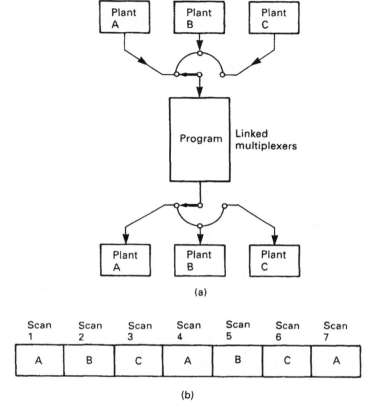

Figure 3.15 *How not to write a PLC program: (a) program layout; (b) program operation*

compare the undocumented program of Figure 8.38 with the annotated version of Figure 8.39.

One of the standard rules for all computer programming is 'when you put some data in memory, record where you have put it'. Failing to do this is like shoving something in a cupboard then being unable to find it a few weeks later. All use of I/O and internal storage should be recorded. Annotation at the programming state helps here; if you have chosen an internal store bit to represent 'Water Overtemperature Alarm' and when you call it up in the program the annotation 'Pump-1 Trip' is attached, you know the same address has possibly been used twice or the memory map is off the rails somewhere.

PLC manufacturers provide store and I/O allocation charts similar to Figure 3.16. These should be used meticulously. MSDOS-based

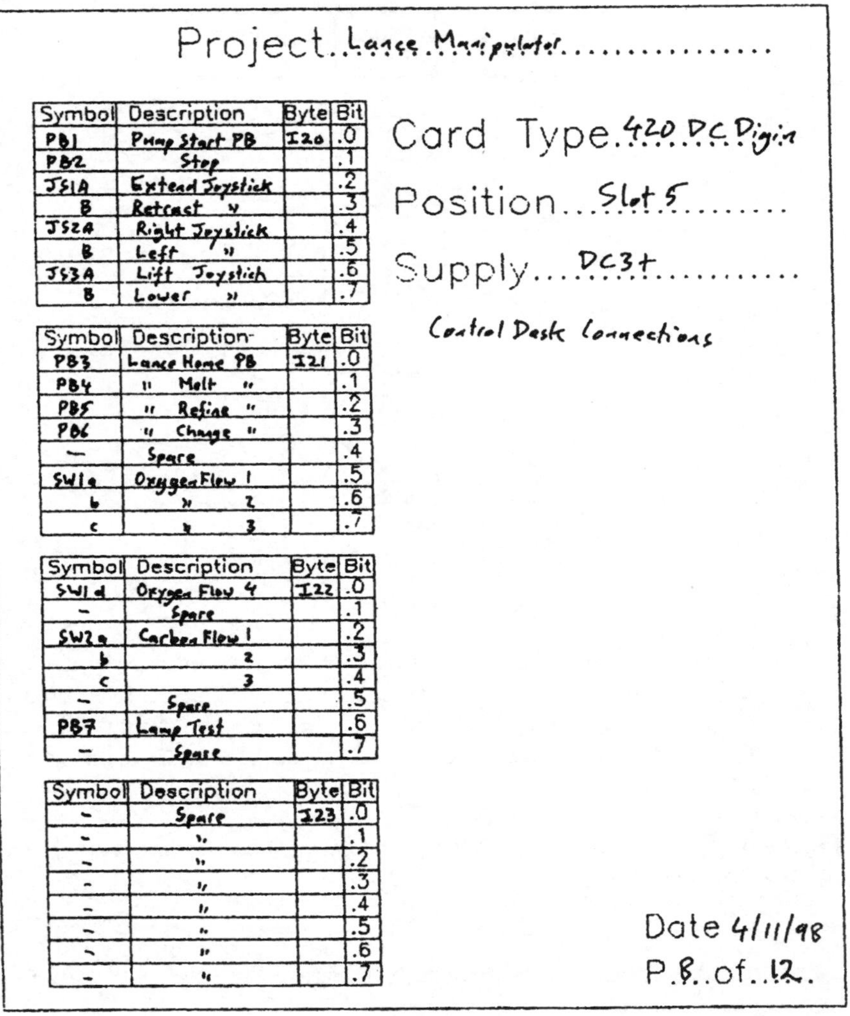

Project..Lance Manipulator.............

Card Type.420 DC Digin

Position...Slot 5........

Supply....DC3+..........

Control Desk Connections

Symbol	Description	Byte	Bit
PB1	Pump Start PB	I20	.0
PB2	Stop		.1
JS1A	Extend Joystick		.2
B	Retract "		.3
JS2A	Right Joystick		.4
B	Left "		.5
JS3A	Lift Joystick		.6
B	Lower "		.7

Symbol	Description	Byte	Bit
PB3	Lance Home PB	I21	.0
PB4	" Melt "		.1
PB5	" Refine "		.2
PB6	" Change "		.3
—	Space		.4
SW1a	Oxygen Flow 1		.5
b	" 2		.6
c	" 3		.7

Symbol	Description	Byte	Bit
SW1d	Oxygen Flow 4	I22	.0
—	Space		.1
SW2a	Carbon Flow 1		.2
b	" 2		.3
c	" 3		.4
—	Space		.5
PB7	Lamp Test		.6
—	Space		.7

Symbol	Description	Byte	Bit
—	Space	I23	.0
—	"		.1
—	"		.2
—	"		.3
—	"		.4
—	"		.5
—	"		.6
—	"		.7

Date 4/11/98
P.8..of..12.

Figure 3.16   *The first stage of a project; hand allocation of I/O*

programming software invariably provides a printout of memory usage, one example of which (for an Allen Bradley PLC-5) is shown in Figure 3.17.

If the interaction of data in a program is complex it is worth producing diagrams of data flow such as that shown in Figure 3.18. Such diagrams help the planning of the software, and assist greatly in fault finding and maintenance.

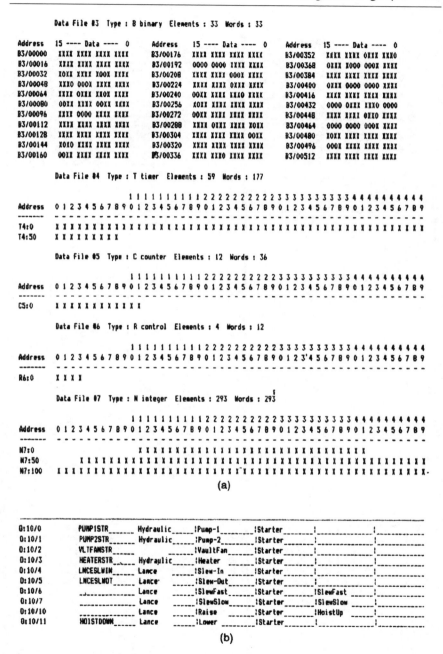

Figure 3.17 *Typical memory usage reports; it can be seen that B3/37 is unused, and the first free timer is T4:59. (a) Database usage; (b) database description*

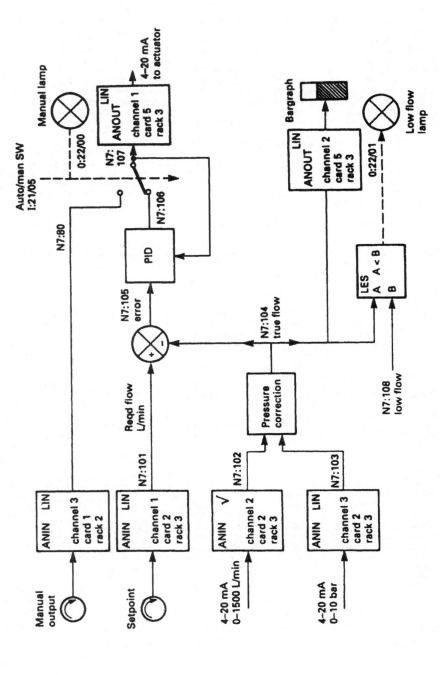

Figure 3.18 PLC program expressed as data flows

It should always be the aim to make the task of the people who inherit the PLC as easy as possible. One way of achieving this is consistency of programming style. If a motor starter has been programmed in a certain way at one point of the program (checking, say, for tripping of the motor protection and operation of the auxiliary contact), this style and method should be repeated for all other motor starters. Particular care over consistency needs to be taken where different parts of a program are being written by different people, and a 'house style' is worth developing. The Ford 'EDDI' concept discussed in Chapter 8 is one example of this approach.

3.6 Speeding up the PLC scan time

A typical PLC scan time will be 10 to 20 ms. This is adequate for most applications linked to contactors, solenoids and similar electromechanical devices. If, however, the PLC is being used with fast moving material the scan time, and more important the variation in the scan time, can be very important. For example, if a PLC is being used to cut material to a set length, and the material is travelling at 10 m/s, a 15 ms variation in scan time will correspond to a 150 mm variation in the length. This section looks at factors which affect the scan time and ways of improving the speed.

Figure 3.19 shows a typical rung. Most PLCs will scan this in the order:

A, B, C, D, E, F, G, H, I,

then update the output. To speed up the scan, though, most PLCs will skip branches once they have found a branch that is true. If signals A and B are true signals C, D, E and F will not be examined and signal G will be tested after signal B.

In the best case, with A, B, G and I all '1', just four contacts have to be examined before the output is energized.

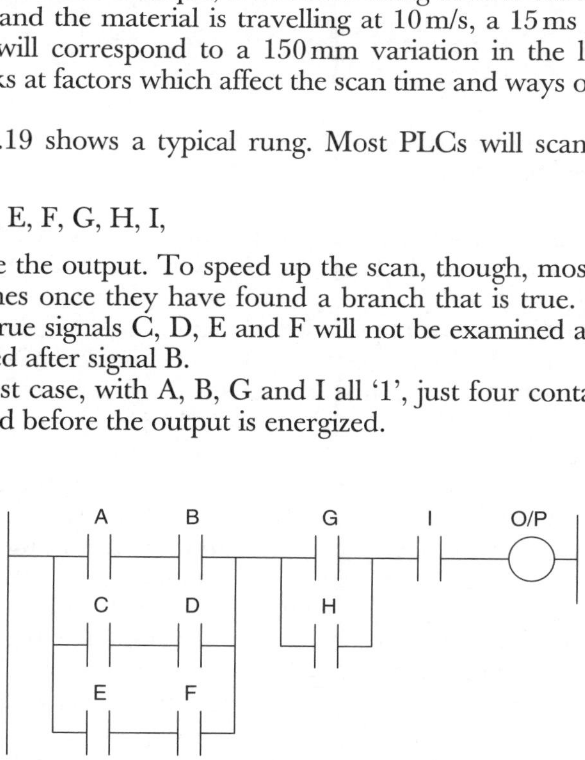

Figure 3.19 *Simple ladder diagram*

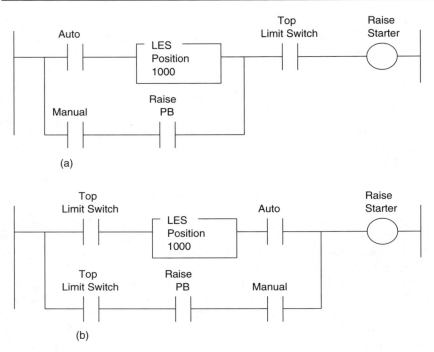

Figure 3.20   *Configuring rungs to improve the scan time: (a) fast operation (b) slow operation*

In the worst case, with A, B, C, D and G all '0' and the remaining signals all '1', all nine contacts have to be examined before the output is energized.

Similar considerations apply when determining if the output is to be de-energized. If A, B, C, D, E and F are all '0', the output can be de-energized immediately without examining the state of G, H or I.

To get fastest speed, therefore, signals which will most probably be '1' or whose presence requires the fastest response should be placed on the top of rungs and nearest the left-hand side. If there is a pair of rungs whose outputs cannot occur together (e.g. Auto Raise and Manual Raise) put the selection signal (e.g. Auto and Manual) as the first contact in each rung so only one contact is examined in the unused rung. Similarly a rung arranged as Figure 3.20(a) will operate faster than Figure 3.20(b). Note that in the slower rung, time has been wasted by repeating the top limit switch signal in each branch.

Repetition of signals is a common error and usually occurs with permissives. The fastest way to handle these is to group all the permissives together once into a permissive store then use this store in each rung with requires the permissive.

Mathematical operations are much much slower than bit operations, and floating point operations are slower than integer operations. If speed is important try to do all mathematical operations using integer numbers, but care obviously has to be taken to avoid rounding errors and overspills. A sixteen bit two's complement number can cover the range −32 768 to +32 767.

Mixed floating point and integer arithmetic should be avoided as these have conversion overheads. For example with the simple arithemetic operation Int2 = Float1 + Int1 the PLC will first convert Int1 to an internal floating point number, add it to Float1 to give a floating result then convert this result back to an integer to go into the integer result Int2.

Many arithmetic operations only have to be done rarely, some only once at power up. A lot of time is wasted if arithmetic instructions are obeyed on every scan. Only do arithmetic operations when they are required.

Many PLCs have control functions which can be used to reduce the scan time. The commonest of these are Jumps which allow unused parts of programs to be skipped as Figure 3.21(a). Subroutine calls can be used in similar way with selected subroutines being called only when required as shown on Figure 3.21(b). Both of these can make major improvements but it should be noted that this can be confusing for people trying to follow the program in the early hours of the morning as rungs which are skipped or live in uncalled subroutines may appear to indicate a PLC fault.

Some PLCs have specific program files which can be allocated a specific scan time. The PLC5 family, for example, can have a Selectable Time Interrupt (STI) program which is obeyed at a fixed time interval which can be as short as 1 ms. With a main program scan of 20 ms, the STI file could operate twenty times for one main program scan. This speed, though, comes at the expense of the main program scan so STI and similar files should be kept as small as possible to avoid dragging the rest of the program files down. Many PLCs, of which the ABB Master is typical, allow scan times to be assigned independently for each and every program file in the processor.

As explained in Section 2.2 and Figure 2.2 a PLC normally goes through a scan which at its simplest level is:

Read Inputs
Obey Program
Update Outputs

Using tricks like STI files is not much use if you have to wait until the end of the main program scan before the outputs are updated. Control functions with names like Immediate Input and Immediate Output are

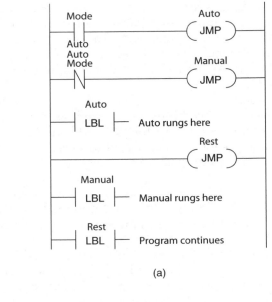

(a)

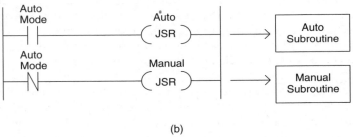

(b)

Figure 3.21   *Methods of sectioning the program to skip unnecessary operations: (a) use of Jumps (JMP) and Labels (LBL); (b) use of Subroutine calls*

therefore often provided which allows an input to be read or an output updated without waiting for the normal beginning and end of the program. An STI or similar file will normally have the structure

Immediate Input <inputs used>
Program
Immediate Output <outputs used>

so the I/O access is as fast as the program scan.

Input and Output signals accessed by serial communication (often called Remote I/O), add further time variations because the program scan and the I/O scan are usually unsynchronized. If speed is of the essence, the I/O cards should be mounted in the same rack as the processor so the access is by the much faster local rack backplane.

Finally, major speed improvements can usually be made by removing junk from the program. Most PLCs contain a vast amount of redundant programs from equipment that has long gone, tests that have been made or simply rungs that access plant devices which are no longer present. These junk rungs take up memory, confuse people and reduce the speed of the processor. It is good practice to go through the program at regular intervals (typical every three months) and remove anything which is not required.

4 Analog signals, closed loop control and intelligent modules

4.1 Introduction

So far we have considered signals that are essentially digital (on/off) in nature plus numerical data from timers and counters. In many systems (and the majority of small systems) these signals are all that is required. Often, though, a PLC will be required to measure, or control, plant signals which can assume any value in some predetermined range. Typical signals of this type are temperatures, flows, pressure, speeds, etc. These are known as analog signals.

In a similar way a PLC may have to produce analog output signals to drive meters and proportional valves, or provide a speed reference for a motor drive controller.

To meet these requirements a PLC needs analog input and output cards. These have somewhat different characteristics to the simple digital cards we have discussed so far. This chapter considers analog signals, the way they are handled and the related topic of other 'smart' PLC modules. First it is useful to briefly review the types of analog signals that are likely to be encountered.

4.2 Common analog signals

4.2.1 Temperature

Measurement of temperature is probably the commonest analog function. Although the simplest measuring device is the common mercury in glass thermometer, it is not readily adaptable to give a remote reading. In industry, remote temperature measurements are primarily made by three methods.

The first of these is the thermocouple, shown in Figure 4.1, where two dissimilar metals are joined together at the point whose temperature is to be measured, and linked to a sensitive voltmeter at some remote location. The voltmeter reading is a function of the two temperatures T_1 and T_2. Variation in the local temperature T_2 will cause errors, so 'cold junction compensation' is normally applied by measuring the local temperature with some separate device and adding a correction signal as in Figure 4.1(b).

Many types of metal combinations can be used for different temperature ranges, and are denoted by a letter. An R-type thermocouple, for example, uses platinum with an alloy of platinum/rhodium and has a range from 0 to 1700 °C; the commoner type-K thermocouple uses chromel/alumel and has a range of 0–1100 °C. The signals from all of these are very small; just 42 μV per °C for a type-K thermocouple.

The next form of thermometer uses the variation of resistance with temperature. A platinum wire constructed with a resistance of 100 Ω at 0 °C will have a resistance of 138.5 Ω at 100 °C. Such devices are known as PT100 sensors (PT for platinum and 100 for the resistance at 0 °C). These can be used over the range −200 °C to +800 °C. To reduce the error from the connecting cable resistance (which can be around 1 Ω) the three- and four-wire connections of Figure 4.2 are used, which allow

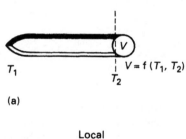

(a)

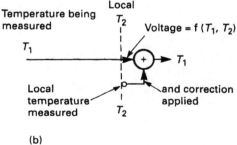

(b)

Figure 4.1 *The thermocouple: (a) principle of operation; (b) cold junction compensation*

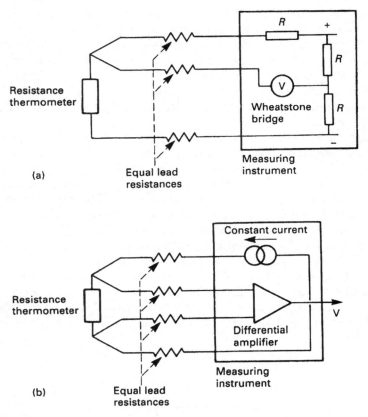

Figure 4.2   *Resistance thermometer circuits: (a) three-wire circuit; (b) four-wire circuit*

the PT100 resistance to be read accurately with a bridge circuit. Variations on the resistance thermometer use semiconductor materials. These devices, called thermistors, exhibit larger, but non-linear, changes in resistance.

The final type of common thermometer is called the pyrometer and measures the infrared radiation emitted from a hot surface. Pyrometers have the advantage that they can be remote from the object being measured, but they can only be used above 500 °C.

4.2.2 Pressure

With pressure measurement, it is important to appreciate that there are three distinct forms of sensor, although all are really variations on the differential pressure transducer of Figure 4.3(a). This gives an output

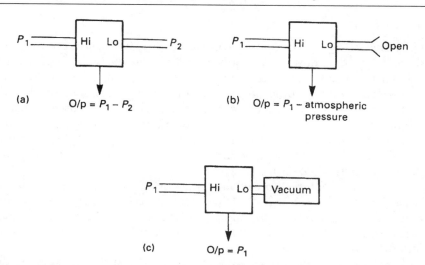

Figure 4.3 *The three forms of pressure measurement: (a) differential; (b) gauge; (c) absolute*

signal proportional to the difference in pressure between its two ports. The gauge pressure transducer measures pressure with respect to atmosphere; easily achieved by leaving one port open as in Figure 4.3(b). Gauge pressures usually have a 'g' suffix, 1.4 psig, for example. The final method connects one port to a vacuum to give absolute pressure as in Figure 4.3(c). With normal atmospheric pressure at about 1 bar, 2.4 bar gauge would be 3.4 bar absolute.

The basic measurement method illustrated in Figure 4.4 applies the two pressures to a diaphragm, and infers the pressure difference by observing the diaphragm deflection or measuring the force needed from an electrical solenoid to keep the diaphragm central. The latter technique is known as the force balance principle.

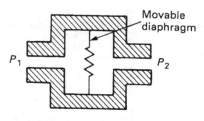

Figure 4.4 *The principle of a pressure transmitter*

4.2.3 Flow

Flow measurement is important in many processes. There are, again, several types of flow. Mass flow refers to the mass of fluid passing a given point per unit of time (e.g. kilograms per minute). Volumetric flow refers to the volume of fluid per unit time (e.g. litres per second). With gases, which are compressible, volumetric flow needs to consider temperature and pressure and refer the fluid back to some standard condition (usually 0 °C and 1 bar, called STP for Standard Temperature and Pressure). Finally we have flow velocity, which is the speed of the fluid (e.g. metres/second).

The commonest method of flow measurement generates a pressure drop across a restriction in the pipe. The simplest of these is the orifice plate of Figure 4.5(a) (with tappings at d and $d/2$ where d is the pipe diameter). Variations on this theme are the venturi tube of Figure 4.5(b) and the annubar of Figure 4.5(c).

There is, however, one disadvantage. The pressure drop (i.e. the differential pressure) is proportional to the square of the flow, i.e.

$$P_d = kF^2$$

where P_d is the differential pressure, F the flow and k a scaling constant. This limits the turndown (the ratio between maximum and minimum flow) to about 4:1, and requires a square root function to be available to convert the differential pressure to linear flow (i.e. $F = B P_d$ where B is a constant). PLCs capable of reading analog inputs invariably include a square root function. Needless to say, floating point (real) numbers are needed if accuracy is not to be lost.

Where a large turndown is needed, the turbine flow meter of Figure 4.6 can be used. The rotation of the turbine blades is sensed by a proximity detector to give a signal which is proportional to flow

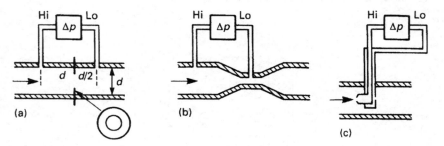

Figure 4.5 *Differential pressure flow measurement: (a) the orifice plate; (b) the venturi; (c) the annubar*

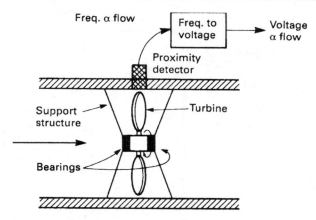

Figure 4.6 *The turbine flow meter*

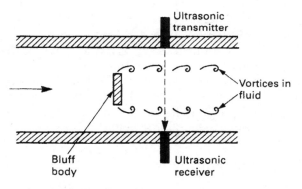

Figure 4.7 *Vortex shedding flow meter*

(i.e. no square root extractor needed) and a large turndown (typically 10:1). The disadvantage is wear in the bearings.

The vortex shedding flow meter of Figure 4.7 also gives a linear signal proportional to flow and measures flow by detecting the small vortices generated downstream of a bluff obstruction in the flow path. (Similar vortices can be seen when moving a hand through water.) In Figure 4.7 an ultrasonic beam is used to detect the vortices.

The final method of Figure 4.8 also uses an ultrasonic beam to measure the flow, with the frequency change caused by Doppler shift (change of frequency with velocity) being measured to indicate flow. This method has the advantage that no intrusion into the pipe is required.

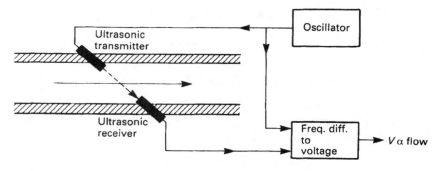

Figure 4.8   *Doppler shift flow meter*

4.2.4   Speed

Speed of motors, pumps, conveyors, etc. is often an indication of process throughput. The commonest way of measuring speed is a DC tachometer, which is simply a DC generator with an output voltage proportional to the rotational speed. A typical device, the GEC BD Tacho, has an output of 100 volts per 1000 rev/min.

Digital pulse tachos use a toothed wheel in front of a proximity detector, or a spoked wheel in front of a photocell, to give a pulse train whose frequency is proportional to speed. A simple electronic circuit can convert the frequency to a linear voltage.

4.2.5   Weighing systems

There are two basic methods of finding the weight of an object; these are known as a strain weigher, or a force balance weigher. In the first type, of which a spring balance is an example, the object to be weighed distorts the support structure, and this distortion is measured to give an indication of the weight. In the second type, of which two-pan kitchen scales to which weights are added or removed is typical, the weight of the object is balanced by some force (electrical, pneumatic or hydraulic) which can then be measured.

Most industrial weighing systems are strain weighers, and use a load cell as the primary measuring element. In its commonest form this is a cylinder to which strain gauges have been attached. These consist of a fine web of thin wires similar to Figure 4.9 which, when subjected to a strain, experience a small change in resistance as their length and cross-section change. Because the gauges are attached to the cylinder of the load cell, their deflection is determined by the changes in the dimension of the cylinder caused by the load.

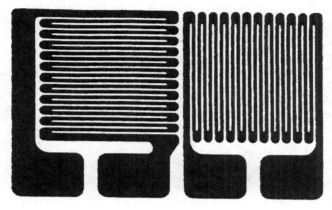

Figure 4.9 *A two-axis strain gauge (Courtesy of Welwyn Strain Measurement)*

The resistance changes are very small (typically less than $0.1\,\Omega$ on a base resistance of $100\,\Omega$) and similar changes can occur from temperature changes (see Section 4.2.1). It is usual, therefore, to arrange four strain gauges per load cell with two loaded and two at 90 degrees unloaded which, in a Wheatstone bridge, will give an output purely dependent on the load and not on temperature.

Weighing is something of a 'black art', and care needs to be taken to ensure that the load is carried solely by the load cells, and jamming or undesirable support by, say, pipes or cables is not occurring. Impact shocks from falling objects should also be considered if damage is not to occur.

4.2.6 Level

Liquid level is needed in many process industries. The simplest, and most reliable, method uses the fact that the gauge pressure in a liquid is directly proportional to the head of liquid above it, as shown in Figure 4.10. The pressure is given by

$P = \rho g h$ for SI units (pascals)

or

$P = \rho h$ for Imperial Units (psi)

where ρ is the liquid density, g is the acceleration due to gravity and h is the liquid head. Care must be taken to ensure that the pressure transducer is compatible with the fluid.

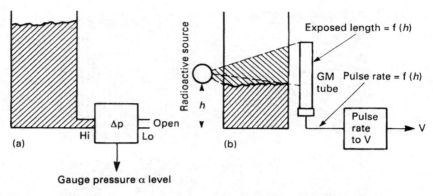

Figure 4.10　*Methods of level measurement: (a) differential pressure;*
(b) radioactive source

Other methods use floats (whose position can be measured) and
techniques using low-level radioactive sources, as in Figure 4.10(b),
where the liquid blocks the radiation, altering the signal from a detector
(such as a Geiger-Muller tube) on the other side of the tank.

4.2.7　Position

Position is often measured not only as a signal in its own right
(measuring the position of a drill head, for example) but also to infer the
value of some other variable. The measurement of level is sometimes
achieved with a float whose position is recorded.

A simple position measurement system can be obtained with a
free-moving linear or rotational potentiometer and a stabilized power
supply to give an output voltage directly related to the slider position
(Figure 4.11).

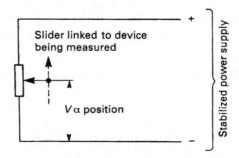

Figure 4.11　*Position measurement with a potentiometer*

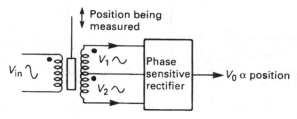

Figure 4.12   *The LVDT*

A more accurate device, with no contact friction and minimal movement force, is the LVDT (for linear variable differential transformer) of Figure 4.12. The AC input signal V_{in} produces voltages V_1 and V_2 in the two secondaries of the transformer whose relative magnitudes depend on the position of the moveable core. A phase-sensitive rectifier produces a DC output signal whose amplitude varies as the core movement. Optical encoders are also often used for accurate position measurement. These are described further in Section 9.10.

4.2.8   Output signals

So far we have considered analog signals which appear as inputs to the PLC. Output signals are also needed, typical examples being to drive meters, control proportional valves or provide required values (setpoints) for other controlling devices. A PLC dealing with analog devices thus needs to be able to handle both input and output signals.

4.3   Signals and standards

It is apparent from the previous sections that the 'raw' signals from the plant sensors are many and varied, ranging from a few millivolts (for a thermocouple) to perhaps over a hundred volts for a tacho, and exhibiting variations in DC volts, AC volts or even resistance. Some form of standardization is obviously desirable if a vast range of analog input cards is not to be required.

The general form of an input signal can therefore be represented by Figure 4.13. The raw signal from the sensor is converted to some standard signal by a local electronics unit, the sensor and the local electronics unit being known as a transmitter or a transducer. The standardized signal representing the plant variable being measured can then be connected to a standard analog input card.

The obvious question is what this standardized signal should be. Analog signals are low level and hence susceptible to electrical interference (or noise as it is more generally known). A signal represented by an

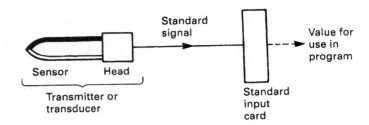

Figure 4.13   *The components of a PLC analog input*

electric current is less affected by noise than a signal represented by a voltage, so a current loop is usually chosen. The transducer and the receiving device are connected as Figure 4.14 with the current signal being locally converted to a voltage by a suitable ballast resistor. A current loop can also be used with several receivers (meter, chart recorder and PLC input, for example) connected in series.

The commonest standard represents an analog signal as a current within the range 4–20 mA, with 4 mA representing the minimum signal level, and 20 mA the maximum. If, for example, a pressure transducer gave a 4–20-mA signal representing a pressure range of 0–10 bar, a pressure of 8 bar would be represented by a current of $8 \times (20 - 4)/10 + 4\,\text{mA} = 16.8\,\text{mA}$. A 4–20-mA signal is often converted to a 1–5-V signal by a local 250-Ω ballast resistor.

The 4-mA 'zero' signal (called the offset) serves two purposes. The first is protection against transducer or cable damage. If the transducer fails, or the signal cable is open or short circuit, the current through the local ballast resistor will be zero, giving a 'negative' signal of 0 V at the receiver. This is easily detected and can be used to give a 'transducer fault' alarm.

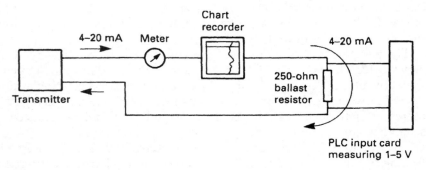

Figure 4.14   *The 4–20 mA current loop*

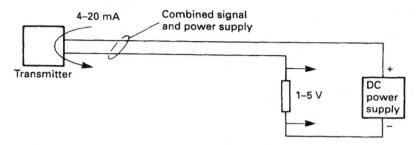

Figure 4.15 *Two-wire 4–20 mA transmitter*

The 4-mA offset current can also simplify installation. In Figure 4.14 we assumed that the transducer was provided with a local power supply and sourced the signal current. Such devices are available, but the arrangement of Figure 4.15 is commoner (and simpler). Here the power supply (usually 24–30 V DC) is mounted local to the receiving device, and the signal lines serve both to power the transducer and transmit the current. The transducer draws current from the power supply in the range 4–20 mA according to the signal being measured. This current is converted to a voltage by a ballast resistor as before.

The 4-mA offset provides the current that the transducer needs to keep working. Obviously a transducer with a signal range of 0–20 mA could not operate in this manner. Transducers similar to Figure 4.15 are commonly called two-wire transducers.

4.4 Analog interfacing

4.4.1 Resolution

An analog interface card converts a continuously varying analog signal to a digital form that can be used inside a PLC program. The analog signal is generally represented, initially at least, as an integer number.

This analog to digital conversion (usually known by the initials ADC) is inherently accompanied by a loss of resolution which depends on the number of bits used. An 8-bit byte, for example, can represent an integer in the range 0–255. If this was used to represent an analog signal measuring a flow with a span (range) from 0 to 1800 l/min, one bit will represent approximately 7 l/min (given by 1800/255). Any control strategy in the program based on finer resolution is meaningless. (Particular care should be taken with comparisons, as some values can never be obtained; a flow of 138 l/min, for example, would never be given by our 8-bit system; it would jump from 134 l/min to 141 l/min.

Table 4.1

No. of bits	Range	Error (%)
8	0–255	0.5
10	0–1023	0.1
12	0–4095	0.025

Comparisons should always be based on (greater than or equal to) or (less than or equal to).

A commoner resolution is 12 bits. This gives a representation as an integer from 0 to 4095. With our flow of 0–1800 l/min, one bit would represent just under 0.5 l/min (1800/4095 = 0.44).

This 'coarseness' is not the problem it might at first appear. Although an analog transducer can give any value in its span, it will have inherent errors. Many first-line transducers are only 2% accurate. If our flow transducer had 2% accuracy, its measurement could be in error by 36 l/min. Alongside this error, the 7-l/min resolution is probably quite reasonable.

It is therefore useful to think of the resolution in terms of an error which is to be added to the error from the transducer itself, as in Table 4.1.

Few industrial transducers have an accuracy better than 0.1%, and a 12-bit conversion will add little error in most applications.

4.4.2 Multiplexed inputs

As we have seen earlier, a PLC generally works with a 16-bit word. If our analog input card occupies one slot in a rack, and reads just one analog input, it will be wasteful of I/O space (and very expensive). For comparison, a normal digital input card reads 16 signals, and costs about a quarter as much as an analog input card.

The cost, and I/O usage, can be reduced by using multiplexing, shown as a block diagram in Figure 4.16. Here four analog input signals, separated from each other by isolation amplifiers, are selected in turn by electronic switches and converted to a digital number by a common ADC. Such cards commonly deal with four, eight or sixteen input signals.

If the card is occupying a single card slot, there obviously has to be some way for the PLC to link the sequential readings from the ADC with the actual input signals. This is a problem we shall return to in Sections 4.4.4 and 4.4.5.

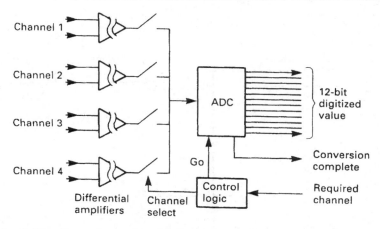

Figure 4.16 *Four-channel multiplexed analog input card*

4.4.3 Conversion times

The conversion from analog to digital signal is not instantaneous (although extremely fast ADCs with conversion times of nanoseconds are used in digital TV systems). In most industrial systems there is a high probability that electrical noise from the local AC mains (50 or 60 Hz, depending on the country) will be present on the signal.

A technique called dual slope integration ties the conversion time to the local mains frequency, giving a high degree of AC mains-related noise rejection. This gives conversion times of 20 ms in the UK (50-Hz supply) and 16.67 ms in the USA (60-Hz supply).

With a four-way multiplexer stepping round each channel in turn, a signal will thus be sampled once every 80 ms. To this must be added the program scan time and the remote scan time if the analog card resides in a remote rack.

An analog input card thus works by taking 'snapshot' samples of the plant signals. A sampled system only knows about the values of its samples. It cannot infer any other information about the signals it is dealing with. Both of the signals of Figure 4.17(a) and (b) would produce the same result of Figure 4.17(c) if sampled at the same rate. An obvious question, therefore, is what sample rate we should choose if our samples are to accurately represent the original signals.

In Figure 4.18(a) a sine wave is being sampled at a relatively fast rate. Intuitively one would assume this sampling rate is adequate. In Figure 4.18(b) the sample rate and the frequency are the same. This is obviously too low as the samples imply a constant unchanging output.

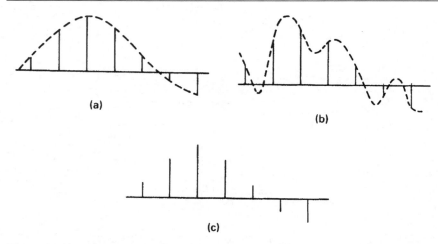

Figure 4.17   *Signals (a) and (b) both produce sample sequence (c). Both are equally valid*

In Figure 4.18(c) the sample rate is lower than the frequency and the sample values are implying a sine wave of much lower frequency than the signal. This latter case is called *aliasing*. A visual effect of aliasing can be seen on cinema screens where moving wheels often appear to go backwards. This effect occurs because the camera samples the world at about 50 times per second.

Any continuous signal will have a frequency range of interest (called its bandwidth). To permit an accurate representation of the original signal to be rebuilt the sampling frequency should be at least twice this frequency bandwidth. This, somewhat simplified, is known as *Shannon's sampling theorem*.

Any real life system will not, however, have a well-defined bandwidth and sharp cut-off point. Noise and similar effects will cause any real signal to have a significant component at higher frequencies. Aliasing may occur with these high-frequency components and cause apparent variations in the frequency band of interest. Before sampling, therefore, any signal should be passed through a low pass *anti-aliasing filter* to ensure only the bandwidth of interest is sampled.

Most industrial control signals have a bandwidth of a few Hz, so sampling within Shannon's limit is not normally a problem. Normally the bandwidth is not known precisely so a sample rate of about 5 to 10 times the envisaged bandwidth is used. For example, with a typical 2 Hz signal a 10 Hz sampling rate would be adequate requiring a sample time of 100 ms.

Surprisingly, this rarely gives problems. Practical industrial systems, dealing with real plant signals concerned with materials with significant

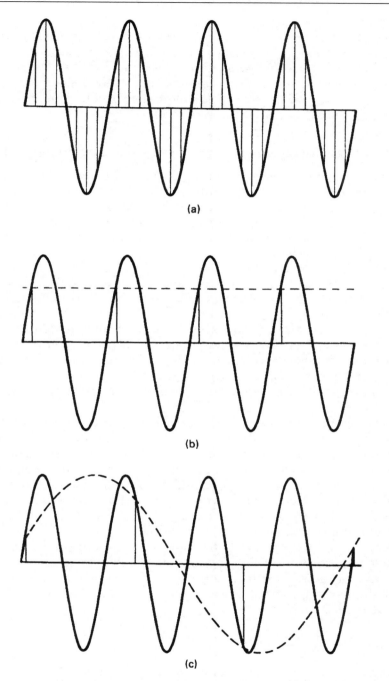

Figure 4.18 *The effect of the sampling rate: (a) good sampling; (b) sample rate too low; (c) aliasing*

mass, rarely have bandwidths greater than 0.5 Hz, and any frequency higher than this can be considered to be extraneous noise and filtered out. Temperature loops, for example, can often be sampled as slowly as once every few minutes without introducing any errors. The designer should, however, always be aware of Shannon's sampling theorem and check that the ADC conversion time is compatible with the signals being measured.

4.4.4 Channel selection and conversion to engineering units

A typical eight way analog input card will provide eight 12-bit signals, each ranging from 0 to 4095 in their 'raw' form. Generally these will need to be accessed via the PLC program and converted to engineering units such as °C, or psi, or l/min. If, for example, the range 0–4095 represents a flow of 0–1000 l/min, a resolution of about 0.25 l/min will theoretically be achievable.

The PLC must therefore address two problems; how to access the multiplexed data via the signal card, and how to use the data in the program. There are essentially two ways of accessing the data, summarized in Figure 4.19.

In Figure 4.19(a) the PLC selects which channel it wants to read by sending a 3- or 4-bit address as an output instruction to the card along with a 'convert' command. The card returns the digitized 12-bit value and a 'done' signal which can be read with a normal word input command. This method has the advantage that the programmer can select different sample rates for different signals.

The commoner method, though, is shown in Figure 4.19(b). A block of storage locations in the PLC store is directly associated with the analog input card. The card 'free runs', writing digitized values into the store from where they can be read by the rest of the program. In Siemens PLCs with fixed slot addressing, for example, the store addresses are determined directly by the analog card position in the rack; a card in slot 2 of the first rack will write its values to a block of stores starting at location 192.

Conversion from a raw 12-bit signal to engineering units can have subtle traps for the unwary. In theory the conversion is simple. If N is the raw signal, HR the high-range signal (corresponding to 4095) and LR the low-range signal (corresponding to zero) then the measured value, MV, is simply

$$MV = \frac{N \times (HR - LR)}{4095} + LR \tag{4.1}$$

If the calculation is done with real (floating point) numbers there should be no problem, and equation (4.1) can be used directly.

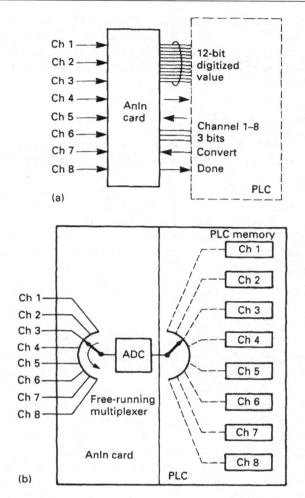

Figure 4.19 *Ways of linking a PLC and an analog input card:*
(a) direct selection; (b) free running

If, however, integer numbers have to be used, great care must be taken. If the multiplication $N\times(HR-LR)$ is performed first, arithmetic overspill is likely unless 32-bit results can be accommodated. If the division $N/4095$ is performed first, the equation will not work as N is always less than 4095, giving an integer result of zero (and an MV of LR). Wherever possible real numbers should be used if equation (4.1) has to be performed.

To avoid this problem, the different manufacturers have devised methods to read analog input signals. In the ABB Master, for example, the database definitions for each signal define HR, LR, the sample rate and a name by which the signal will be referred to in the program. There are, obviously, detail differences, so in the next section we will look, by way of an example, at the way analog signals are read by an Allen Bradley PLC-5.

4.4.5   Analog input cards

The Allen Bradley PLC-5 reads analog signals with an analog input card (1771-IFE) which can, in its simplest form, read eight analog inputs.

The PLC communicates with the card via instructions called 'block transfers' which transfer data to (or from) a block of store locations. Data transfers from the PLC to a card are called 'block transfer writes' (BTW) and, not surprisingly, transfers from a card to the store are 'block transfer reads' (BTR). For each type of instruction, somewhat simplified, the programmer states:

(a)   the direction of transfer (BTW or BTR)
(b)   the card address (rack, slot and slot half, left or right)
(c)   the store location address
(d)   the number of 16-bit words to be transferred

The analog input card uses both BTW and BTR instructions, the BTW being used once, after power up, to configure the module and the BTRs subsequently to read the data as summarized in Figure 4.20.

The post-power-up BTW in Figure 4.21 sets how the module is to behave; whether it gives data in binary or BCD, whether the module uses eight differential signals or 16 signals referenced to a common 0 V, and the maximum and minimum values for the input range (HR and LR in equation (4.1)) on each channel. The card uses these to return readings in engineering units (in 12-bit binary integer, two's complement format or 12-bit BCD).

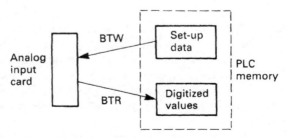

Figure 4.20   *The Allen Bradley BTW and BTR instructions*

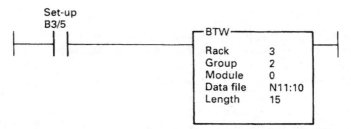

Figure 4.21 *The BTW instruction slightly simplified. This rung sends 15 words of data starting at N11:10 to the card in the left-hand position of slot 2 in rack 3 whenever B3/5 makes*

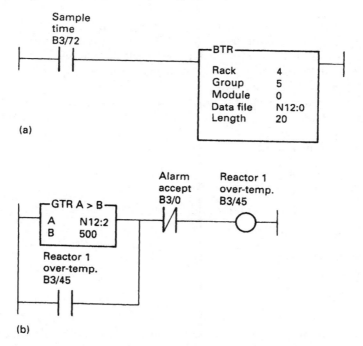

Figure 4.22 *The BTR instruction. (a) Again slightly simplified, this rung reads 20 data values from the card in the left-hand position of slot 5 in rack 4 whenever B3/72 makes. The values (which could include digitized analog signals and error diagnostics) are stored from memory location N12:0. (b) Range checking the third value from (a)*

Once set up, values can be read at the required time intervals with a BTR as in Figure 4.22(a). This gives signal values in the specified store locations along with over-range and similar alarms. The values can then be used in the program; in Figure 4.22(b) an over-temperature check is being made on the third analog signal from Figure 4.22(a).

4.4.6 Filtering

The set-up data in Figure 4.21 include a filter time constant to provide a
first-order filter on the input signal. This helps to remove any noise on
the signal. If the analog readings on another system are unfiltered, or an
unusually long time constant is needed (the 1771-IFE of Figure 4.20 can
provide a time constant of up to about one second), the programmer
can provide a separate filter routine.

A first-order filter can be represented by the simple differential
equation

$$T\frac{\mathrm{d}y}{\mathrm{d}t}+y=x \tag{4.2}$$

where x is the input signal (corresponding to the raw value from the
input card), y is the filtered signal and T is the time constant.

In a PLC system we do not have continuous values for y and x, but
sampled values of y and x taken at intervals Δt (the update time of the
card, or the rate at which samples are initiated).

We thus have a snapshot input sequence x_n, x_{n-1}, x_{n-2}, where x_n is
the most recent, and x_{n-1} the previous, and a similar output sequence
y_n, y_{n-1}, etc.

We can approximate:

$$\frac{\mathrm{d}y}{\mathrm{d}t}\simeq\frac{\Delta y}{\Delta t}\simeq\frac{y_n-y_{n-1}}{\Delta t} \tag{4.3}$$

Substituting back into equation (4.2) gives

$$Ty_n-Ty_{n-1}+y_n\Delta t=x_n\Delta t \tag{4.4}$$

Solving for the filtered value y_n gives

$$y_n=\frac{y_{n-1}+(\Delta t/T)x_n}{1+(\Delta t/T)} \tag{4.5}$$

This requires one store location to hold the last value of y (denoted by
y_{n-1}) and can be easily performed by the simple sequence of Figure 4.23.
Note that for equation (4.5) and Figure 4.23 to work, the update time
Δt must be consistent and known.

4.5 Analog output signals

PLCs are often required to drive analog output signals as well as read
analog inputs. Common applications are driving analog meters or chart
recorders and providing reference signals such as the desired speed for
a thyristor drive. Like analog inputs, these signals have standard voltage
ranges of 1–5 V or 0–10 V or the current range of 4–20 mA.

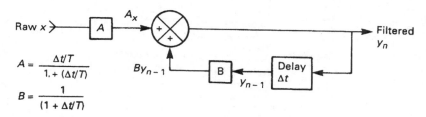

Figure 4.23 *A simple first-order filter. If $T \gg \Delta t$ (as is usually the case), $A = \Delta t/T$ and $B = 1$*

A typical analog output card, the Allen Bradley 1771-OFE, has four output channels, each turning a 12-bit (0–4095) digital signal into an analog output. Isolation amplifiers are used on the outputs to reduce the effects of noise and allow the signals to connect into external devices fed from different electrical supplies. The digital signals come from storage locations inside the PLC as shown in Figure 4.24. This action is known as a digital to analog conversion, or DAC.

For best resolution the PLC should use the full 0–4095 range, but this is frequently impossible. If the PLC, for example, is setting the speed range of a motor from 0 to 1350 rev/min, it will need to convert 0–1350 into the range 4–20 mA. Equation (4.1) can be rearranged as

$$V_{DAC} = \frac{4095 \times (N-LR)}{(HR-LR)} \tag{4.6}$$

where V_{DAC} is the value passed to the DAC (in the range 0–4095). N is the output number in engineering units, and HR/LR are the high- and low-range values. As before, great care must be taken with equation (4.6) to avoid overspill or loss of resolution.

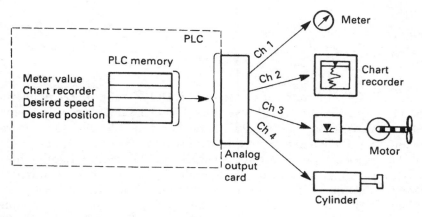

Figure 4.24 *Analog output signals*

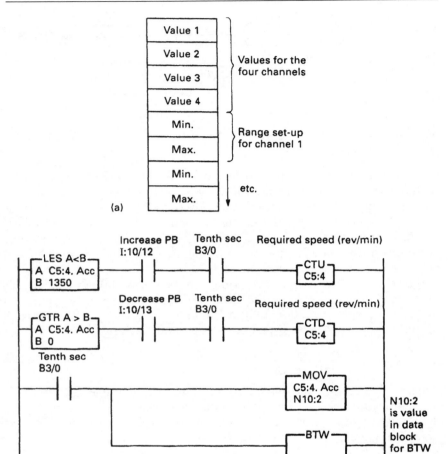

Figure 4.25   *Using an analog output card: (a) the data block (slightly simplified) for an Allen Bradley analog output card; (b) setting speed for a motor with a counter and an analog output card*

The PLC-5 communicates with the 1771-OFE with the BTW instruction described in the previous section. The programmer sets up a block of 12 words as in Figure 4.25(a), the first four of which contain the values, and the balance the set-up data such as HR and LR. The block of data is then written to the card with a BTW. Figure 4.25(b) shows a typical example where an analog speed reference can be raised or lowered by operator-controlled pushbuttons. Note the use of Greater Than (GTR) and Less Than (LES) instructions to confine the counter value within the allowed range of 0–1350 rev/min.

Ranging as in Figure 4.25 allows engineering units to be used inside the program; the counter in Figure 4.25, for example, holds the speed directly in rev/min, but this is accompanied by a loss of resolution, as explained earlier. For the range 0–1350 rev/min, we have a resolution of about 0.1%, compared with the theoretical 0.025% resolution available from the card.

4.6 Analog-related program functions

There are other operations that can be performed on analog signals. A typical list, for the GEM-80, is

SQRT   Square root, mainly used with differential pressure flow measuring devices (such as orifice plates)

LINCON   Performs $X*(A/B)+C$ with limiting

FGEN   Multipoint straightline function generator used for linearizing and similar functions (see Figure 4.26(a))

LIMIT   Performs limiting of signals as shown in Figure 4.26(b)

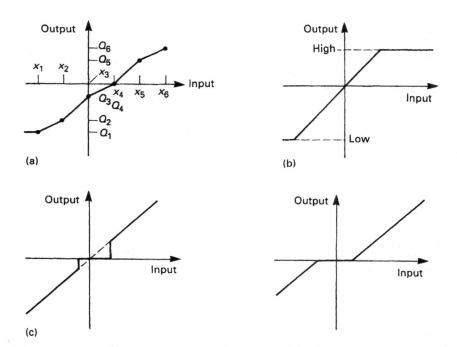

(a)

(b)

(c)

Figure 4.26 *GEM-80 special functions useful for analog signals. (a) FGEN function, with N points at equal intervals Δx. (b) LIMIT. High and low limits do not have to have the same values. (c) DEDBAND without and with offset*

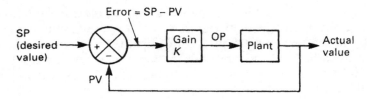

Figure 4.27   *Simple proportional closed loop control*

RAMP   Rate limiting (with different rise and fall rates).
DEDBAND   Deadband functions as in Figure 4.26(c). Useful for preventing 'dither' in closed loop control when PV and SP are close.
ANALAG   First-order lag. Used for filtering (see also Section 4.4.6).

4.7   Closed loop control

4.7.1   Introduction to control theory

Many industrial processes require some plant variable (temperature, pressure or flow, for example) to be kept at a fixed value or to follow some profile. These schemes are normally based on the block diagram of Figure 4.27 where the actual value is fed back from the plant and compared with the desired value. This is known as feedback or closed loop control.

The required value, denoted by SP for Setpoint, is compared with the actual value PV, for process variable, to give an error E, which is simply

$$E = SP - PV \tag{4.7}$$

This is multiplied by a gain K to give an output OP from the control mechanism where

$$OP = KE = K(SP - PV) \tag{4.8}$$

This output causes a change in the plant, giving the output PV. In a well-ordered plant, PV will be directly related to OP, allowing us to write

$$PV = A \times OP \tag{4.9}$$

where A is a simple gain factor.

Combining equations (4.7) to (4.9) allows us to write

$$PV = \frac{AK}{(1 + AK)} \times SP \tag{4.10}$$

i.e. the plant signal PV will follow the SP multiplied by a scaling factor $AK/(1+AK)$. The term AK is known as the open loop gain, and is often denoted by G, allowing us to write

$$PV = \frac{G}{1+G} \times SP \tag{4.11}$$

It can be seen that for large values of G, the error between PV and SP will be small. For $G=10$, for example,

$$PV = 0.91SP$$

A large value of G can be obtained by using a large value of gain, K. Unfortunately, in practical systems this often leads to instability, a topic discussed in the following section.

Figure 4.28 shows a modified type of control strategy, where the output signal is the sum of the error *plus* the time integral of the error, i.e.

$$OP = K(E + M\int E\, dt) \tag{4.12}$$

This is known, for obvious reasons, as PI control, for proportional plus integral control.

The integral term will cause OP to change as long as there is an error and OP will only be constant when the error is zero and SP = PV. (The full analysis of a PI controller requires a knowledge of calculus. Interested readers are referred to control theory textbooks such as the author's *Industrial Control Handbook* published by Butterworth-Heinemann.) A PI controller thus provides zero error in the steady state without the need for a high gain.

In practical controllers, the term M in equation (4.12) is replaced by $1/T_i$ giving

$$OP = K\left(E + \frac{1}{T_i}\int E\, dt\right) \tag{4.13}$$

where T_i is known as the integral time. The reason for this change arises out of the underlying mathematics.

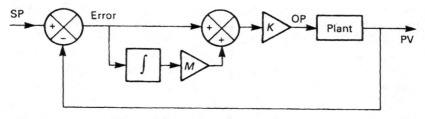

Figure 4.28 *Proportional plus integral (PI) control*

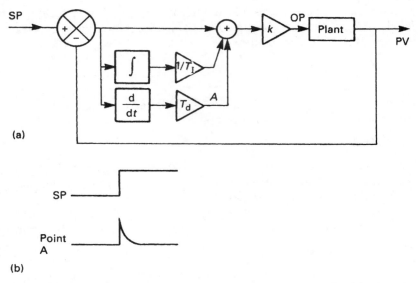

Figure 4.29   *The PID or three-term controller: (a) controller block diagram; (b) effect of derivative term*

A further modification is shown in Figure 4.29. Here a time derivative (rate of change) term has been added, giving

$$OP = K\left(E + \frac{1}{T_i}\int E\, dt + T_d\frac{dE}{dt}\right) \qquad (4.14)$$

Not surprisingly this is known as a three-term or PID (for proportional plus integral plus derivative) controller. The multiplier T_d is known as the derivative time.

The derivative term brings two benefits. Because it responds to the rate of change of error it will 'kick' the output as in Figure 4.29(b), when the setpoint changes rapidly. The derivative term can also make a system more stable and reduce overshoot.

So far we have considered a plant signal following a setpoint. Closed loop control is also useful where a plant is subject to disturbances from the outside world as summarized in Figure 4.30. A level control system, for example, could be affected by changes in the outflow rate, a temperature control system will be affected by changes in ambient temperature and the temperature of the material it is heating, and a flow control system will be sensitive to changes in source pressure. Because all of these will produce a change in PV, the controller will detect them and modify OP to make PV=SP again, and remove the effect of the disturbance.

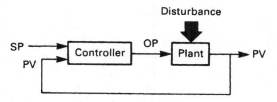

Figure 4.30 *The effect of a disturbance*

4.7.2 Stability and loop tuning

Ideally we want PV to equal SP at all times. A real plant has finite reaction times and non-linearities, and the ideal response is impossible. Figure 4.31 shows some possible responses to a step change in SP (similar curves could be obtained for a step disturbance).

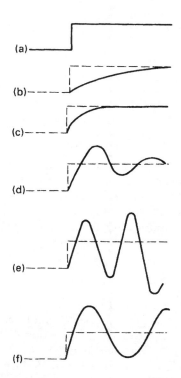

Figure 4.31 *Forms of step response: (a) required response; (b) overdamped; (c) critical damping; (d) underdamped; (e) unstable; (f) constant amplitude*

An achievable practical response is usually similar to the somewhat underdamped Figure 4.31(d). A common standard, known as quarter-amplitude damping, aims for each overshoot to be 25% of the previous.

The engineer has control over the response by adjustment of the gain K, the integral time T_i and the derivative time T_d. These can be set by trial and error (a good starting point is often $K=0.5$, $T_i=20$ s, $T_d=5$ s) or a good deal of time can be spent analysing the mathematics governing the plant (again see the author's *Industrial Control Handbook*).

The required values can also be determined by experiment (although the reader should be aware that the effects on the plant can be severe and safety implications should be considered). The experimental method described below, known as the Ziegler–Nichols method, should give a quarter-amplitude response.

The controller is initially set up as a proportional-only controller (T_i=infinity, and T_d=zero). Varying the gain K will change the step response from the underdamped Figure 4.31(d) to the unstable Figure 4.31(e). There will be a critical value of gain K_c where constant oscillation will occur as in Figure 4.31(f). The period of these oscillations, T_c, is then measured. The required controller settings are then

PI
$$K = 0.45 K_c$$
$$T_i = 0.8 T_c$$
PID
$$K = 0.6 K_c$$
$$T_i = 0.5 T_c$$
$$T_d = 0.12 T_c$$

These values are best viewed as initial settings, which can be tuned for best response. Increasing the gain, or decreasing T_i, makes a system respond faster but decreases the stability. A good rule of thumb is to make $T_d = T_i/4$ when derivative action is used.

Further tuning techniques are given in the author's *Industrial Control Handbook*.

4.7.3   Closed loop control and PLCs

A closed loop system based on PLCs will be similar to Figure 4.32. The plant variable, PV, is read by an analog input card, and the output OP provided by analog output cards. The setpoint, SP, is provided by the operator (via a graphics terminal in Figure 4.32) or by some program sequence. The PID algorithm is then provided by the program.

It is possible to write PID algorithms with four-function $(+-*/)$ mathematics, but it needs great care. The program scan time must be known for the integral and derivative routines, and protection against output

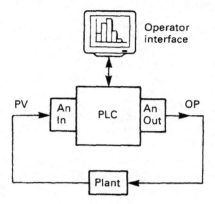

Figure 4.32 *Closed loop control on a PLC*

actuator saturation must be built in to overcome an effect called integral wind-up.

In most, if not all, PLCs which can support analog input and output cards, the manufacturer supplies a PID function in the program library. In this section we will look at the way the GEM-80 handles closed loop control.

The GEM-80 gets a value from the outside world with an AND address instruction, and sends values to the outside world with OUT addresses as in Figure 4.33(a) (from Sections 4.4.5 and 4.5 it will be apparent that this is a simplification of what is really going on). Input

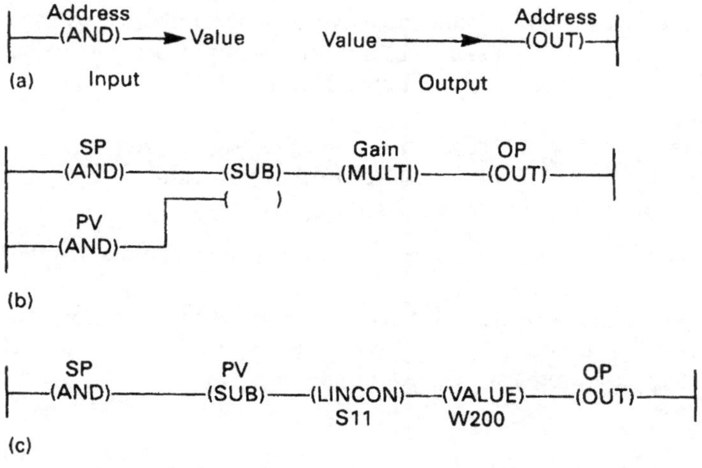

Figure 4.33 *Simple control with a GEM-80: (a) analog input and output; (b) integer proportional control; (c) proportional control with LINCON*

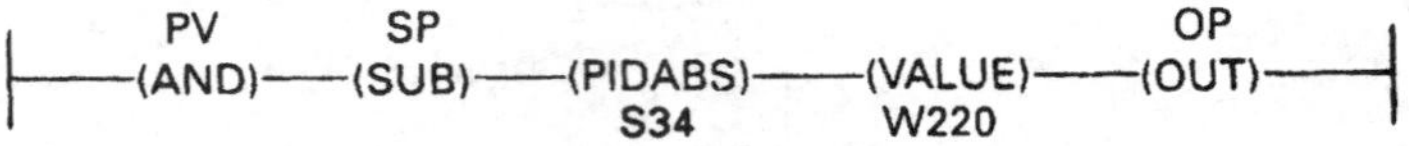

```
       PV        SP                              OP
├───(AND)───(SUB)───(PIDABS)───(VALUE)───(OUT)───┤
                      S34        W220
```

Figure 4.34   *Three-term control with a GEM-80*

addresses have the form A*n* (e.g. A4) and output addresses the form B*n* (e.g. B3). The GEM-80 can also store numbers in W locations (e.g. a value can be stored in W112).

Let us first consider a simple proportional-only controller, where $E = (SP - PV)$. We could write $OP = K(SP - PV)$ and perform the operation with integer math functions as in Figure 4.33(b). This, however, has the disadvantage that the output will change in large steps. If K, for example, has the value 4, the output will change in steps of 4.

The GEM-80 has many built-in functions working with real numbers. One of these is a linear function $Y = AX + B$, denoted by LINCON-S11. We can use this as a gain function by setting $B = 0$. Combining this with a more succinct way of writing the subtraction function gives us the single rung of Figure 4.33(c). The A and B values for the LINCON are stored in the addresses defined in the VALUE function.

A full PID block, PIDABS-S34, is available and is used in its simplest form as in Figure 4.34. The addresses for K, T_i, T_d are specified in the VALUE block following the PIDABS block. As written, these start at address W220. Altogether, there are 15 storage locations concerned with PIDABS. These include

Fault Code

Settings for K, T_i, T_d Changeable by the programmer or by the program itself to allow, for example, settings to be used in different circumstances

Output limits

Rate limits   Maximum rate of change of output

Hold mode   Drives output to a fixed value

Suicide mode   Drives output to zero

These are all accessible by the program, and can be used to make complex control schemes.

A close relation of the PIDABS is the PIDINC function used when the controller drives a motorized valve as in Figure 4.35. Here the motor acts as an integrator itself, since

$$V_{pos} = \int OP \, dt \tag{4.15}$$

If a conventional PID controller is used here, instability will result. The PIDINC is the derivative of PIDABs and has the function

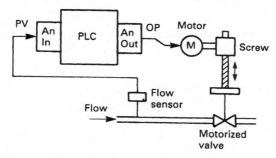

Figure 4.35 *The need for incremental PID control*

$$\mathrm{OP} = K\frac{\mathrm{d}}{\mathrm{d}t}\left(E + \frac{1}{T_i}\int E\,\mathrm{d}t + T_d\frac{\mathrm{d}E}{\mathrm{d}t}\right) \tag{4.16}$$

When combined with the integral action of the motorized valve, a normal PID action results.

So far we have been concerned with purely automatic actions. Often a manual mode is required, with an automatic changeover selection as shown in Figure 4.36. There is a hidden problem here.

When manual is selected the PID function is still active, and as it is highly unlikely that PV will equal SP in manual, the integral term in the PID controller will cause the output from the PID function to rise to full output or drop to zero. When auto is reselected, PV will swing wildly for some time until the PID function regains control.

What is needed is a bumpless transfer which matches automatic and manual values on changeover. In a GEM-80 this can be achieved with the three rungs of Figure 4.37. The operating mode, automatic or manual, is selected by a switch connected to digital input A2.0, which is energized for automatic. (Note the good design practice; a lost supply to the switch will make the control change to manual and will hold the last output value.)

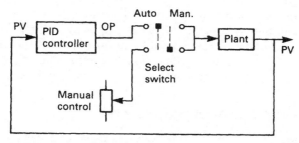

Figure 4.36 *Auto/manual selection*

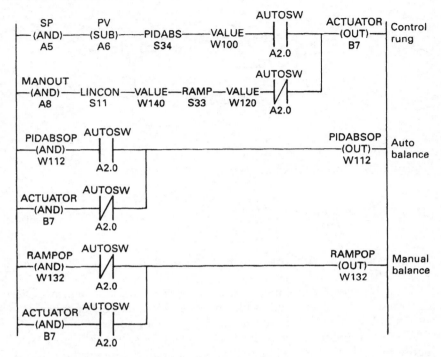

Figure 4.37   *Bumpless transfer on a GEM-80*

The first rung selects the output from the PIDABS function (for automatic) or from the manual control (in manual). The RAMP function limits the rate of change of the manual signal.

The second and third rungs perform the balance. W112 holds the output value of the PIDABS function (with block starting at address W100) and W132 holds the output value of the ramp function (block starting at address W120).

In automatic, rung 2 has no effect, but rung 3 writes the output value B7 back to the ramp output W132. When a change takes place from automatic to manual, the ramp will start at the last automatic value.

Similarly, in manual, rung 2 writes the output B7 to the PIDABS output W112, so on a change from manual to automatic, the PID function starts at the last manual value. Rung 3 has no effect in manual.

A bumpless transfer is thus achieved in both directions.

4.8 Specialist control processors

PLCs are not, of course, the only devices capable of performing closed loop control. Three-term controllers are readily available (even as plug-in modules for a PLC rack) and controller manufacturers overlap with PLCs by providing programmable analog controllers.

Typical of these is the TCS Tactician, which uses a graphical programming method to link together standard signal-processing blocks. The 'program' is built on the screen with a mouse.

Figure 4.38 shows a control scheme for temperature control of an oil-fired furnace. There are three PID controllers, one for the temperature loop, and one each for the air and oil flow. The correct ratio is maintained between the oil and air flows, with the air leading for increasing heat and the oil leading for decreasing heat (known as lead/lag control). This structure is literally drawn on a VDU screen.

Schemes such as Figure 4.38 are much easier to program in control-based schemes such as the Tactician, but general arithmetical and sequencing schemes are simpler in PLCs. Which is used in a specific scheme is a matter of judgement for the project engineer.

4.9 Bar codes

All articles bought in supermarkets are labelled with a bar code which identifies the country of origin, manufacturer and the item itself. These bar codes are read at the checkout by a scanner, and the price found automatically from the store's computer and added to the customer's

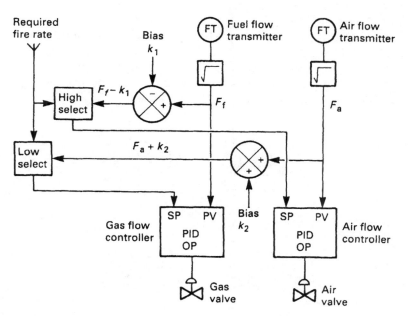

Figure 4.38 *A more complex system; lead/lag burner control*

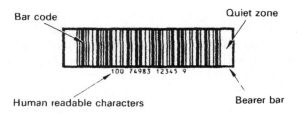

Bar code

Quiet zone

100 74983 12345 9

Human readable characters

Bearer bar

Figure 4.39 *The structure of a bar code*

bill. This is a typical application of bar code technology which is being widely introduced in industry and commerce as a way of tracking and keeping inventory control of items with minimal human intervention. For example, the ISBN numbers used to identify books are now read by bar code readers in libraries and used to keep track of library loans for the UK Public Lending Right (PLR) scheme.

The structure of a bar code is shown in Figure 4.39. It consists of a series of bars and spaces used to encode alphanumeric symbols in a machine readable format. At either end is a quiet zone to allow the reading system to sense the start and finish of the bar code. A quiet zone is needed because not all bar code representations are of a consistent length. The bearer bars are added when there is a danger that a misaligned read scan may not catch all of the bar code data. The bearer bars will then give a broad pulse which the reader will detect and cause the read to be rejected.

There is not a single universal bar code, and many different codings are in use throughout the world. Some of the commoner ones are shown in Figure 4.40. Code 39 uses nine bar elements per digit, three of which are wide (hence the name). The nine elements are always made up of four spaces and five bars.

1234567890

Interleaved 2-of-5

100864-216736

UPC

ABCDEF

Code 39

4014561780123

EAN

Figure 4.40 *Common bar codes*

EAN stands for European Article Number and was derived from the earlier American UPC or Universal Product Code. Both use characters constructed from two bars and two spaces occupying seven positions. Only numeric data can be represented. EAN and UPC have long data streams which can be subdivided into subsets by longer twin bars. Supermarkets in the UK use EAN product coding, with three groups denoting country of manufacture (UK is 5, France is 3 and so on), the manufacturer (012427 is the Scottish soup maker Baxter) and the product itself (020108 is Royal Game Soup). EAN is also used for ISBN book markings.

Interleaved 2 of 5 is again a numeric only coding and it represents characters in pairs, one by the bars and one by the spaces. Each character uses five positions, two of which are wide.

The amount of data that can be held in a linear bar code is determined by the length. Where a large amount of data is required, two dimensional bar codes may be used. Some of the commonest 2D codes are shown in Figure 4.41. These allow over 3 kbytes of data to be encoded. Typically 1800 characters can be held in a 50 mm by 50 mm square. All have extensive error detection and error correction built into the codes.

Figure 4.41 *Examples of two-dimensional bar codes*

There are two methods of encoding data in two dimensions. The first, called stacked codes, is essentially multiple rows of linear bar codes. The commonest of these are PDF-417 (widely used in the automotive industry) and Code 16K. PDF-417 can have from 3 to 90 rows, and Code 16K up to 16 rows.

The second encoding method, matrix codes, uses an array of squares (e.g. data matrix) or hexagons (e.g. Maxicode) to encode the data and some method of providing a position datum. The centre datum circles used with Maxicode can be clearly seen.

Stacked codes can be read with normal scanner systems, but need careful alignment. Matrix codes need to be viewed as a whole, and they are read by a charge coupled device (CCD) TV camera and the digitized image analysed by a computer. This allows the reading to be largely independent of rotational errors. The use of a CCD camera removes the need for scanning; all that is required is a fairly high and even level of illumination with no glare or reflections. For this reason it is likely that CCD reading will become more common for linear bar codes as well.

Considerable self-checking is built into these codings. The structure itself has a definite machine readable format which is easily checked by a machine (e.g. the 3/9 relationship in Code 39). In addition the last digit in the code is usually a check digit which is formed using ideas similar to the CRC method described in Section 5.2.7.

Industrial systems, unlike supermarkets, will normally use automatic bar code marking systems and unattended readers. A bar code is read by scanning a light beam across the code and detecting the reflection. Visible or infrared light can be used with LEDs or low powered lasers as the source. Infrared is attractive for industry because reads can be made through oil and grease coatings. Usually the scanning is performed by a vibrating mirror.

The reflection can be specular (as occurs with a mirror) or diffuse (as occurs from a sheet of paper). Bar code readers rely solely on diffuse reflection. A bar code reading system can thus be represented by Figure 4.42. The light beam should not strike the bar code at 90° as might be first thought, as the resultant specular reflection could dazzle the receiver. Usually angles between 60° and 80° are used.

A bar code is scanned continuously, not just once, and a good read is declared when the same information has been received several times. The number of identical reads needed is set by the designer, but five is a typical number.

The physical relationship between the scan and the bar code determines how many reads can be attempted. A bar code can be arranged vertically (called ladder orientation) or horizontally (called picket fence).

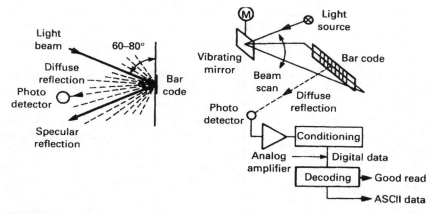

Figure 4.42 *Operation of a bar code reader*

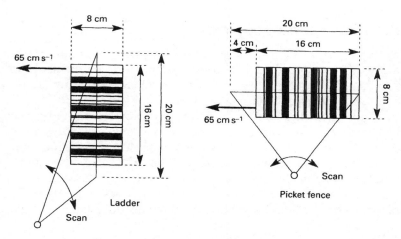

Figure 4.43 *Timings for a bar code read for ladder and picket fence orientation*

A reader at a given distance from a target bar code will have a fixed scan length. In Figure 4.43, a reader with a scan length of 20 cm and a scan time of 5 ms (200 scans per second) is being used to read a bar code sized 16 cm by 8 cm. The bar code is moving transversely at 65 cm/s. In the ladder orientation it will remain in view for 8/65 = 0.123 s which will allow 24 reads (each taking 5 ms). With the picket fence arrangement it can only travel 4 cm whilst remaining fully in view, which, by a similar calculation, will only allow 12 reads. The chance of getting (say) five identical reads increases with the number of reads, so for this application the ladder arrangement is obviously the correct choice.

The reliability of a bar code system is measured by the first read rate (FRR) which is how many times the first scan gives the correct data. The FRR determines how many identical readings are needed before a good read is declared. Typical FRRs are 90% for which three to five identical reads would be needed. Systems can be made to operate at much lower FRRs by increasing the number of identical reads required.

Industrial systems requiring product identification will generally use a bar code reader driven by a PLC. The Allen Bradley PLC family, for example, includes bar code readers which can be directly connected to a PLC rack via a bar code interface card. This uses block transfer read/ write (see Section 4.4.5) to transfer set-up data to the reader, trigger the read, and receive bar code data back from the product. This section is based on material supplied by Allen Bradley.

4.10   High-speed counters

We saw in Section 2.2 that the scan time limits the maximum count rate of a PLC to about 10 Hz. High-speed counter cards are available for use where higher count speeds are needed, or the program scan time introduces an unacceptable random error.

In Figure 4.44, the counter is driven by a directional pulse encoder (which produces two offset pulse trains as shown, allowing the count direction to be observed). The counter value can be loaded from the PLC, and read back when needed. The PLC can also download a preset value, and the counter card drives these outputs showing the relationship between the count and the preset. These outputs are DC to minimize delays (a 50-Hz signal has a 10–20-ms uncertainty).

4.11   Intelligent modules

Most PLCs can be fitted with a wide range of intelligent modules. As well as the bar code readers and high-speed counters described above, intelligent modules can include vision systems for pattern recognition (useful in quality control), position control systems for CNC machines and robotics plus sensor modules for thermocouples and PT100 temperature transducers. All minimize the programming effort needed in the main program. There are also add-on processor modules which allow complex mathematical codes to be written in high-level languages such as Basic or C and linked into the PLC program.

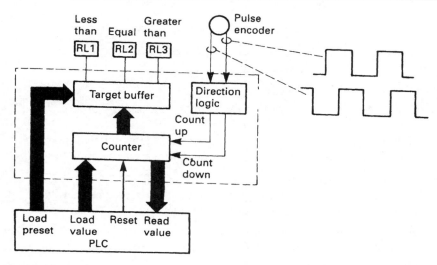

Figure 4.44 *An intelligent module; a high-speed counter*

4.12 Installation notes

Analog systems are generally based on low voltages and are consequently vulnerable to electrical noise. In most plants, a PLC may be controlling 415-V high-power motors at 100 A, and reading thermocouple signals of a few millivolts. Great care must be taken to avoid interference from the high-voltage signals.

The first precaution is to adopt a sensible earthing layout. A badly laid out system, as in Figure 4.45, will have common return paths, and currents from the high-powered load returning through the common

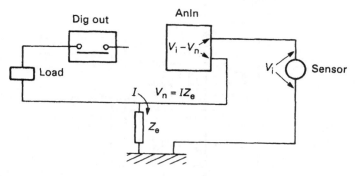

Figure 4.45 *Interference from poor earthing*

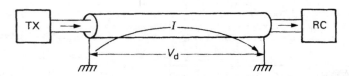

Figure 4.46 *Earth loop formed by multiple earthing of screen*

impedance Z_e will induce error voltages into the low-level analog circuit. It should be realized that there are three distinct 'earths' in a system:

(a) A safety earth (used for doors, frames, etc.)
(b) A dirty earth (used for high-voltage/high-current signals)
(c) A clean earth (for low-voltage analog signals)

These should meet at one point and one point only (which implies that all analog signals should return, and hence be referenced to, the same point).

Screened cable is needed for all analog signals, with foil screening to be used in preference to braided screen. The screen should *not* be earthed at both ends as any difference in earth potential between the two points will cause current to flow in the screen as in Figure 4.46, and induce noise onto the signal lines. A screen must be earthed at one point only, ideally the receiving end. When a screened cable goes through intermediate junction boxes, screen continuity must be maintained, *and* the screen must be sleeved to prevent it from touching the frame of the junction boxes. In the author's experience, this needs almost personal involvement as contract electricians seem almost brainwashed to earth screens everywhere despite written instructions to the contrary! Earthing faults in screened cables can cause very elusive problems.

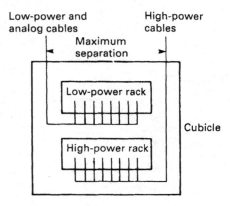

Figure 4.47 *Cable separation*

High-voltage and low-voltage cable should be well separated; most manufacturers suggest at least 1 m between 415 V and low-voltage cables but this can be difficult to achieve in practice. In any case, separation can only be achieved until some other person, not knowing the system well, straps a 415-V cable to the same cable tray as a multicore thermocouple cable. The author tends to use trunking or conduit for low-voltage signals as a way of identifying low-voltage cables for future installers. Some people achieve the same result by using cables with different-coloured PVC sleeves.

In an ideal world, separate cubicles should be provided for 110-V/ high-current signals and low-voltage signals, but this is rarely cost-effective. Where both types of signals have to share a cubicle the cables should take separate, well-separated routes, and the cards be separated as far as possible, as summarized in Figure 4.47.

5 Distributed systems

5.1 Parallel and serial communications

Cabling is one of the most costly parts of any control scheme. There is the cost of the cable itself, the support structure and cable tray, plus the labour costs of pulling cable, ferruling and terminating the ends. If, in the course of commissioning, it is discovered that some extra signals are needed and there are insufficient spare cores, another expensive cable will have to be pulled, with all the attendant costs and time delay.

Figure 5.1 shows two PLC systems that need to exchange data. As shown there are eight signals one way, 12 signals the other at 110 V AC, and two 16-bit numbers at 24 V DC. Along with supplies, neutrals and DC returns this represents 56 cores needing, probably, one 27 core and one 37 core steel wire armoured cable, 3 off 110-V 8-bit digital output cards, 3 off 110-V 8-bit digital input cards, 2 off 24-V digital output cards and 2 off 24-V digital input cards. All the cards require labour to terminate them inside the cubicles at each end. All told, it is not a cheap exercise.

Examples similar to Figure 5.1 are common. At my plant there are arc furnaces (each controlled by PLCs) whose fume extraction is handled by baghouses (each controlled by a separate PLC). The two PLCs for each furnace need to exchange information so the fume extraction can set its fan speeds, suction level, etc. to the furnace performances, and to ensure that each furnace is interlocked with the baghouse operation.

In Chapter 1 we described how remote I/O can be used to reduce cabling costs. In this chapter we will see how similar ideas can be developed to provide communication between PLCs, computers and intelligent instruments.

Figure 5.1 is a form of parallel transmission; all the data to be sent are passed simultaneously. This method is widely used (at lower voltages) to connect computers to printers and for bus-based computer instrumentation schemes such as the IEEE-488 bus described in Chapter 7.

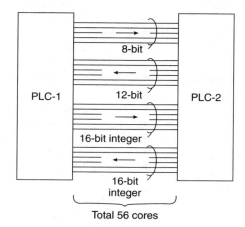

Figure 5.1   *Parallel data transfer*

In Figure 5.2 a single data line (plus a return) connects the transmitter and the receiver, and the data are transmitted as a serial string of bits. Since computers, peripherals, PLCs, etc. all work internally in parallel for speed, parallel to serial conversion is required at the transmitter and serial to parallel conversion at the receiver. The simplest way of achieving this is to use shift registers as shown in Figure 5.2, into which data can be loaded in parallel and shifted out one bit at a time. Specialist integrated circuits called UARTs (universal asynchronous receivers–transmitters) are used to provide this conversion and the control functions. Not surprisingly, this is known as serial transmission.

The advantages of serial transmission arise from cost and flexibility. All that needs to be installed for bidirectional communication is a small, cheap, usually four-core (two pair) screened cable, although the signal

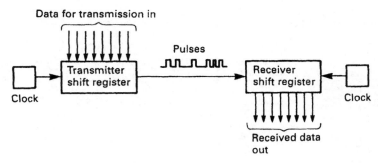

Figure 5.2   *Serial data transmission*

levels are small and there is usually a cost penalty in that trunking or conduit needs to be used for protection.

Once installed, a serial communication system is not really constrained in the amount of data that can be passed (although there will be a time penalty for large amounts of data). Additional data items can be added with no installation costs.

The disadvantages are speed, noise immunity, safety and program comprehensibility. Serial communication is obviously slower than parallel transmission (by a factor equal to the number of parallel lines). This is generally not a problem; on a dedicated PLC communication system a response time of 0.5 s is easily achievable (and remote I/O systems normally achieve around 30 ms). Response times can be longer on commercial systems such as Ethernet, but these are generally not interfacing directly with human beings or a plant in time-critical applications.

The voltages in serial transmission are low, usually of the order of 10 V, and hence prone to noise. Care needs to be taken in the installation (conduit or trunking, separation and screening are advisable) and proprietary systems include methods for error detection and repetition of faulty messages.

Despite these error-detecting correction schemes, a serial communication system should never be considered totally secure, and must *not* be used for purely safety functions such as emergency stops. These must always be hardwired. We will return to safety considerations in Chapter 8.

Finally we have program comprehensibility. The idea of serial communications can be difficult to comprehend in the middle of a fault at 3.00 a.m. Essentially, what we are achieving is to link two areas of memory in separate PLCs as shown in Figure 5.3. This added complexity can bring great confusion if it is not supported by good plant documentation.

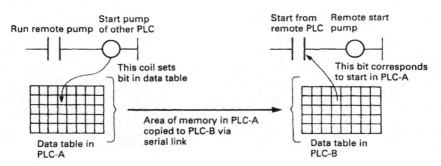

Figure 5.3   *Inter-PLC serial communication*

5.2 Serial standards

5.2.1 Introduction

For a serial communication system to work, there needs to be a consistency between the transmitter and the receiver. There must be definition of:

1 Signal voltage levels.
2 The transmission code (what the bit patterns mean and how the message is built up).
3 Transmission rates (the speed at which the bit pattern is sent).
4 Synchronization. In Figure 5.2 we showed clocks at both ends of the link. If these have a small difference in frequency (as they inevitably will) the receiver will get out of alignment with the transmitter. Some method must be provided to give synchronization between transmitter and receiver.
5 Protocols. Apart from the data, there will need to be some method for the transmitter and receiver to interchange control signals such as 'I am unable to receive a message at present'.
6 Error-checking methods and recovery procedures ('that last message didn't make sense, please send it again').

Getting equipment from different manufacturers to work together over a serial link can sometimes be very difficult. The problems usually arise out of differences in one (or more) of the above points.

5.2.2 Synchronization

The theoretically simplest way to achieve synchronization is to have a common clock for both the transmitter and the receiver, as the two can never, in theory, get out of alignment. This is known as synchronous transmission.

Most systems, however, are asynchronous, and use separate clocks, as in Figure 5.4. The messages are broken down into characters

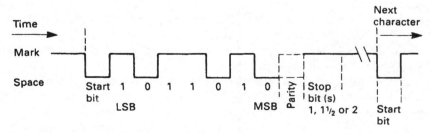

Figure 5.4 *An asynchronous data character*

(typically 5–8 bits in length) and the two clocks are synchronized at the start of each character.

The idle state of the line is a '1' signal (called a 'mark' in telecommunications). The character starts with a '0' signal (called a space) followed by the data bits, usually with least significant bit first. An error-correcting bit (called the parity bit) is sometimes added after the data bits. This is discussed in Section 5.2.7. Finally, the signal returns to the idle mark state for a time before the next character can be sent. This is known as the stop bit and can be 1, 1.5 or 2 bits in width, depending on the system. The next character can follow a random time after the stop bit. The transmitter and receiver clocks are synchronized at the start bit, and only have to stay aligned for the 10 or so bits needed to send a character.

It may be thought that, with noise, mark to space transitions in the data could be mistaken for start bits. In practice, the link will pull itself back into synchronization in a few characters as shown in Figure 5.5. A framing error is signalled by the UART when it receives a zero where it would expect a stop bit.

5.2.3   Character codes

Many types of character code have evolved over the years, but now the almost universal standard is the ASCII code (American Standard Code for Information Interchange, also known as ISO 646) shown in Table 5.1. Variations on this are the CCITT alphabet No. 5, and national options such as the £ symbol in the UK.

ASCII is a 7-bit code giving 128 different combinations covering full upper/lower case alphanumeric characters along with punctuation and 32 control characters that we will return to in Section 5.2.6.

5.2.4   Transmission rates

The transmission signalling rate is expressed in baud, which is the number of signal transitions per second. For the majority of serial links that we shall consider, with two signalling states (0 and 1), the baud rate and the bits/s are identical. For linking a PLC with an instrument, a rate of 1200 baud might be typical. For proprietary PLC to PLC or remote I/O links, with high-quality communication cable, rates as high as 115 kilobaud will be used.

This should not be interpreted as an ability to send 115 000 bits of data down the cable in 1 s. We have already seen in Figure 5.4 that splitting the data into characters with start/stop bits involves some overheads, which increase with the error checking when full messages are sent.

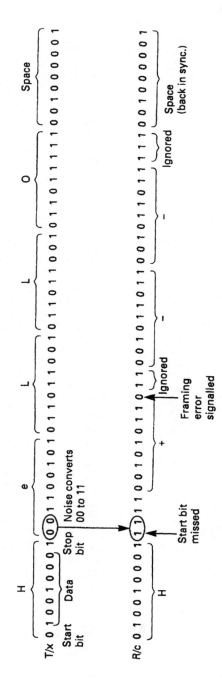

Figure 5.5 Framing errors and the ability of an asynchronous transmission to recover from a fault; 7 data bits and a single stop bit used, ASCII coding, parity not used

Table 5.1 ASCII codes

		Control characters			
Decimal	*Hex*	*Char*	*Decimal*	*Hex*	*Char*
0	00	NUL	14	0E	SO
1	01	SOH	15	0F	SI
2	02	STX	16	10	DLE
3	03	ETX	17	11	DC1
4	04	EOT	18	12	DC2
5	05	ENQ	19	13	DC3
6	06	ACK	20	14	DC4
7	07	BEL	21	15	NAK
8	08	BS	22	16	SYN
9	09	HT	23	17	ETB
10	0A	LF	24	18	CAN
11	0B	VT	25	19	EM
12	0C	FF	26	1A	SUB
13	0D	CR	27	1B	ESC

Control characters can be obtained via the use of the CONTROL key and the character in the right-hand column. Backspace (BS) for example is ctrl-H.

		Printable characters			
Decimal	*Hex*	*Char*	*Decimal*	*Hex*	*Char*
28	1C	FS	48	30	0
29	1D	GS	49	31	1
30	1E	RS	50	32	2
31	1F	US	51	33	3
32	20	space	52	34	4
33	21	!	53	35	5
34	22	"	54	36	6
35	23	#	55	37	7
36	24	$	56	38	8
37	25	%	57	39	9
38	26	&	58	3A	:
39	27	'	59	3B	;
40	28	(	60	3C	<
41	29	)	61	3D	=
42	2A	*	62	3E	>
43	2B	+	63	3F	?
44	2C	'	64	40	@
45	2D	–	65	41	A
46	2E	.	66	42	B
47	2F	/	67	43	C

Table 5.1   (*cont.*)   *Printable characters*

Decimal	Hex	Char	Decimal	Hex	Char
68	44	D	98	62	b
69	45	E	99	63	c
70	46	F	100	64	d
71	47	G	101	65	e
72	48	H	102	66	f
73	49	I	103	67	g
74	4A	J	104	68	h
75	4B	K	105	69	i
76	4C	L	106	6A	j
77	4D	M	107	6B	k
78	4E	N	108	6C	l
79	4F	O	109	6D	m
80	50	P	110	6E	n
81	51	Q	111	6F	o
82	52	R	112	70	p
83	53	S	113	71	q
84	54	T	114	72	r
85	55	U	115	73	s
86	56	V	116	74	t
87	57	W	117	75	u
88	58	X	118	76	v
89	59	Y	119	77	w
90	5A	Z	120	78	x
91	5B	[	121	79	y
92	5C	\	122	7A	z
93	5D	]	123	7B	{
94	5E	^	124	7C	|
95	5F	–	125	7D	}
96	60	'	126	7E	~
97	61	a	127	7F	DEL

5.2.5 *Modulation of digital signals*

So far we have considered a serial link transmitting digital data in its 'raw' form, i.e. as a series of voltage levels directly representing the bit pattern we wish to send. This is known as the baseband transmission.

A digital signal has a bandwidth from 0 Hz (DC corresponding to a string of continuous zeros or ones) to at least half the bit rate. Many transmission media, such as radio telemetry and the telephone network, have inherent low-frequency limitations and cannot handle baseband signals.

The data are therefore modulated onto a carrier wave. There are three different ways of achieving this: amplitude shift keying (ASK),

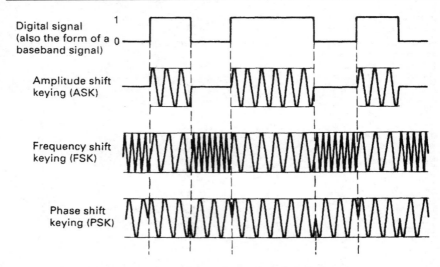

Figure 5.6   *Various forms of modulation for digital signals*

frequency shift keying (FSK) and phase shift keying (PSK). All are summarized in Figure 5.6. One advantage of modulation is that it allows several independent signals, modulated onto different carrier frequencies, to be carried on the same cable. A modulated digital signal is said to be using broad band or carrier band transmission. Often the term 'carrier band' is used to imply FSK with one signal on the cable, and 'broad band' is used where several signals share the cable.

Broad band and carrier band both require devices to interface the digital signals at the receiver and transmitter to the transmission media. These modulate the signal at the transmitter, and demodulate it again at the receiver. Such devices are known as modulators/demodulators or modems.

Figure 5.7 shows a typical two-way arrangement using the public telephone network and FSK. 'Originator' refers to the station which originally established the link; subsequent communications are bidirectional.

In Figure 5.7 there are two types of equipment whose names, and more commonly abbreviations, appear widely in data transmission and are the source of much confusion. The equipment at the transmitting and receiving ends is known as data terminal equipment (DTE). This covers computers, PLCs, printers, terminals, VDUs, graphics displays, etc. The communication equipment (i.e. the modems) is known as data communication equipment (DCE).

The confusion arises because communication standards and protocols are concerned with connecting a DTE and a DCE. When we link a PLC

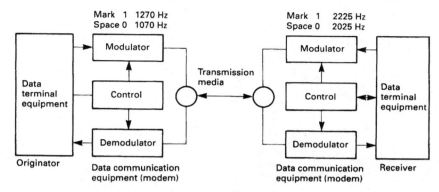

Figure 5.7 *A typical digital transmission system using modems*

and a printer we are linking two DTEs, and will probably have difficulties. We will return to this problem in the following section.

5.2.6 Standards and protocols

RS232E

For successful communications to take place, a set of rules must exist to govern the transmission of data. These rules can be split into standards, which govern voltage levels, the connection and control of DTE–DCE interface, and protocols, which determine the content and control of the message itself.

Much of the early work on data transmission was done by the Bell Telephone company in the USA, and the result of their work was formalized by the Electrical Industries Association (EIA) into 'A standard for the interface between DTEs and DCEs employing serial binary interchange'. This standard is known as RS232 and is currently at revision E.

Worldwide standards are set by the Comite Consultatif International Telephonique et Telegraphique (CCITT), which is a part of the United Nations International Telegraph Union. The CCITT publishes standards and recommendations, those for data transmission being prefixed by letters V or X. Standard V24 is, for all practical purposes, identical to RS232.

Signal levels defined for RS232 and V24 are +6 V to +12 V at the source for a space (zero) and −6 V to −12 V for a mark (one). These are allowed to degenerate to +3 V and −3 V at the receiver. Other characteristics such as line capacitance and edge speeds are also defined. The connections are made with a 25-pin D-type connector. Figure 5.8 summarizes the main connections between a DTE and a DCE and the

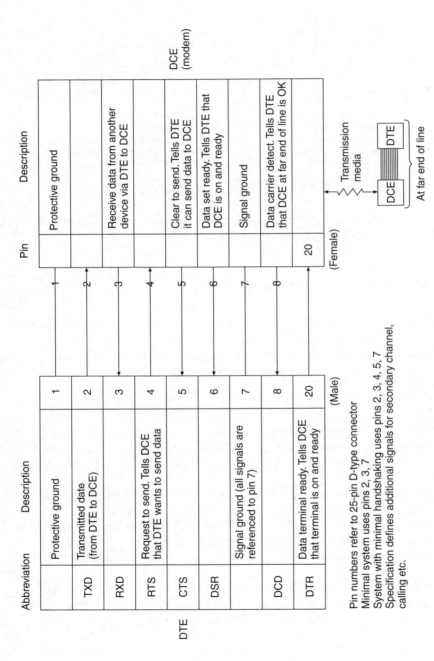

Figure 5.8 *Connections between DTE and DCE as specified by RS232*

Pin numbers refer to 25-pin D-type connector
Minimal system uses pins 2, 3, 7
System with minimal handshaking uses pins 2, 3, 4, 5, 7
Specification defines additional signals for secondary channel, calling etc.

meanings attached to each. These are only a subset of the full specification (which is a rather lengthy document and very heavy going).

There are many common sources of trouble with 'standard RS232'. The standard covers the connection of a DTE and a DCE. Connecting a PLC to a printer is linking two DTEs. Theoretically, a 'null modem cable' which crosses signals such as pins 2 and 3 (data transmit and receive) should work, but usually does not. Manufacturers usually assign their own, often peculiar, ideas to the pin allocation. Many printer manufacturers, for example, use pin 2 to receive data and make the printer a DCE. Even IBM use a 9-pin D-type connector (rather than the standard 25 pin) for RS232 connections on their AT range of computers. Acorn (for reasons known only to themselves) used a 4-pin DIN plug on their BBC computer, and this has proliferated into other equipment.

Allied with this is an almost random interpretation of the use of the control signals. It is not unknown for an 'RS232-compatible instrument' to have just two connections (corresponding to pins 2 and 7 on the DTE in Figure 5.8). Such a device can have no data flow control at all.

'RS232 compatible' thus nearly always means an extended period with a breakout box or line analyser (both essential equipment for use with serial links) and a collection of crimp plugs/sockets and D-type shells. We will return to this problem in Section 5.2.8.

RS422 and RS423

RS232 was designed for a short-haul link between a DTE and a DCE, usually within the same room. If RS232 is used at high speeds over long distances (greater than a few metres), problems will occur.

The EIA have acknowledged the limitations of RS232 for DTE/DTE communications, and have issued two other standards, illustrated in Figure 5.9. One major problem with RS232 is the referencing of signals to a common ground (pin 7 in Figure 5.8) as in Figure 5.9(a). RS423 and RS422 (Figure 5.9(b) and (c)) use differential receivers to dispense with the ground connection and overcome common mode noise.

Nominal transmitter voltages are $\pm 6\,V$ with the signal sense being determined by the relative polarity. Connection A is negative with respect to B for a mark (one) and vice versa for a space.

RS423 uses a single-ended transmitter and a differential receiver, allowing a standard RS232 transmitter to be used provided the difference in ground potentials does not exceed $4\,V$. RS422 uses both a differential transmitter and receiver. The mechanical (37-pin connector) details are defined in RS449.

The 20 mA loop

An unofficial early 'standard' is the 20-mA loop. This originated with the early electromechanical teleprinters, but is still found in many

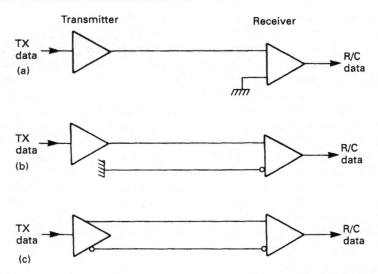

Figure 5.9   *Common data transmission standards: (a) RS232; (b) RS423; (c) RS422*

applications. It consists of a switch driven by data at the transmitter, a current source, and a current sensor at the receiver. Presence of current is a mark (one) and absence a space (zero).

The current loop, isolated from earth, gives good common mode noise immunity, and overcomes differences between ground potentials at either end of the loop. This is the main reason for its continued use.

Unfortunately there are no common standards for control, or even which end of the loop provides the current source. Figure 5.10(a) is known as active transmit, passive receive, and Figure 5.10(b) is passive transmit, active receive. Little communication can take place between a passive transmitter and a passive receiver.

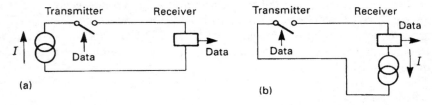

Figure 5.10   *The two forms of 20-mA data transmission: (a) active transmitter, passive receiver; (b) passive transmitter, active receiver*

Message protocols

The standards described above cover the 'mechanics' of data transmission. The message content is defined by the protocol used. In addition to defining the form of a message (i.e. what group of bits form characters, and what groups of characters form a message), the protocol must define how communication is initiated and terminated, and what actions must be taken if the link is broken during a message. The protocol must also cover how errors are detected, and what action is then to be taken.

There are essentially three types of protocol in use, as shown in Figure 5.11. Character-based protocols (Figure 5.11(a)) use control characters from the ASCII set of Table 5.1 to format the message. Most character-based protocols are based, to some extent, on IBM's BISYNC standard.

Bit pattern protocols, such as IBM's SDLC and ISO's HDLC and CCITT X 25, are based on Figure 5.11(b). Flag characters define the start and end of the message, with the end flag being preceded by some form of error control.

The final type of protocol uses a byte count. The start of the message is signalled by a start flag character followed by a count showing the total number of characters in the message. The receiver counts in the

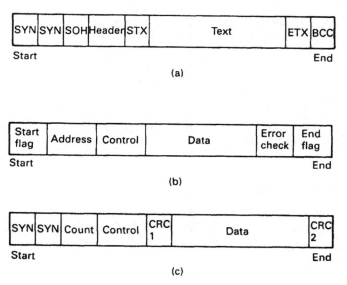

Figure 5.11 *The three types of protocol commonly used in data communications. (a) The basic form of a character-based protocol. (b) The basic form of a bit-pattern-based protocol. (c) The basic form of a byte count protocol. Note that two error checks are used; one for the header (count and control) and one for the data*

message characters and then validates the message with the error-checking data. A common example of this type of protocol is DEC's DDCMP.

Of these, character-based protocols which are variations on BISYNC are probably most commonly used (sometimes called BSC for binary synchronous protocols). They are easy to implement and have the advantage that they can be monitored with a simple terminal across the signal lines.

The control characters from the ASCII set commonly used are:

- Hex 04 EOT End of transmission (often used as a reset to clear the line).
- Hex 16 SYN Synchronizing character, establishes synchronization (i.e. start) and sometimes used as a fill character.
- Hex 05 ENQ Enquiry, used to bid for the line in a multidrop system (see Section 5.3.4).
- Hex 02 STX Start of Text. What follows is the message.
- Hex 01 SOH Start of header. What follows is header information, e.g. message type.
- Hex 17 ETB End of transmission block. Data commenced with STX or SOH is complete.
- Hex 03 ETX End of text. Data commenced with STX or SOH is complete and the end of a sequence block. ETX is normally followed by some form of error-checking information, which is validated by the receiver which replies with either:
- Hex 06 ACK Acknowledgement. Message received error free and I am ready for more data. Also used to acknowledge selection on a multidrop system (see later), or:
- Hex 15 NAK Negative acknowledgement. Message received with errors, please retransmit. Also used to say 'not available' when selected on a multidrop system.

5.2.7 Error control

The addition of noise to a digital signal does not necessarily result in corruption. The original signal can be regenerated at the receiving end providing the noise has not been sufficiently severe to turn a '1' into a '0' or vice versa.

Noise generally has a power density distribution similar to that of Figure 5.12, with zero mean and tails going off to infinity. If the digital signal has voltage levels $+V$ and $0\,V$, noise in region A will corrupt a '0' to a '1', and noise in region B will corrupt a '1' to a '0'. The probability of error is thus the sum of areas A and B divided by the total area under the curve.

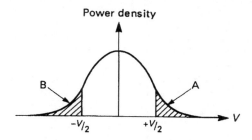

Figure 5.12 *Power density spectrum of a noise signal. Noise in regions A, B will corrupt a digital signal with V volts between a '1' and a '0'*

This probability depends on the ratio between the magnitude of the signal and the noise. The signal to noise ratio, SNR, is defined as:

$$SNR = \frac{\text{Mean square value of signal}}{\text{Mean square value of noise}} \tag{5.1}$$

An SNR of 20 is normally achievable. There is, however, no 'cut-off' value for noise, and there is a possibility of error whatever the value of SNR. This probability can be calculated (using statistical mathematics) and has the form of Figure 5.13. From this graph, a link with an SNR of 20 will have an error rate of 10^{-5}. This sounds good, but it represents some 30 corrupt bits in the transmission of a 360 kbyte floppy disk (which contains 2.88 Mbits).

Rather interestingly, as the signal gets swamped by noise (SNR<1) the error rate does not tend to 1 (as might be first thought) but 0.5. What will be received will be a random stream of '1's and '0's, half of which will, on average, be correct by chance.

With even higher SNRs, 100% reception cannot be guaranteed. A single bit in error can have severe results, changing the sign of a number, or turning an 'open' command to a 'close' command, so some form of error control is generally needed.

An error rate of 1 in 10^5 implies a single error bit followed by 99 999 correct bits. This is not a true picture. Anyone who has used a phone will be aware that interference normally has the form of 'clicks' or 'pops' introduced by the switching of inductive loads local to the line. This is similar to the noise found on data transmission lines. A click of 0.05 s is ignored in speech, but represents the demise of 60 bits of data at 1200 baud. Noise, therefore, tends to introduce short error bursts separated by periods of error-free transmission, and the error rate represents the average over an extended period of time.

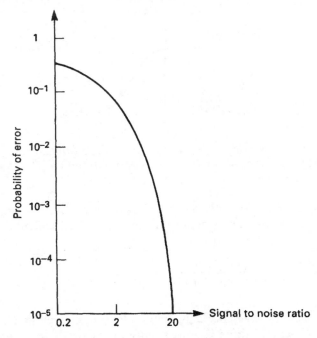

Figure 5.13   *Relationship between error probability and SNR*

There are generally two ways of handling error control. The simplest, used in almost all industrial systems, detects that an error has occurred, and the receiving station asks for a retransmission. This is known as automatic transmission on request, or ARQ. The ASCII characters ACK (received OK) and NAK (received with errors, please send again) are used for handshaking and control.

The second method attempts to detect and correct any errors by adding redundant characters into the message. This is known as forward error control (FEC). The English language contains a lot of redundancy (allowing communication by speech in difficult circumstances). Given the sentence:

Tod?? t?e w?a??er ?s su??y

which has an error rate of 40%, and the fact that it is a statement about the weather, it is quite straightforward to fill in the missing characters to give 'Today the weather is sunny'.

FEC is needed for radio links to and from satellites and is also used for the page addressing on Teletext (which uses a technique called the

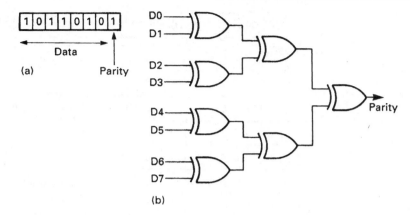

Figure 5.14 *Error checking with a parity bit. (a) The parity bit makes the number of bits in the word an odd number; (b) parity circuit for an 8-bit word using exclusive OR (XOR) gates*

Hamming code). It adds significantly to the message length, and is consequently not widely used in industrial networks.

The simplest form of error detection is the parity bit. This is an extra bit added to ensure that the number of bits in a single character or byte is always odd, as shown in Figure 5.14(a). This is known as odd parity; even parity (parity bit added to make number of bits in each character even) is equally feasible, but odd parity is more commonly used. An ASCII character has 7 bits, so the addition of a parity bit increases the length to 8 bits.

Parity is easily calculated with exclusive OR gates as shown for an 8-bit character in Figure 5.14(b). Parity-calculating ICs are readily available, such as the TTL 74180 and the CMOS 4531.

Parity (or, to give it its full title, vertical parity check) can detect single (or 3, 5, 7) error bits, but will be defeated by an even number (2, 4, 6) of error bits.

Additional protection can be provided by breaking the message down into blocks, each character of which is protected by a parity bit, and following the block with a block check character (BCC) which contains a single parity bit for each column position as shown in Figure 5.15. Normally, even parity is used for the column parity bits. This is known as longitudinal parity checking. The BCC character has its own odd parity bit which is calculated from the BCC character, not the parity bits in the message. The initial STX or SOH are excluded from the BCC calculations, but the terminating ETB or ETX are included.

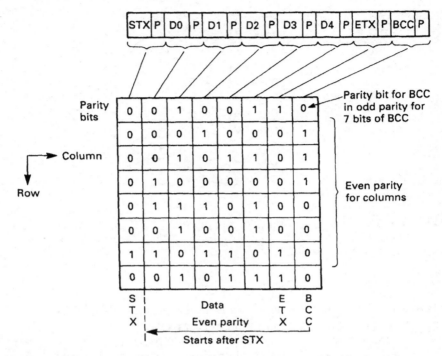

Figure 5.15 *Block check character used for error checking*

BCC can detect all odd numbers of errors, and many multiple-bit combinations. It is defeated by an even number of errors spaced symmetrically around the block.

The most powerful error detection method is known as the cyclic redundancy code (or CRC). Like the BCC method, this splits the message into blocks. Each block is then treated like a (large) binary number which is divided by a predetermined number. The remainder from this division, called the CRC, is sent as a 16-bit number (two characters of 8 bits) after the message. The same calculation is performed at the receiver, errors being detected by differences in the CRC.

The calculation of the CRC is performed with a shift register and exclusive OR gates, a typical example being the CRC-CCITT circuit of Figure 5.16.

A similar scheme, used on GEM-80 links, detects:

- all single-bit errors
- any odd number of errors
- all single and double errors in the GEM message format
- any two burst errors of two bits in the GEM message format
- any single burst of 16 bits or less

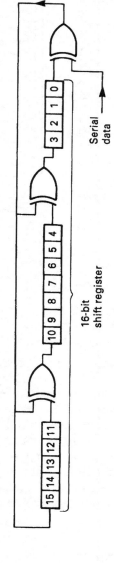

Figure 5.16 Error checking with CRC-CCITT circuit

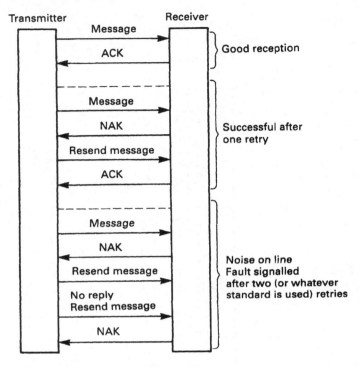

Figure 5.17   *Retries with an ARQ system*

- all but 1 in 32 768 bursts of exactly 17 bits
- all but 1 in 65 536 bursts of greater than 17 bits

The use of CRC-checked blocks greatly improves the error rate. Typical improvements of the order of 10^5 are achieved, giving an undetected error rate of 1 in 10^{10} for a circuit with a basic error rate of 1 in 10^5.

Normally, ARQ systems provide an acknowledgement or an error signal to the initiating device or procedure at the transmitting end, good reception being determined by the reception of the ACK from the receiver. On receipt of a NAK (or no reception of ACK or NAK within a predetermined time) the transmitter will resend the message. To stop a line being clogged with retries, it is usual to set a limit on the number of retries (often three or five) before an error is declared. This procedure is summarized in Figure 5.17.

5.2.8   Point to point communication

A PLC is often required to establish a simple serial link with a device. Typical applications are reading data from an instrument or a bar code reader, or sending data as a setpoint to an instrument or producing a report on a printer. In this section we will look at how this can be achieved with a typical device, an Allen Bradley 1771-DA ASCII module which communicates with the PLC-5 via the BTW and BTR instructions discussed in Section 4.4.5. Other manufacturers' PLCs operate in similar (but of course not identical) manner. It should be appreciated that the description below is a vast simplification of the actual operation, and serves only to outline the principles (the manual is 150 pages long).

Point to point links are usually simple, employing, at most, parity checks for error control. Where data are being read from an instrument, the port on the instrument was probably designed to be connected to a printer, and few, if any, of the control signals on Figure 5.8 will be used. The first step for reading or writing data is therefore to determine:

(a)   the connections on the instrument/device
(b)   the baud rate
(c)   the data format (ASCII, number of bits, parity used, number of stop bits)
(d)   the way the control signals are used
(e)   how message transfer is initiated (when reading data)
(f)   the form of the message

Fortunately (and rarely!) the ASCII module is pinned, and can behave as a pure DTE. Its operating parameters (baud rate, etc.) are set up with data sent from the PLC via a BTW instruction.

The module operation is summarized in Figure 5.18. Data to/from the outside world are buffered. The total size of these buffers is 2K bytes, the split between input and output sizes being set as part of the BTW configuration.

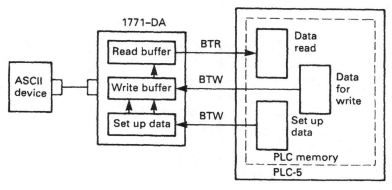

Figure 5.18   *Operation of ASCII module with block transfer instructions*

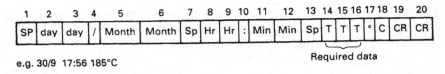

1	2	3	4	5	6	7	8	9	10	11	12	13	14	15	16	17	18	19	20
SP	day	day	/	Month	Month	Sp	Hr	Hr	:	Min	Min	Sp	T	T	T	°	C	CR	CR

e.g. 30/9 17:56 185°C

Required data

Figure 5.19　*A typical ASCII string from a transducer*

Data from the outside world come into the input buffer, and are passed to a block of store in the PLC-5 with a BTR instruction. Data to the outside world are written to the output buffer with a BTW instruction.

A typical input message could come from a temperature transducer with the form of Figure 5.19. We need to know when a message has been received. The message will be read into the input buffer from the instrument, and the PLC allowed to perform a BTR in two circumstances:

1　When the buffer is full (i.e. the buffer has been sized exactly to the size of the data message), or

2　A character predefined as a 'terminator' by the BTW configuration has been received. This is commonly a carriage return < CR > or < ETX > or even a character specific to the application. 'C' or a < CR > could be used for Figure 5.19.

Once in the memory, the data must be converted to numeric form. From Table 5.1, it can be seen that hex 30 (decimal 48) must be subtracted from an ASCII code to give a number. A procedure similar to that in Figure 5.20

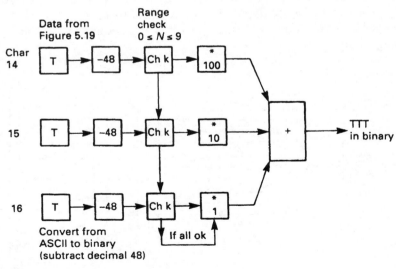

Figure 5.20　*Conversion from ASCII digits to binary*

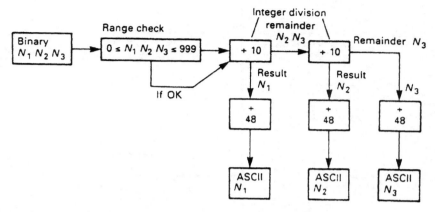

Figure 5.21 *Conversion from binary to ASCII characters*

must be used (with suitable range checking to prevent garbage being accepted).

Writing data is similar, except that the ASCII string must be built up inside the PLC. This requires breaking a number down into a byte for each digit as summarized in Figure 5.21. The data are then sent to the buffer with a BTW instruction.

5.3 Area networks

5.3.1 Introduction

So far we have considered point to point links. For a true distributed control system we need a method whereby several PLCs or computers can be linked together to allow communication to freely take place between any members of the system.

To achieve this we need to establish a connection topology, some way of sharing the common network that prevents time-wasting contention and an address system that allows messages to be sent from one member to another. Such systems are known as local area networks (LANs) or wide area networks (WANs), depending on the size of the area and the number of stations.

5.3.2 Transmission lines

Any network will be based, to some extent, on cable, and at the high speeds used there are aspects of transmission line theory that need to be considered.

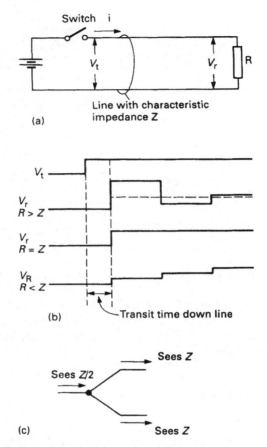

Figure 5.22   *Transmission lines and characteristic impedance: (a) a transmission line; (b) effect of terminating resistor; (c) effect of a branch*

Consider the simple circuit of Figure 5.22. At the instant that the switch closes, the source voltage does not know the value of the load at the far end of the line. The initial current step, i, is therefore determined not by the load, but by the characteristics of the cable (dependent on the inductance and capacitance per unit length). A line has a characteristic impedance, typically $75\,\Omega$ or $50\,\Omega$ for coaxial, and 120–$150\,\Omega$ for biaxial or screened twisted pair. The initial current step will therefore be V/Z where Z is the characteristic impedance.

After a finite time, this current step reaches the load R, and produces a voltage step iR. If R is not the same as Z, this voltage step will not be the same as V, and a reflection will result. Typical results are shown in Figure 5.22(b).

This effect occurs on all cables and is normally of no concern as the reflections only persist for a short time. If, however, the propagation delay down the line is similar to the maximum frequency rate of the signal, the reflections can cause problems. It follows that a transmission line should be terminated by a resistance equal to the characteristic impedance of the line. Normally, devices for connecting onto a transmission line have a high input impedance to allow them to tap in anywhere, with terminating resistors being used at the ends of the line.

A side effect of this is that T connections, or spurs, are not allowed (unless the length of the spur is short). In Figure 5.22(c) a T has been formed. To the signal, coming from the left, the two legs appear in parallel, giving an apparent impedance of $Z/2$ and a reflection.

5.3.3 Network topologies

From the previous section it should be apparent that any network can sensibly only be based on a ring (which needs no terminating resistors) or a line (with a terminating resistor at each end).

Figure 5.23 is a master/slave system where a common master wishes to receive or send data from/to slave devices, but the slaves never wish to talk to each other. All the slaves have addresses, which allows the master to issue commands such as 'Station 3; give me the value of analog input 4' or 'Station 14; your setpoint is 751.2'. Such systems are often based on RS422.

The star network of Figure 5.24 is again based on a master with a point to point link to individual stations. This arrangement is commonly used for high level computer systems. Communication control is performed by the master station. Station to station communication is possible via, and with the co-operation of, the master.

In Figure 5.25 all the stations have been connected in a ring. There is no master, and all stations can talk to any other station and all have equal right of access. The term 'peer to peer link' is often used for this arrangement. With Figures 5.23 and 5.24 control was firmly in the hands of the master. With the ring, some technique is needed to avoid

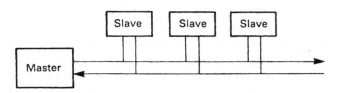

Figure 5.23   *Master/slave network*

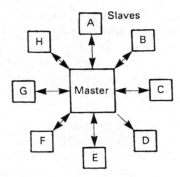

Figure 5.24 *A star network*

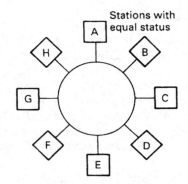

Figure 5.25 *Masterless peer to peer link or ring*

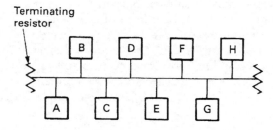

Figure 5.26 *Peer to peer link arranged as a single highway with terminating resistors*

clashes when two stations wish to use the line at the same time. We will discuss this in the following section.

Figure 5.26 is probably the commonest type of network used by PLCs. It is a single line with terminating resistors and, like the ring, is a peer to peer link where all stations have equal standing.

5.3.4 Network sharing

A peer to peer link allows many stations to use the same network. Inevitably two stations will want to communicate at the same time. If no precautions are taken, the result will be chaos. Various methods are used to govern access to the network.

One idea is to allocate time slots into which each station can put its messages. This is known as time division multiplexing, or TDM. Whilst it prevents clashes, it can be inefficient, as a station will have to wait for its time slot even if no other station has a message to send. To some extent a mismatch between the frequency of messages from different stations can be overcome by giving more slots to a hardworking station. With a five-station network and stations labelled A to E, if A has a high workload an order ABACADAEAB, etc. might be adopted. This is sometimes known as statistical TDM.

The empty time slot of Figure 5.27 uses a packet which continuously circulates around the ring. When a station wishes to send a message it waits for the empty slot to come round, when it adds its message. In Figure 5.27, station A wishes to send a message to station D. It waits until the empty packet comes round. Then it puts its message onto the network along with the destination address D. Stations B and C pass the message but ignore it because it is not for their address. Station D matches the address and reads the contents (and appends that it has received the message). Stations E–H ignore it, but pass it on. Station A receives the message back again, sees the acknowledgement and removes its message, leading the empty packet circulating the ring again.

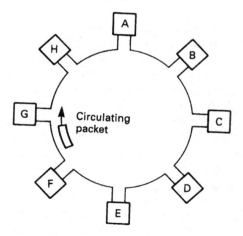

Figure 5.27 *Empty slot and token passing*

A similar idea is a token passing, where a 'permit to send' token circulates round the network. A station can only transmit when it is in possession of the token, which is released when the acknowledgement that the message arrived is received.

Both empty slot and token passing require some way of reinstating the packet or token if the network is corrupted by noise or broken. This is usually provided by a master station, or monitor station, but it should be noted this is not fulfilling the same role as the masters in Figures 5.23 and 5.24.

Empty slot and token passing are usually associated with rings, although they can be used with a bus-based system if the stations are arranged as a logical ring.

Bus systems usually employ a method where a station wishing to send a message listens to the network to see if it is in use. If it is, the station waits. If the network is free, the station sends its message (thereby locking out any other station until the message ends). This is known as carrier sense multiple access (CSMA).

Situations can still arise, however, where two stations simultaneously start to send a message, and a collision (and garbage) results.

This situation can easily be detected, and both stations then stop and wait for a random time before trying again. A random time is used to stop the two stations clashing again. This is known as carrier sense multiple access with collision detection (CSMA/CD).

There is a fundamental difference between TDM, empty slot, and token passing as one group, and CSMA. With the former there is a certain amount of time wasting, but every station is guaranteed access within a specified time. With CSMA there is a little time wasting, but a station can, in theory, suffer repeated collisions and never get access at all.

A useful analogy, for which the author is indebted to Allan Roworth of Siemens, is to consider traffic control. TDM/token passing approximates to traffic lights, and CSMA to roundabouts. In heavy traffic the best solution is traffic lights; everyone gets through and the waiting is shared evenly. Roundabouts can 'lock out' one road when the traffic flow is heavy and uneven from one direction. In light traffic, however, roundabouts keep the traffic flowing smoothly; and there are few things more annoying than being brought to a halt by a red light, then have nothing go past in the other direction.

5.3.5 A communication hierarchy

Early process control systems tended to be based on a single large computer or PLC. The advent of cheap PLCs with good communications has led to the development of a hierarchy of machines which split the

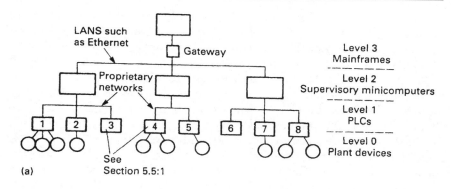

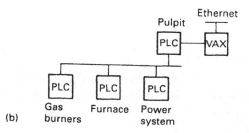

Figure 5.28 *Communication hierarchies: (a) a company-wide network; (b) a real system*

tasks between them. This is generally arranged as in Figure 5.28(a) with a hierarchy split into four levels.

Level 0 is the actual plant, with devices linking to the next level by direct wiring or simple RS232/422 serial links.

Level 1 is the level the majority of this book is concerned with, consisting of PLCs and small computers directly controlling the plant.

Level 2 is microcomputers, such as the DEC VAX, acting as supervisors for large areas of plants.

Level 3 is the large company mainframes, such as IBM's AS400.

Usually the layout is not as clearcut as Figure 5.28 implies. There are also differences between different companies; some number the layers from top to bottom and some ignore level 0. Normally there will be a split of responsibility in the hierarchy; at the author's plant, engineering is responsible for levels 0 and 1 and data processing for levels 2 and 3.

There are many advantages to distributed systems. The resulting tree is conceptually simple, and as such is easy to design, commission, maintain and modify.

A correctly designed system will be, for short periods, fault tolerant and can cope in a limited mode with the failure of individual stations.

At the time of writing the author is concerned with the design of a new arc furnace which employs four PLCs and a VAX computer arranged as in Figure 5.28(b). This split allows individual parts to be designed and commissioned separately, and allows the plant to be put into a safe state if any PLC fails or the communications are lost.

A distributed system can also bring about an increase in performance as lower level machines take the work off higher level machines. In Figure 5.28(b), the pulpit PLC issues broad commands to the lower level PLCs, and concerns itself mainly with data gathering for the VAX system. The lower level machines concern themselves with running the plant and monitoring for alarm conditions, passing any information the operator should be aware of back to the pulpit PLC for display on VDU screens.

5.4   The ISO/OSI model

Neat as Figure 5.28 is, the interconnection between different machines can bring even more problems than linking two 'RS232-compatible devices'. Common problems are different baud rates, flow control, routing and protocols.

In 1977 the International Standards Organization (ISO) started work on standards to try to ensure compatibility between different manufacturer's equipment. This is known as the open systems interconnection (OSI) model, and is primarily concerned with communication between level 2/3 systems in Figure 5.28(a).

It consists of definitions for the seven layers of Figure 5.29. Each layer at the transmission end has a direct relationship with the same layer at the receiving end. The function of each layer is, from the bottom:

1   The physical link layer – concerned with the coding and physical transmission of the message. Requirements such as transmission speed are covered.
2   Data link layer – controls error detection and correction. It ensures integrity within the network and controls access to it by CSMA/CD or token passing.
3   Network layer – performs switching and makes connection between modes.
4   Transport layer – provides error detection and correction for the whole message by ARQ, and controls message flow to prevent overrun at the receiver.
5   Session layer – provides the function to set up, maintain and disconnect a link, and the methods used to re-establish communication if there are problems with the link.

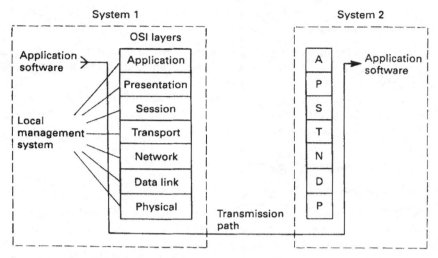

Figure 5.29 *The OSI model*

6 Presentation layer – provides the data in a standard format (which may require the data to be converted from their original form in the initiating application).

7 Application layer – links the user program into the communication process and determines what functions it requires.

As a very rough analogy, consider the placing of a verbal order by telephone. This analogy is based on Siemens material published in their brochure 'Communications Setting the Pace in Automation'.

1 Physical link layer – the phone is lifted and connected to the telephone network. A dialling tone is heard.

2 Error detection and control – it is a good line with no noise.

3 Network layer – the number is dialled, 9 for an outside line and then the number. The phone rings at the other end.

4 Transport layer – the telephone is lifted at the receiving end. 'This is ACME products, could you hold on please, I'm handling another call. OK, go ahead now. Sorry, I didn't get that, could you repeat please?'

5 Session layer – 'This is Aphrodite Glue Works, I have a verbal order for you, number CAP4057, my account number is 7322D.' The receiver makes a note of these details in case the call is broken prematurely.

6 Presentation layer – 'I am using order number from the June 1998 catalogue.'

7 Application layer – 'I require 100 off 302-706 and 50 off 209-417, delivery by datapost.' 'OK, 100 off 302-706 and 50 off 209-417 will

be despatched by datapost this afternoon. Total cost £147.20, invoice to follow.'

At any stage, the lower layers can interact. A burst of noise on the line, for example, will cause the transport layer to ask for a repeat of the last message.

It can be seen that layers (1) to (4) are concerned with the communication and layers (5) to (7) are concerned with processing functions for the particular applications.

5.5   Proprietary systems

5.5.1   Introduction

The ISO/OSI model is mainly concerned with higher level communications such as linking minicomputers. At the level this book is concerned with, we are primarily interested in linking PLCs. Each manufacturer has tended to have its own standard (Modicon's MOD-BUS, Texas Instrument's TIWAY, CEGELEC's ESP) and these link their own equipment in a straightforward manner. If, in Figure 5.28, PLCs 1, 2 and 3 were Allen Bradley, 4 and 5 were GEM-80s and 6, 7 and 8 were Siemens there would be no real problems in linking similar PLCs. Allen Bradley Data Highway would be used for the first three, CEGELEC's CORONET for the GEMs, and Siemens SINEC L1 or L2 for units 6, 7 and 8. Each of these is simple to use and, in the author's experience, very reliable. Linking between the different systems, however, is another story.

In this section we will look at various proprietary systems and at the tentative steps taken to provide standards that allow linking between different manufacturers. All are similar in principle, and provide useful internal diagnostics for fault finding, and less useful green and red communication LEDs on the cards.

For reasons of space we shall consider how machine to machine links are achieved in Allen Bradley, CEGELEC and Siemens PLCs. Each PLC manufacturer has his own proprietary link, some of which are (in no particular order):

- ABB   Masternet and Master Fieldbus
- Gould/Modicon   Modbus
- General Electric   GENET
- Mitsubishi   MelsecNET
- Square D   SYNET
- Texas Instruments   TIWAY

All use similar ideas and are often tantalizingly close; this is a topic that we shall return to in Section 5.5.6.

5.5.2 Allen Bradley Data Highway

PLC-5s communicate with each other on a peer to peer (no master) token passing highway based on twinaxial cable and operating at 57.6 kbaud. Their trade name is Data Highway Plus (an earlier version called Data Highway linked the predecessor of the PLC-5, the PLC-2 range). The PLC station addresses are set on DIP switches on each PLC, and up to 64 stations can exist on one line with octal addresses 0–77.

Communication is established with a single message (MSG) instruction. This can be set up to read or write a block of data, the programmer specifying:

(a)   the start address at the local end
(b)   the start address at the target end
(c)   the length of the block to be transferred (in words)
(d)   the station address at the remote end

In Figure 5.30(a), station 5 is performing an MSG write, sending six words starting from N10:40 to a block from N7:10 at station 12. In Figure 5.30(b), station 7 is performing a MSG read, taking eight words starting from N10:0 at station 12 and copying them into a block starting at its own N7:32.

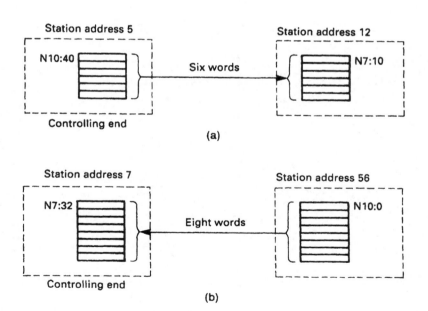

Figure 5.30   *The Allen Bradley message (MSG) instruction: (a) write message; (b) read message*

The MSG instruction appears in a program as in Figure 5.31(a), the transfer being initiated every time the rung goes true. The DoNe bit goes true when it has been successfully completed. The ERRor flag goes true when an error occurs. Common errors are a line fault, a non-existent address at the far end, or the PLC at the far end shut down. The cause of the fault is given in flags set in the message control word. Link statistics (e.g. number of retries) are kept in the processor for diagnostic purposes.

The details of the MSG instruction are set up by the programmer via the screen of Figure 5.31(b). These are mostly self-explanatory, with the possible exception of the remote link, which is concerned with sending data via a gateway module to a different highway, possibly of a different type.

The data highway is also used by the programming terminal, so a programmer can connect anywhere onto the data highway and link into any machine on the network.

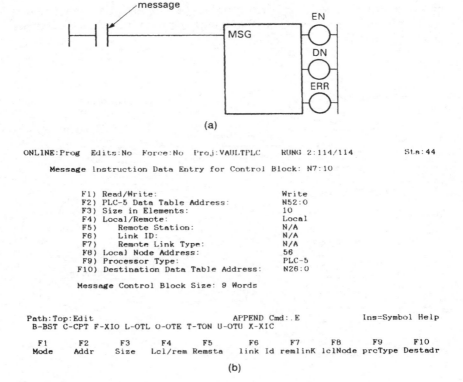

Figure 5.31 *(a) Using the MSG instruction; (b) Set-up data for a message instruction as seen on the programming terminal*

Allen Bradley have followed the hierarchy of Figure 5.28, and one of their products is the Pyramid Integrator, which combines a PLC-5 (the 5-250) and a MicroVax computer in a single rack, providing a direct (backplane) link between the PLCs data table and the VAX program.

5.5.3   Gem-80 Starnet, ESP and CORONET

The GEM-80 has two forms of inter-machine serial communication. The first, and simplest, method, known as Starnet, provides point to point of master/slave communications (similar to Figure 5.23) with a 20-mA current loop and a protocol known as ESP (extended simple protocol). This only provides master/slave communication, but slave/slave communication is possible using the master to relay messages.

The very flexible GEM data table was discussed in Section 2.3.4. Serial communication uses the J and K tables and the P (preset table). The basic form of the mechanism is shown in Figure 5.32.

Addresses in the K table are used to hold data for serial output, and those in the J table for serial input. The P table is used to set up the presets for the link, such as the baud rate, whether a given machine is a master or slave (GEMs use the term 'control/tributary'), the size of the blocks to be transferred and whether the transfers free-run continuously or are initiated by the program.

In free-running mode, the operation of the link is totally invisible to the user; data written into K7 in the master machine, for example, will appear, automatically, in J7 in the tributary port with address 0. The actual operation is more flexible than this simple description would imply; in practice, flexible allocation of the data table gives more control and greater speed.

The second form of GEM communication is a masterless peer to peer link called CORONET. This operates on a screened twisted pair cable (RG108AU) at 9.6K baud to RS485 signal standards. A line length of 4 km is possible with up to 32 stations in the basic form.

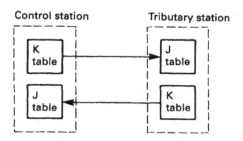

Figure 5.32   *The GEM-80 J and K tables*

The J and K tables hold the Output/Input data and the P table is used to preset each station on the link as before. The link does not free-run, and each transfer has to be initiated by program. This is performed by the I table, to which is written the target address and a send bit to start the transfer. Check bits to say the message has been received are provided in the I table.

Both forms of communication set flags in the F (fault) table, and link statistics showing the number of retries and failed messages are kept.

5.5.4 Siemens SINEC

Siemens PLCs have access to four forms of communication network under the common name SINEC (Siemens Network Architecture for Automation and Engineering). Two of these, prefix L, are low level networks, and two, prefix H, are high level networks.

SINECLI is a master/slave network covering a single master and up to 30 slaves. It operates to RS485 standards on twin twisted pair. The programmer can define the polling order, including repeats as described in Section 5.3.4. The network basically provides master/slave communication, although, like CEGELEC's Starnet, slave/slave communication is possible by having the master act as a repeater.

SINECL2, known as Fieldbus, is a peer to peer link using token passing. This uses an open architecture, and is a possible candidate for a future standard, as discussed in Section 5.5.6.

SINELECH1 is a baseband network operating at 10 Mbaud on coaxial cable with a maximum length of 2.5 km and implementing the first four layers of the OSI model discussed in Section 5.4. It operates to standard IEEE 802.3, better known as thick-wire Ethernet (see also Section 5.5.5). Up to 1024 stations can be supported, with CSMA/CD being used for access control.

SINELECH2B is a broadband network also operating at 10 Mbaud on coaxial cable. It is based on standards IEE 802.4 and IEE 802.7, conforming to MAP 3.0 (see Section 5.5.6). Access control, as required for MAP, is by token passing. With the data being modulated onto a carrier, the cable can also carry other services such as telephones and closed circuit TV.

5.5.5 Ethernet

Ethernet is a very popular bus-based LAN originated by DEC, Xerox and Intel and commonly used to link the computers at level 2 in Figure 5.28. It uses $50\,\Omega$ coaxial cable, with a maximum cable length of 500 m (although this can be extended with repeaters). Up to 1024 stations can be accommodated, although in practical systems the number is far lower. Baseband signalling is used with CSMA/CD access control. The raw data rate is 10 Mbaud, giving very fast response at loading levels up to about 20–30% of the theoretical maximum. Beyond this, collisions start to occur.

Stations are connected onto the cables by transducers known as nodes on the network. Commonly, 'vampire technology' is used for these transceivers, as shown in Figure 5.33(a). The transceiver clamps onto the cable, with a sharp pin piercing the cable and contacting the centre conductor. The arrangement of the pin shrouding prevents it from contacting the screen. This approach allows transceivers to be added, or removed, without disturbing the rest of the network. I must admit to being more than a little apprehensive about vampire technology, but it does seem to work. To avoid reflections (as discussed in Section 5.3.2) a minimum spacing of 2.5 m must be maintained between nodes. To assist the user, Ethernet cable has 'tap-in' points marked on its sleeving. Where a large number of nodes are to be connected locally, the cabling arrangement of Figure 5.33(b) is used.

An alternative to the vampire transceivers is the plug-in transceiver using coaxial cable plugs, as in Figure 5.33(c). These are obviously more secure, but have the disadvantage that the network is disrupted if a node is added or removed.

The transceivers are connected to a local controller which performs the access control. Ethernet has three layers, shown in Figure 5.34, which approximate to the functions performed by the same layers in the OSI model discussed in Section 5.4.

Ethernet is possibly the most successful and widely used LAN. Both ABB Masterview and Siemens SINECH1 are essentially Ethernet (although this is not specifically stated as such in their material).

5.5.6 Towards standardization

We have already discussed the difficulties of linking different equipment. There are normally few problems in linking PLC networks to higher level computers. PLC manufacturers publish their message format and protocols, and interfacing software (called 'drivers') has been written for all common computers and PLCs. The difficulty comes when you want to link two machines at level 1 in Figure 5.28. In many cases, the only economical solution is to do it through the computers and the higher level link.

General Motors (GM) in the USA were faced with this problem and attempted to specify a LAN for industrial control. This was called MAP (Manufacturing Automation Protocol). A similar office-based LAN called TOP (Technical Office Protocol) was conceived at the same time. With GM's purchasing muscle, it involved several automation equipment

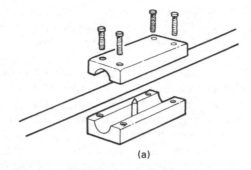

(a)

(b)

Figure 5.33 *Ethernet connections: (a) vampire connector; (b) Ethernet cable arranged in loops to provide connection separation*

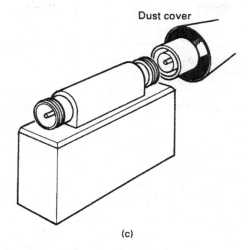

Figure 5.33 *(cont.) (c) break the line screw connector*

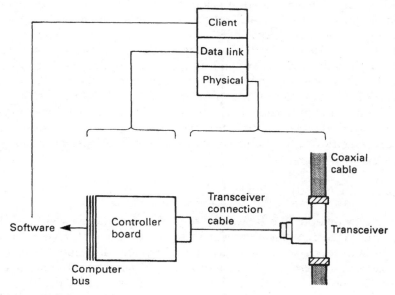

Figure 5.34 *Ethernet architecture*

manufacturers. A firm commitment to the OSI model was made, and the network based on broadband token bus as specified in IEEE 802.4 (compare Ethernet; baseband CSMA/CD, to IEEE 802.3). The token bus was chosen as it is deterministic; the response time can be predicted (see discussion on roundabouts and traffic lights in Section 5.3.4).

MAP (currently at version 3.0 at the time of writing) seems to have gone slightly off the rails. The draft title for this chapter was 'whatever happened to MAP?'. In the course of the research for this book, major PLC manufacturers were visited. Each could interface MAP, but for each (with the exception of Siemens SINECH2B) it seemed to be an expensive add-on which the customer could 'have if he really wanted'.

There appear to be several reasons for this distinct lack of enthusiasm. The first is a bureaucratic organization and a changing specification. The term 'moving target' was used independently on several occasions by different manufacturers. The second reason is cost; MAP links often cost more than the PLC to which they are connected. The expression 'Designed by big organizations, for big organizations' was used, and seems apt. The third reason is speed; by using token passing MAP is slow by comparison with Ethernet and the OSI model is not really designed for time-critical applications. The non-deterministic nature of CSMA/CD does not seem to cause any problems up to about 30% of the theoretical maximum loading, and real systems normally operate below 10% loading. The final, and perhaps most crucial, fact is that MAP seems to have settled at a level where it is in direct competition with established LANs such as Ethernet rather than the proprietary systems at level 1 of Figure 5.28.

In the mid-1980s MAP was going to be the common standard of industrial control. MAP systems have been installed, both in Europe and the USA, but it has not yet achieved anything like acceptance.

A typical example of the problems that a fieldbus system may encounter is the introduction of new ideas. All the communications systems described so far are based on what is called the source/destination model. If station A has information for station B, a message is sent with the format:

Source A | Destination B | Data | CRC

If this information is to be sent to several stations, each will need their own message. In applications where multiple setpoints have to be sent to multiple controllers, the delay caused by the time shift between the messages can cause problems, although this can be overcome to some extent by the use of group or global addresses as used by Profibus.

In addition, if station A needs information from station B (the state of an interlock for example), station A must perform a read on each occasion the data is required.

A recent development, called the producer/consumer model, uses a different approach. Here data is placed onto the network with no indication as to who it is for. The format is now simply:

Identifier | Data | CRC

All stations using this data accept it at the same time, eliminating the need for multiple messages. This significantly reduces the number of messages and hence increases the network speed.

The placement of data onto the network can be done in two ways. The first, and fastest, is 'notify on change'. Here a station only places information on the network when a new value is different to the old. Stations with an interest in this data assume that the status or value remains the same until notified otherwise. There are obvious dangers in this, and a regular pre-defined 'heartbeat' (similar in principle to the later Figure 5.37) is included to say a station is active on the network.

The second approach updates on a time basis, each data item having its own, or a global, update time.

At the time of writing, Foundation Fieldbus is the only producer/consumer fieldbus network, and Rockwell (Allen Bradley) have also adopted the method for their ControlNet. The latter is interesting as it combines the ideas of their remote I/O and Data Highway onto one system and allows PLC racks (and their data) to be shared equally amongst several processors and not dedicated to one as before. There are also European attempts at standardization. In conjunction with the Instrument Society of America, specifications for a low cost (twisted pair) low level network called Fieldbus has emerged. Its full specification was due to be completed by 1992, but (inevitably) has been delayed by commercial and political infighting. It is available (Siemens SINECL2 is one example) and demonstrations linking different manufacturers' equipment can be seen at most automation and control exhibitions. It could, perhaps, fulfil the role that MAP was publicized for.

Other possible contenders are the, originally German, Profibus described below which is again supported by several companies, and the French FIP. Both of these are similar to, but not identical with, Fieldbus.

5.5.7 Profibus

Profibus is one of the more common fieldbus contenders at present, largely because it has been adopted by Siemens and many other German electrical companies. There are three versions of Profibus designed for three different application areas. All use token passing and are based on the ISO/OSI model.

The first, called Profibus-DP, for Decentralized Periphery, is by far the commonest and is designed to link intelligent masters (e.g. a PLC), to slave devices such as sensors, drives or actuators. Profibus only uses levels 1 and 2 of the ISO/OSI model. Twisted pair RS485 or fibre optics are used for transmission.

The second, Profibus-FMS, for Field Message Specification, is designed for the higher level with multiple masters and allowing peer to peer communication. Levels 1, 2 and 7 of the ISO/OSI model are used and RS485 or fibre optics for transmission.

Both DP and FMS share the same transmission standards and can consequently work together on the same network.

The final form, designed for Process Automation in hazardous areas, is Profibus-PA which permits the construction of an intrinsically safe network. Profibus-PA uses slightly different standards to DP and FMS, but can be linked by a segment coupler device.

All are linear bus systems, i.e. a straight line. Transmission speeds from 9.6 kbit/s (up to 1200 m) to 12 Mbit/s (up to 100 m) can be used. Screened twisted pair is used, with terminating resistors at each end of the bus. Up to 32 stations can be used in each segment, each with a unique station address. Segments can be coupled with segment repeaters, allowing a total of 127 stations to be addressed. Addresses are assigned for global or group data reducing the number of messages and time lag problems when data for several devices are to be changed together.

Connections to masters or slaves are made via standard 9-pin D-type connectors, as shown in Figure 5.35(a). Terminating resistors are either switched in internally at the end stations or connected inside the final plugs. Note that the terminating resistors require power, this normally comes from the end stations themselves.

The manufacturer of each device on the network, e.g. a VF drive, provides a disc file, called the GSD, which is a description of the data exchange the device can support (e.g. accepting speed reference and run command and providing load current and drive state, etc.) plus operating parameters such as supported transmission speeds. Included in the GSD file is a unique Identification Number assigned by the Profibus User Organization. The GSD files for all the devices on the network are used along with the station addresses to build a network description which is held in the master.

Because Profibus DP only uses levels 1 and 2, the data exchange maps onto pre-determined areas in the master controller (usually a PLC) as shown in Figure 5.35(b). To change the speed of the drive, the user simply writes the new speed into the mapped area, and the data is transferred with no further action. In a similar manner, slave data and status is automatically read from the mapped area. A Profibus DP network is thus totally transparent to the user.

5.6 Safety and practical considerations

Figure 5.36 shows a fairly common situation where a switch connected to one PLC is, via a serial link, causing a motor to run in another. Supposing the motor is started and the link is severed. The bit corresponding to

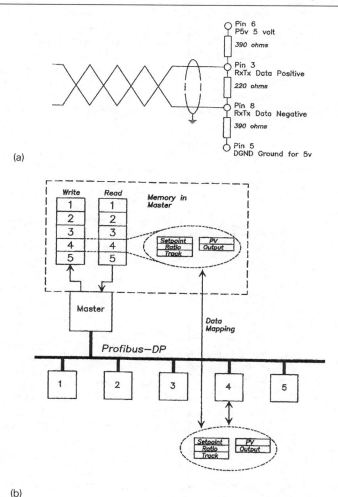

(a)

(b)

Figure 5.35 *Profibus-DP network. (a) Termination and network connections. Non-terminating devices only use pins 3 and 8. A 24 V DC supply for external devices is often provided on pins 2 (–ve) and 7 (+ve). Pin 1 is for the shield. (b) Memory mapping between a device and the master*

'motor run' which is set inside PLCB will not be cleared by the link failure, and PLCA will be unable to stop the motor.

When the switch is turned off, the serial link control in PLCA will signal an error, but this is of no use to PLCB which does not know that PLCA is trying to communicate with it. This may, or may not, be

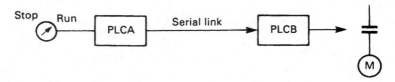

Figure 5.36   *Safety considerations with a serial link*

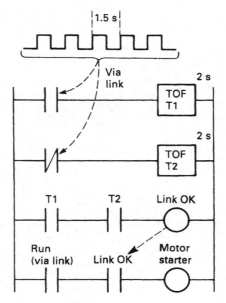

Figure 5.37   *A way of checking a serial link*

a problem, depending on the application, but it is obvious that there are implications that need to be considered.

One approach is to define how long an output driven by the link is allowed to be uncontrolled, say 2 s. The originating PLC then sends a toggling signal via the link at a slightly shorter period, say 1.5 s, as in Figure 5.37. Inside PLCB, the true and complement forms of this signal trigger two TOFs (delay off) set for 2 s. With the link healthy coil energized, the link-driven outputs can be energized. If the link fails, one TOF will de-energize (and one stay energized), causing the 'link healthy' signal to de-energize and all link-controlled outputs to go to a safe state.

A network introduces extra delays into the system. These delays obviously depend on the loading on the network, but are typically of the order of 0.2–0.5 s on proprietary networks, and a bit slower on Ethernet and MAP.

Noise is a major source of problems, and normally manifests itself as an increase in the delay time introduced by the network (caused by a large number of retries). Because of checking and CRC discussed earlier, noise rarely causes operational problems, and when it does (in really severe cases) the effect is almost always something not working when requested (rather than something starting unexpectedly). Noise prevents signals getting through; it does not usually cause faulty signals to be accepted.

Obvious precautions against noise are separation from power cables, and the use of conduit or trunking (mainly to identify low-signal-level cables). Cable screens should be continuous and earthed in one, and only one, place. Great care should be taken to prevent screens accidentally grounding inside junction boxes.

Most proprietary networks have monitoring facilities. Figure 5.38 shows the diagnostics available for the Allen Bradley Data Highway Plus. Some errors are inevitable on all systems (see Figure 5.13) and it is worth logging the rate when a network is first commissioned. This allows checks to be made at a later date, and any deterioration noticed early before problems start to arise.

Fibre optics, discussed in the next section, give almost total freedom from interference.

5.7 Fibre optics

When light passes from one medium to another, the beam is bent as in Figure 5.39(a). This is known as refraction, and is the cause of water appearing shallower than it really is. If the angle of incidence, α, is greater than a certain critical angle θ_c, the light beam does not emerge from the surface, but is reflected internally as in Figure 5.39(b). This is known as total internal reflection, and it can be shown that

$\sin \theta_c = 1/\mu$

where μ is the refractive index for the two materials.

In Figure 5.39(c), a small-diameter tube of glass has been constructed. Light entering at a shallow angle will be conveyed down the tube with little loss by repetitive total internal reflection. This principle, known as fibre optics, is the basis of an interference-free form of data communication.

The principle is very simple. Data at the transmitter are converted into light pulses which are conveyed down the fibre optic cable and detected by a photosensor at the receiver. Fibre optic cable has a very large bandwidth, so modulation or signal multiplexing allows several high-speed serial channels to be carried down one cable.

There are many advantages to fibre optic cables. The transmission is totally free from problems caused by noise, crosstalk, and ground loops

```
┌─────────────────────────────────────────────────────────────────────────┐
│                 Who Active - Active Station Status                        │
│ ┌─────────────────────────────────────────────────────────────────────┐ │
│ │                                        Ack Timeouts        0          │ │
│ │                         DH+                Nak No Mem Rcvd     0       │ │
│ │   Messages Transmitted  27518          Claim Tokens        1          │ │
│ │   Messages Received     27519          Nak No Mem Sent     0          │ │
│ │   Commands Generated    0              CRC Errors          0          │ │
│ │   Requests Executed     27519          Duplicate Packets   0          │ │
│ │   Reply Sent            27518          Token Timeouts      18         │ │
│ │   PLC 5/25  Series:A  Revision:H       Retries             0          │ │
│ │   Operational Mode:Run                 Adapter Timeouts    0          │ │
│ │                                        Undeliverable       0          │ │
│ │   Rack Errors   1    2    3    4    5    6    7                        │ │
│ │   Timeouts    104  100   92  153  116   0  156   Memory: 13824 Words   │ │
│ │   CRC           0    0    0    0    0   0    0    Memory Unprotected    │ │
│ │   Block Transfers 0  0    0    0    0   0    0    53 Data Files        │ │
│ │   Retries       4   4▼   4    6   28   0    4    4 Program Files       │ │
│ │                                                  No Forces             │ │
│ └─────────────────────────────────────────────────────────────────────┘ │
│   F2 - Clear Counters                                                     │
│   F3 - Freeze/Un-Freeze Counters                                          │
│   Esc - Exit Diagnostic Display                                           │
└─────────────────────────────────────────────────────────────────────────┘
```

(a)

```
ONLINE:Run   Edits:No  Force:No  Proj:VAULTPLC    RUNG 2:114/115          Sta:44
   Message Instruction Data Entry for Control Block: N7:10

Read/Write:                      Write       Ignore if Timed-Out:  0 TO
PLC-5 Data Table Address:        N52:0              To be Retried:  0 NR
Size in Elements:                10          Awaiting Execution:   0 EW
Local/Remote:                    Local              Continuous:    0 CO
Remote Station:                  N/A                     Error:    1 ER
Link ID:                         N/A             Message Done:     0 DN
Remote Link Type:                N/A       Message Transmitting:   0 ST
Local Node Address:              56            Message Enabled:    0 EN
Processor Type:                  PLC-5
Destination Data Table Address:  N26:0

Error Code: 131

Message Control Block Size:      9 Words

Sym:              Des:
N7:10/8 =

  F1      F2      F3      F4      F5       F6      F7      F8      F9     F10
                                          Des    Next    Prev  neWaddr  Help
```

(b)

Figure 5.38 *LAN diagnostics from a programming terminal. (a) Diagnosis for a LAN station. (b) Diagnostic page for a MSG instruction. The destination station (56) has gone off line, causing a fault bit to be set at the transmitter*

and gives total isolation between transmitter and receiver. It can also pass through explosive atmospheres with total safety, as a cable breakage cannot result in sparks.

There are two basic types of fibre. Step index fibre operates as in Figure 5.39(c), with reflections occurring at the fibre wall. Graded index fibre has a non-uniform refractive index, causing the light beam to follow a gentler curve as shown in Figure 5.39(d). Graded index fibres have lower losses.

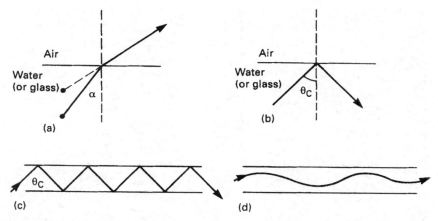

Figure 5.39 *Fibre optic data transmission: (a) refraction; (b) total internal reflection; (c) a fibre optic cable; (d) graded index fibre*

The optical signal is attenuated as it passes down the cable; these losses are usually given in dB/km (typically 5–20 dB/km). Further losses occur at curves (the minimum bending radius is usually related to losses rather than mechanical damage) and at the couplings at each end of the cable. A typical link can operate for 1–2 km without repeaters.

A data transmission cable will usually consist of two fibres (one for each direction) of 200 μm diameter loose inside a protective sheath. Loose sheathing reduces the chance of impact damage. At each end, the sheathing has to be removed, and protective sleeves put onto the individual fibre optic cores as shown in Figure 5.40.

The commonest connector is the SMA connector used in Figure 5.40. This allows fibre optic cables to be disconnected and reconnected like a normal signal cable. Fitting these connectors to the cable is, however, a skilled job. The best method with least subsequent signal attenuation is time consuming, using epoxy resin and a laborious polishing routine. A simpler (and commoner) method uses a crimping tool and specialist cutter. Both require the termination to be checked with a microscope viewing a light source sent down the fibre from the other end.

There are several disadvantages to fibre-optic-based links. The first is that the link is strictly point to point. Topologies such as Ts, multidrops or buses can only be achieved by the use of (expensive) repeaters at each node.

Fibre optic cables are also vulnerable to damage. The cable is not only less robust than conventional cable, but it cannot be easily (or quickly) jointed. With normal coaxial cable, a damaged length can be quite readily cut out and a new length spliced in with through connectors

(a)

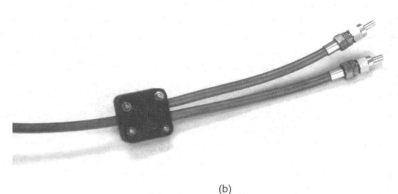

(b)

Figure 5.40   *(a) Fibre optic cable used for communication between programmable controllers; (b) SMA fibre optic connector. Note the cable split and minimum bend radius apparent in both photographs*

and little, or no, ill-effects. Through connectors are possible with fibre optic cables, but they introduce high losses into the link and may even prevent it from working. It is not unknown for a new run of cable to be needed, or a repeater introduced, as a result of a single cable break.

Fibre optic cable should always be well protected with conduit or robust trunking to minimize damage. Although there is no technical reason why fibre optic links should not share cable tray with 33-kV cables, it is not good practice as they will probably be damaged if any more power cables are added.

A final important point is safety. Most fibre optic links use high-power optical sources, sometimes lasers. Never look down a cable 'to see if the transmitter is working'. If there is any doubt about cable continuity, disconnect the cable at both ends (taking care to ensure that the right cable has been disconnected in multicable applications) and use low-power incandescent sources for testing. In many cases, anyway, the source used for data transmission is outside the visible range, and cannot be seen. It can still, however, cause damage to the eye.

6 The man–machine interface

6.1 Introduction

So far we have discussed connecting a PLC to the plant, and the ways in which the control is achieved. A PLC also has to 'connect' to the human operators, accepting commands from them and displaying the status of the plant in a form that can readily and easily be understood. This is known as the man–machine interface, or MMI, and can be summarized by Figure 6.1.

The study and design of this interface is known as ergonomics, and tries to ensure that operators can perform their duties efficiently, in comfort, and with minimum error.

The most important aspect is probably the worker's immediate workspace and environment. Reliable error-free performance cannot reasonably be expected from an operator who has a headache or a sore back within an hour of starting work. Factors such as noise, dust, smell, vibration, humidity, temperature (and temperature changes), lighting levels (and glare) all contribute to a worker's ability to concentrate.

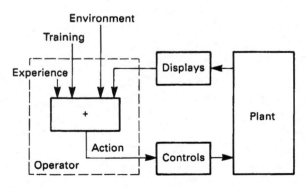

Figure 6.1 *The operator as part of the control loop*

Psychological factors such as stress and the degree of concentration needed are important, as is the ability to mentally rest and 'coast' for a short period from time to time. Directive 90/270/EEC and the HSE booklet *Working with VDUs* cover legislation regarding display screens.

The layout of controls, displays and seating are important. Many desks the author has seen have been laid out for workers 1.5 m tall with a 3 m armspan. Figure 6.2 shows comfortable working positions for seated and standing operators, and Figure 6.3 the boundaries of human perception.

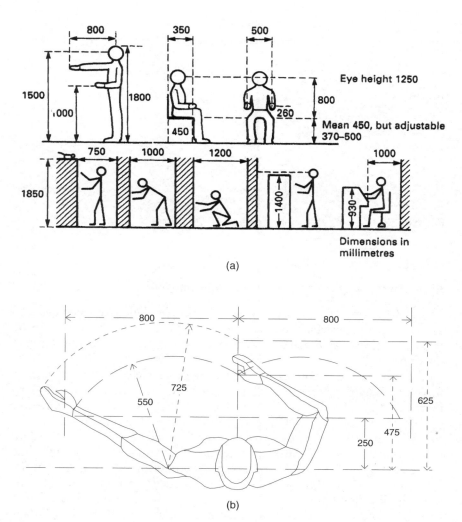

Figure 6.2 *Comfortable working positions*

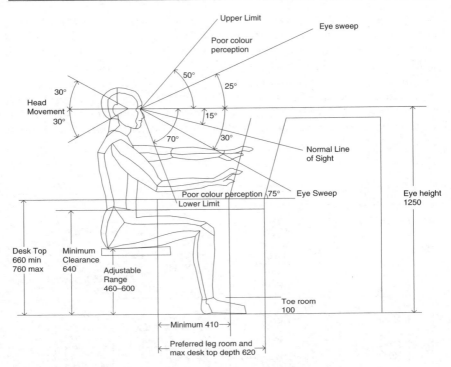

Figure 6.3 *The seated operator and limits of perception*

6.2 Simple digital control and indicators

Most of an operator's controls will be simple digital devices such as switches, pushbuttons, joysticks and indicator lamps. These should be laid out within easy reach and view of the operator, as shown in Figures 6.2 and 6.3.

The function of controls should be made as clear and instinctive as possible. Useful techniques are grouping by function (with boundary marks on the desk surface) or grouping by different device manufacturer (e.g. Siemens controls on the clamp and Telemecanique controls on the press). One of the worst desks the author has seen had 14 visually identical joysticks (but with different motion axis) arranged in a straight line along a (single-operator) 3 m desk top. At a nuclear power station in the USA the operators broke up a similar layout by crimping drink cans over the joystick handles; the Coke can was the long travel, the Fanta can the cross travel, the Pepsi can the raise lower and so on. This is actually a sensible idea if implemented with more style!

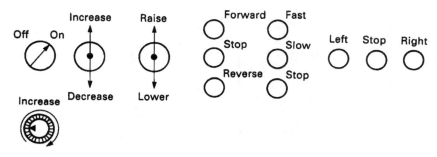

Figure 6.4 *Intuitive control actions*

Desk layouts should be consistent, particularly where operators change jobs at regular intervals (a good practice; it maintains interest, prevents boredom and helps with training). On desks the author has designed, for example, emergency stop is usually top right-hand side, lamp test is bottom left-hand side, and fault indication/alarm is top left-hand side. This layout is not important, but the consistency is. For the emergency stop, the top right-hand side is most convenient for right-handed operators, and minimizes accidental presses.

Consistency of operation is equally important. A clockwise switch for start in one location (the intuitive operation) and an anticlockwise switch somewhere else is confusing. Figure 6.4 shows the expected response for common controls. Where a control causes a plant motion (e.g. long travel, cross travel) the controls should mimic the plant.

Pushbutton colours aid clarity. Recommended colours (in BS-2771) are:

Red   Stop, Off, Emergency Action
Green   Start
Black   Other functions (e.g. jog, reset, test)
Yellow   Intervention (e.g. continue after a fault)

White/grey/blue can be used for 'other functions', but in practical industrial applications dirt degrades the colour clarity. The use of push-on/push-off buttons should be avoided, particularly without a separate state (run/stop) indication.

Similar recommendations exist for the colour of indicator lamps:

Red   Fault, danger, warning, action needed
Amber   Caution, warning, operator should be aware of, deviation from normal, overload
Green   Healthy, sequence running normally, ready
Blue/white Informative, e.g. speed selections

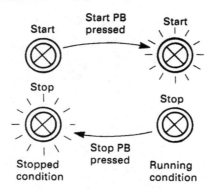

Figure 6.5 *Normal action of illuminated pushbuttons. The extinguished button is pressed, and then lights*

A human being is very good at recognizing visual patterns, and can easily pick out a change. In normal operation a desk should be 'green' with perhaps the odd passing blue or white.

Ambers and reds suggest that action is needed. Wherever a lamp exists, a lamp test pushbutton should be provided.

Illuminated pushbuttons are commonly used to reduce desk space. The intuitive method is to press the extinguished button, which then lights (and the other goes out) as in Figure 6.5. Running illuminated buttons should be green (and stopped buttons red) to give the green desk.

It improves operator confidence if every action is confirmed in some way. In many cases this confirmation is automatic when the result of an action can be seen or heard. If there is no direct feedback (for a remote lubrication unit, say) a confirmatory indicator (probably from an auxiliary contact on the starter) should be provided. Operators intuitively expect a response in a time shorter than one second, after which there is a growing sense of unease.

A plant should not, however, be over indicated. An operator can easily be swamped by a visual overkill. Above all avoid multiple flashing flickering lamp patterns. A flashing light demands action *now* and continual presence means a serious design error.

6.3 Numerical outputs and inputs

6.3.1 Numerical outputs

An operator's desk will often have to display numerical data; hours run, position, temperature and so on. Usually these data will be held inside store locations in the PLC. Most digital displays operate in BCD (see Appendix), so a four-digit display (capable of showing 0000 to 9999)

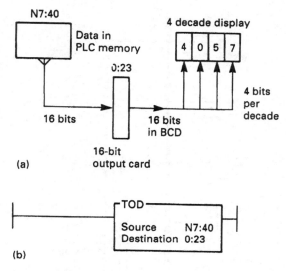

Figure 6.6 *Driving a BCD display: (a) physical connection; (b) TOD (TO Decimal) instruction*

will require 16 output signals from the PLC, usually at 12 or 24 V. Figure 6.6 summarizes this operation for a PLC-5, where the number in store location N7:40 is sent to the digital displays connected to a 16-bit output card in slot 3 of rack 2. One small complication is that, internally, PLCs work in binary. The output to the display is in BCD. All PLCs capable of handling numerical data have a simple instruction to convert from binary to BCD. For a PLC-5 (and the example of Figure 6.6(a)) this is the TOD (TO Decimal instruction) of Figure 6.6(b).

6.3.2 Multiplexed outputs

Figure 6.6 is a reasonable solution for one display, but one display/one output card becomes expensive and wasteful of I/O and cable where many displays are needed. A very economical solution is to use a multi-plexed output. The basic idea is shown in Figure 6.7(a) for four displays (although the idea can be extended almost indefinitely).

The digital displays have 16 data lines as before plus an additional strobe line. If the strobe line is high, the display reads in a number from the data lines. If the strobe is low the display memorizes (and shows) the last data. The four displays share the same output word which cycles through the data to be displayed. Strobe pulses are generated in the centre of each data word as shown in Figure 6.7(b).

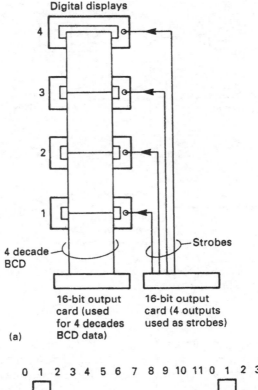

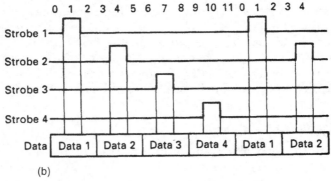

(b)

Figure 6.7 *Multiplexed displays: (a) physical connection; (b) operation*

The basis of the program to achieve this is shown in Figure 6.8. A counter acts as the multiplexer 'clock' driven by pulses from elsewhere in the PLC program (a GEM-80, for example, has clock flags in the E data table). Three clock pulses are allocated per display; the first puts the 16-bit data onto the output word, the second energizes the strobe, and the third does nothing in the program, but has the effect of

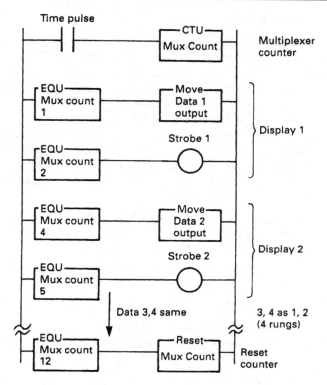

Figure 6.8 *Ladder program to drive the four displays of Figure 6.7. EQU tests for equality of the two items (e.g. the value in the Mux counter and a number) and Move transfers data to the output card*

removing the strobe whilst leaving the data on the output. (Once data have been written to an output, they stay there until the next data are written, whereas the coil will de-energize when its input conditions go false.) The program segment thus causes the strobe to energize in the centre of the data as in Figure 6.7(b).

The one disadvantage is speed. With eight displays, 24 'clock' pulses will be needed per cycle. A typical clock pulse would be 30 ms, giving an update time of just over 0.7 s. This will normally be acceptable, the one possible exception being where the display is used to enter numbers from a keypad (similar to data entry on a hand-held calculator).

The idea can be taken further, down to multiplexing individual digits. An eight-digit display can be driven with just four data lines, three-digit select lines and a strobe. Eight displays of this type can be driven with just 11 bits; four data lines, three digit lines, three display select lines and one strobe. The program, though, would be lengthy (but simple) and the response rather slow.

6.3.3 Leading zero suppression

If we display 25 on a four-digit display, we want it to appear as 25 and not 0025. This is called leading zero suppression. A digital display driven in BCD normally uses one of the unused binary codes 1010 to 1111 to generate a blank display (usually 1111, 'F' in hex). Other unused codes often show + − and a decimal point.

We can therefore provide leading zero suppression for a four-digit display with the three rungs of Figure 6.9. These simply detect 0, 00 or 000 and write hex F, FF or FFF (or equivalent) to the display. It is usual not to blank the bottom digit, so 0 appears as 0. Figure 6.9 works equally well with multiplexed displays. The 'write F' instruction is simply the data OR'd with hex F000, e.g.

Data	0000	0100	1001	0111	(0497 in BCD)
	1111	0000	0000	0000	(F000 in hex)
OR	1111	0100	1001	0111	(F497 in BCD,
					top digit blank)

6.3.4 Numerical inputs

Simple digital inputs (pushbuttons, joysticks, switches) are normally allocated one input per motion, and require nothing particularly special.

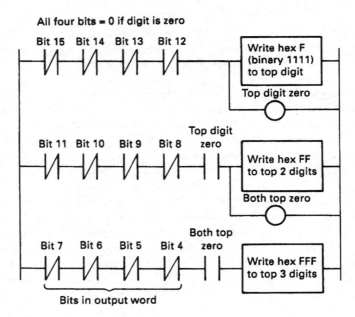

Figure 6.9 *Leading zero suppression*

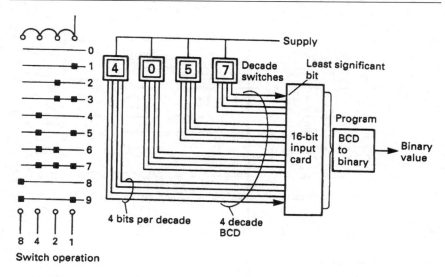

Figure 6.10 *Reading decade switches*

A simplification is possible for multiple position selectors (an eight position rotary switch, for example) which can use binary coded contacts; position 6, for example, being represented by 110 on three input lines.

Where numbers are to be entered, however, there are two basic entry schemes. The first is BCD-coded decade switches. Each switch has 10 positions, and gives a 4-bit BCD output. To read a four-digit number thus requires a 16-bit input card as shown in Figure 6.10.

Where more than one value has to be entered, multiplexing can again be used to reduce the number of inputs required. The principle, for three four-decade switches, is shown in Figure 6.11. The diodes on each switch output prevent sneak paths through unstrobed switches. Normally these diodes are part of the switch construction and all the designer has to do is specify the signal polarity.

The multiplexing is controlled by a software counter as summarized in Figure 6.12. As with multiplexed outputs, the major disadvantage is a lower update speed, and a decrease in program comprehensibility.

The techniques of multiplexing can also be used to minimize pushbutton/ switch cabling. In Figure 6.13 sixteen inputs are being read with four outputs, four inputs and an eight-core cable. Again, diodes are needed to prevent sneak paths. Speed of response and program comprehensibility are disadvantages as before.

The second approach is similar to a calculator, with ten number pushbuttons, Enter and Cancel pushbuttons and a digital display (driven as described in the previous section). The entered number is built up in

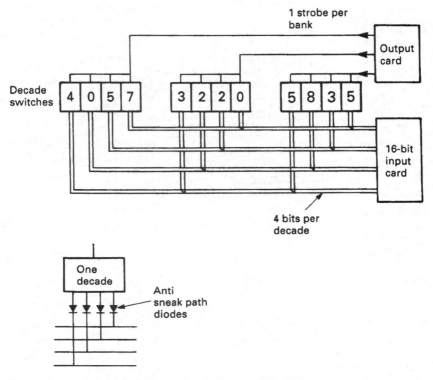

Figure 6.11 *Multiplexed reading of decade switches*

a buffer digit by digit by multiplying the buffer by ten and adding in a number each time a button is pressed. Figure 6.14 summarizes the operation.

6.4 Alarm annunciation

Faults inevitably occur in all plants, and as a result all bar the simplest system will incorporate an alarm system to draw the operators' attention to developing problems. This alarm system can be as simple as a lamp saying 'Pump Tripped' or as complex as a large SCADA system which can generate thousands of alarm banners on computer screens.

For simple systems all that is required is a lamp for each alarm which operates as Figure 6.15. When the alarm occurs the light flashes. When the alarm is acknowledged (accepted) by the operator the lamp goes solid if the alarm is still present, or goes out if the alarm was a transient event which has passed. Normally an audible alarm is sounded until the alarm is accepted. It is good practice to have two bulbs for each alarm and provide a lamp test button on an alarm annunciator.

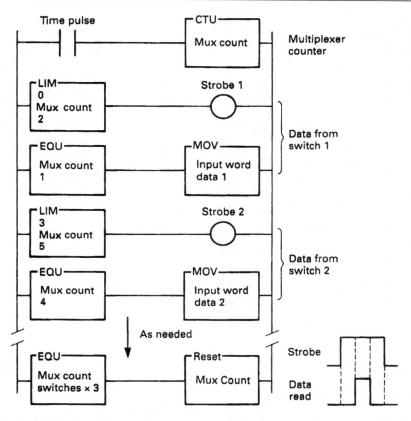

Figure 6.12 *Sample program for multiplexed input. The instruction LIM (for limit) checks that the value is in the range of the outer numbers. The top LIM, for example, gives a true output for MuxCount 0, 1 or 2*

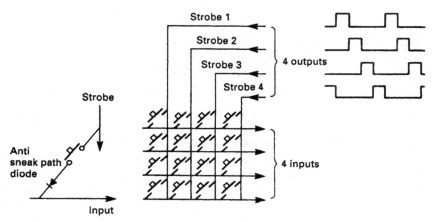

Figure 6.13 *Reading input contacts with a multiplexer*

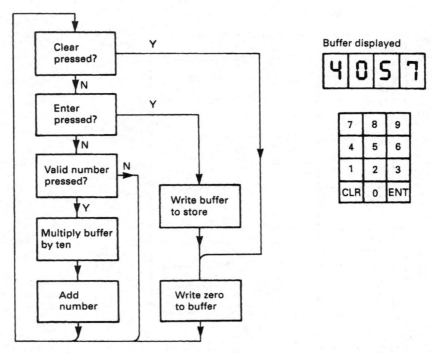

Figure 6.14   *Flow chart for keypad operation*

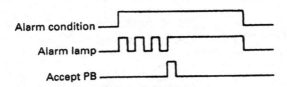

Figure 6.15   *Operation of a simple alarm annunciator. When an alarm occurs the lamp flashes, and an audible alarm sounds, until the operator accepts the alarm. This simple system, though, is vulnerable to alarm floods which can overwhelm the operator*

Alarm systems are commonly installed without much thought, and this can cause problems. The alarm system at Three Mile Island gave wrong information to the operators. The alarm system at Milford Haven oil refinery produced an overkill of information which overwhelmed the operators. During the Channel Tunnel fire, the alarm system gave the wrong impression of the seriousness of the situation and delayed rescue operations. At Chernobyl the alarm system itself was over-run and alarms were being automatically removed before they were presented to the operators.

An alarm system should make operators aware of a developing problem in time for appropriate action to be taken, and the operators should know what this action should be. It is worth listing the requirements for a good alarm system:

The alarms should be clear, unique, concise and unambiguous.
If an alarm appears saying *FT205 Low*, does the operator know what this means? The author has seen alarms where even the designers could not explain the meaning.

The alarms should have a defined operator response.
Every alarm should be followed by an operator action. This could be starting another pump, putting the plant into a holding state or calling for help from the maintenance crew. If the operator does not know what to do the alarm is not achieving anything useful.

Alarms should not cry wolf.
If they do, the level of confidence in *all* alarms will be reduced and possibly important alarms will be ignored.

Alarms should be implicitly trusted.
If they are not, the plant will run with standing alarms, a potentially dangerous situation.

Alarms should only tell the operators what they need to know.
If they flood the operators with irrelevant information important alarms may be missed. For example, if an operator stops a fan this should not trigger any alarms unless the plant responds in an unexpected way.

Alarms should only occur when the operator has time to respond.
Giving the operator one second notice of a problem requiring deep thought and complex action is pointless.

One of the biggest problems with alarm system is a single problem which can result in a deluge of alarms. In the five hours before the accident at Milford Haven oil refinery in 1994 the operators were presented with a new alarm every two to three seconds. They barely had time to accept the alarms let alone understand them or take appropriate action.

Alarm floods can be reduced by grouping of alarms so only first up in a group is given. A two pump hydraulic system operating a cylinder, for example, could have the following alarms:

Pump Stopped
Low Pressure
Standby pump failed to start

246 *Programmable Controllers*

LowLow oil level
Cylinder stroke not achieved in five seconds

If there is an oil leak, LowLow oil level will occur, the pump will stop automatically (to avoid pump damage), the system pressure will fall, the standby pump will be inhibited from starting and the cylinder will not move. One fault could thus cause five alarms. A first-up alarm system will inhibit alarms which will naturally follow from the first event.

Alarm floods can also be caused by repeating alarms. These are commonly caused by an analog signal which is just wandering either side of an alarm trigger point These can be reduced by *hysteresis* with the alarm event and alarm clear occurring at two different setting as shown later in Figure 9.10. A simple way of achieving this is shown in Figure 9.11 in Section 9.4. An alternative approach for reducing repeating alarms is to give the operator the ability to *Shelve* alarms; i.e. provide a function which says 'Don't tell me about this again for (say) thirty minutes'.

Alarms should have different priority levels. The priority will be a function of importance and the speed with which an operator must respond. At the highest level will be problems related to personnel safety, followed by problems which require immediate action to avoid expensive plant damage or extended loss of production. Lower level alarms are less important or require less urgent action. Commonly three alarms priority levels are used. Alarm floods can be reduced by allowing the operator to 'shelve' the lower level, and hence less important, alarms.

SCADA systems invariably include an alarm system and these usually display alarms which appear as a banner on the screen which must be acknowledged by the operator. It is usual to store these alarm banners in an alarm history file which shows the time at which the alarm occurred and the time the alarm was accepted. These alarm histories, commonly called an Alarm Log, can be very usual for fault finding or subsequent post mortems. They are essential if the alarm system does have a 'first-up' system to filter alarms.

Alarm systems frequently have too many alarms. Great care should be taken to ensure that the alarms are really alarms. If a combustion air fan trips this is almost certainly an alarm event. If the operator stops the fan for production reasons this is probably not an alarm event. It is very useful, both for maintenance and production, to keep a log of plant events, but the difference between events (which should be logged without operator action) and alarms (which require operator action) should be clearly understood.

Most alarm systems attract the operators' attention with audible alarms. These certainly demand attention but continuing alarms can be very stressful and can distract an operator when careful thought is required. An alarm flood with each alarm accompanied by intrusive sounds can be totally self defeating. If alarms can occur frequently or in a flood the operator should be given the opportunity to mute, or at least reduce the volume of, the alarms. Spoken alarms, commonly used on aircraft for collision avoidance and ground warning systems, are very effective if applied sensibly. Soft female voices suggesting the correct action are recommended. Most modern SCADA systems permit sound files to be attached to alarms.

Alarm duplication commonly occurs where simple lamp based alarm annunicators are used along with banner alarms on computer or PLC driven screens. If these are not planned with care the operator may end up having to accept an alarm on the annunciator then accept the identical alarm on the computer system. This wastes valuable time and increases operator workload and stress. Even worse it can cause considerable confusion if the identical alarms have different descriptions (e.g. *LE205 level* on the annunciator and *Imminent Sump Overflow* on the computer banner).

6.5 Analog indication

Where a PLC is concerned with analog signals, flows, temperatures, pressures, etc. these will usually be displayed to the operator. Section 6.3 discussed the display of information with numerical digital displays. In this section we will consider analog displays, i.e. meters and bargraphs.

Analog meters and digital displays can both be used to display varying signals, so it is best first to consider their good and bad points. A digital display can display a value to any achievable resolution and accuracy; a four-digit display has a resolution of 0.01% of full scale. An analog meter, however, can only be read to a resolution of about 1% of full scale, regardless of the accuracy of the signal. If high accuracy is needed, a digital display is best.

An analog meter, though, is best used where an operator is required to pick up a general pattern or impression without a great need for accuracy. Digital meters need to be read individually. The rogue temperature on the bar graphs of Figure 6.16 can be seen instantly; the same data on digital displays are far less obvious.

Another consideration is the speed of change of signal. Digital displays need time for the human mind to assimilate the information. With fast-changing data on a four-digit display, all the eye sees is 8888. In situations where the operator is required to handle fast-changing signals, an analog meter is preferred.

There are also implications for the cost of the system and the ease of maintenance, which will depend on the specific application. A three-digit

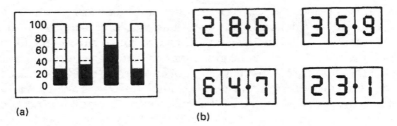

Figure 6.16   *The same data displayed in analog and digital forms: (a) analog; (b) digital*

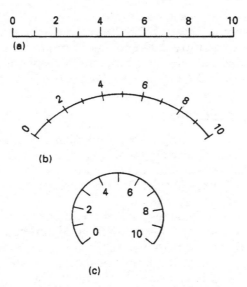

Figure 6.17   *Different meter scales, all having equal scale length: (a) linear; (b) 100°; (c) 270°*

digital display requires around 14 signal lines, plus a supply. An analog meter requires just two (but needs one output from a relatively expensive analog output card). Digital meters have no moving parts whereas analog meters are less robust and can be damaged by sharp blows.

A meter scale should be chosen to be easy to read at the normal viewing distance. A useful rule of thumb is a scale length about 1/15th of the viewing distance (e.g. 20 cm scale length for a 3 m viewing distance). Figure 6.17 shows different meters with identical scale length. Normally 20 scale divisions are used. With a correct choice of viewing distance/scale length, an observer can interpolate to one-fifth of a scale

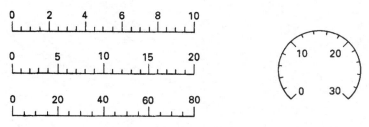

Figure 6.18 *Scale markings readable to about 1% resolution*

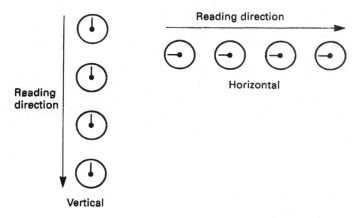

Figure 6.19 *Grouping of meters which are scanned for deviations rather than precisely. Normal indication arranged as shown*

division to give a resolution of 1%. Perhaps surprisingly, more than 20 scale divisions can decrease the resolution by making the scale look cluttered and difficult to read. Figure 6.18 shows typical scales readable to 1%. Meters to higher accuracy are available (BS89 defines nine accuracy ranges from 0.05% to 5%), but they have little practical application in industry.

Meter ranges should be chosen so that a normal reading is between 40% and 60% of full scale. If a block of meters is to be quickly scanned for an anomalous reading, the 'normal' value should be at 9 o'clock for a horizontal row of meters, or at 12 o'clock for a vertical column, as shown in Figure 6.19.

Where a signal (shown on a meter) is to be controlled, it is important to have the correct link between the signal movement on the meter and the operator's expectation. Everyone expects 'increase' to equate to 'up' or 'clockwise', giving the relationships of Figure 6.20.

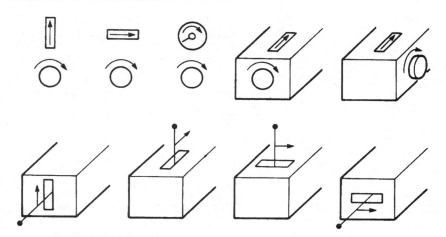

Figure 6.20   *Human expectations of controls and indications. Arrows show direction for increase (of flow, power, speed, temperature, etc.). It is odd to note that UK light and power outlet switches are reversed from normal expectations. Controls relating to motions (e.g. crane controls) should follow plant movement*

6.6   Computer graphics

6.6.1   Introduction

So far we have discussed individual operator devices, pushbuttons, switches, indicators, analog meters and digital displays. Increasingly all of these functions are being provided by computer graphics screens. This can be a display device designed specifically for a particular range of PLCs (the Allen Bradley Panelview and the CEGELEC Imagem which we will discuss in the following section are of this type), general purpose graphic display devices (such as ABB/ASEA's excellent Tesselator) or graphics software running on conventional industrial computers.

It is useful, as ever, to first consider the merits and disadvantages of using computer displays. If everyone was totally honest, often the main reason for the choice is that they look good and impress visitors. Too often the result is stunning colourful flashy graphics that are impossible to view for more than a few minutes without acquiring a headache, have to be searched for useful information and have an update speed of several seconds. At my site several plants are controlled by screen, and there are usually spectacular 'visitor's screens' and more mundane, restful (and useful) working screens.

The major advantages are simplicity of installation and flexibility. A graphics terminal has just two connections to the outside world, a serial link connection (see Chapter 5) and a power supply. If it is used to replace a desk full of switches and indicators there are obvious cost savings. A good quality switch occupying about 60×40 mm of desk space costs about £20 at the time of writing, to which must be added 1/16th of the cost of an input card, a share of a PLC rack, about three connectors, one core in a multicore cable plus labour for building the desk and PLC cubicle, pulling the cable, and ferrulling the cable cores. A single device can be very expensive when all the costs are considered. There are software costs and a large capital cost for a graphics terminal, but these generally work out significantly cheaper.

The designer of desks or control stations often has to deal with changes and modification (another example of the 'didn't we tell you' syndrome which usually manifests itself as a retrofitted 30-mm push-button with dymotape label in a desk originally fitted with 20-mm controls). Constructing a desk is always a fine balance of time, choosing between waiting until all the requirements are clear, and the minimum time needed to make it. Modifications at the commissioning stage rarely look neat.

The displays on a graphics terminal can be modified relatively easily, and, more importantly, the modifications leave no scars. If the design of a normal desk can only start when the desk contents are 95% finalized (which is about right) a graphics screen can be started at 75% finalized. This flexibility is of great assistance as no job is ever right first time.

There are disadvantages, though. The most important of these is the limited amount of information that can be displayed on a single screen. It is very easy to overcrowd a screen (giving a screen similar to a page full of text on a word processor), making it difficult for the operator to identify critical items. A useful rule of thumb is not to use more than 25–30% of the screen. For a typical 80×25-character screen this means about 500 available positions, which include both identifying text and data. 'Motor Speed NNNrpm', for example, uses 16 characters.

The effect of this is often a need to build up a hierarchy of screens, the top screen showing an overview, lower screens showing more and more detail. The problem with this is the time delay needed to shift through the screens. Direct screen to screen movement is possible by calling for a page number (which needs a good human operator memory, or a directory piece of paper, or wasted screen space) or by making all screen changes via an intermediate directory page (with additional delay). These time delays are small (less than a second typically) but the cumulative annoyance is large.

The time taken to update screen data can also be problematical, particularly where a machine to machine link is involved. Again a response

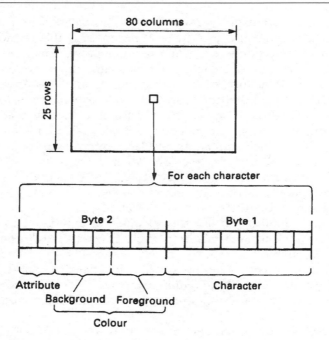

Figure 6.21 *Block graphic memory allocation*

time of around one second is typical, but several seconds is by no means uncommon. The use of a graphics terminal for fault finding on a fast-moving plant is not really feasible.

There are generally two types of graphics terminal. The simplest, known as block graphics, has one store location for each character position on the screen. An 80×25-character display will thus have 2000 store locations. Each location will commonly have two bytes (one 16-bit word) arranged as in Figure 6.21. The first of these holds the character to be displayed, a single byte giving 256 possibilities. Standard ASCII (see Table 5.1) provides 128 alphanumeric characters, the other 128 being assigned to useful semigraphics characters. Figure 6.24 shows some of the block graphic symbols available on IBM PC clones, and the Allen Bradley Panelview. The second byte determines the colour, using 3 bits for foreground colour (giving eight colours) and 3 bits for background colour (again 3 bits), leaving 2 bits for functions such as flash, double height or bright/dim.

The second type of display deals not with individual characters, but with individual points on the screen called 'pixels'. Characters are built from pixels, typically 8 wide by 14 high (for EGA on an IBM-PC, giving a total of 112 pixels per character), the pattern for each character being stored in a read only memory (ROM) as in Figure 6.22.

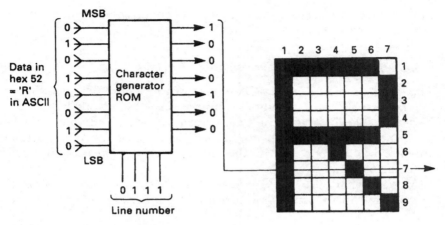

Figure 6.22 *Generating block graphics*

A typical medium-resolution screen will have 640 (horizontal) by 350 (vertical) pixels, a total of 224 000 points. High-resolution screens for computer aided design (CAD) use even more pixels. Each of these can be accessed individually, allowing lines to be drawn at any angle, fill patterns of any type to be used and trend graphs of plant variables to be displayed. Each individual pixel can have its own colour (from over 256 possible colours in some displays) and intensity. The result is an almost photographic resolution.

There are additional costs, the most obvious of which is a large store requirement. The system hardware and software are more complex (and hence more expensive) but, perhaps surprisingly, this is not apparent to the user; pixel graphics displays are often easier to program than block graphics units. Recent home computers (from the era of the BBC and the Spectrum) have all used pixel graphics.

The next two subsections look at examples of block graphics and pixel graphic displays. The first example (the Allen Bradley Panelview) is a relatively simple block graphics display designed to replace simple desk controls (pushbutton, indicators, bar graphs, digital displays and simple mimics). The second, the GEM-80 Imagem, is a more sophisticated display system using pixel graphics and a powerful graphics language.

All PLC manufacturers have graphics packages, usually both block and pixel. Allen Bradley, for example, have the pixel-based Controlview, and the GEM-80 has a simple block graphics video processor. Siemens have the SCADIX family, ABB the Tesselator, and so on. There are differences in the progamming methods used by each (although the basic ideas are similar). To cover all would be time consuming, tedious and confusing. The two examples below were therefore chosen from equipment I have ready access to, not by way of any recommendation over other manufacturers.

6.6.2 The Allen Bradley Panelview

The Allen Bradley Panelview family shown in Figure 6.23 is designed to replace switches, pushbuttons, numeric displays and similar devices. It is available in keypad or touchscreen versions, with both having monochrome or colour displays. In the keypad version, operator actions are linked to function keys, in the touch screen version 120 touch cells are available.

All Panelviews use a VGA screen with 640×480 resolution which can display full uppercase/lowercase alphanumerics (for normal text)

Figure 6.23 *Members of the Panelview family. Picture courtesy of Rockwell Automation*

plus built in ISA (Instrument Society of America) symbols for motors, pumps, etc. Bitmaps can also be imported to provide user designed graphic symbols or backgrounds.

The unit can have several 'pages' of displays, the limit (set by the memory) is around 40 pages of reasonable complexity. These are programmed with a standard IBM PC clone.

Any display contains two distinct types of object. Static objects are fixed and do not change. These are used for fixed text, fixed titles, and unchanging graphics. More interesting are dynamic objects. These are linked to inputs, outputs and numerical data in the controlling PLC.

Panelviews use various forms of serial communication to link to the PLC, including DeviceNet, ControlNet, Data Highway and RS485. The version we shall discuss as an example in this section connects to the PLC via the normal remote I/O cable and looks, to the PLC, like one rack of standard I/O cards. The form of communication and the rack number is part of the initial configuration of the Panelview. An indicator on a Panelview screen, say, could be allocated to bit 3 of card 5 in rack 7, and be driven by address 0:75/03 in the PLC program. Similarly a button on the screen could be assigned to bit 14 of card 2 also in rack 7 and read by the program as I:73/14.

Dynamic objects are added by selection from pulldown menus. These include:

Displays to the operator:
Multistate indicator (Running, Stopped, Tripped, Isolated, Fault, etc.)
Numeric Displays with fixed or floating decimal point (Tank Contents 4057 litres)
Bar graphs and scales (vertical or horizontal)
Circular meters and scales
Alarm messages (appearing as windows and can be stored on an alarm history page)
Message displays (predefined messages with stored ASCII strings such as '*Drill Sequence now completed*'). These strings can also include embedded data from the PLC ('*Warning: motor current is NNN amps*').

Data inputs from the operator:
Pushbuttons. A selection of five different types, including normally open, normally closed, latched, maintained and interlocked (select one and only one from a group). Pushbuttons can control a single bit or write a numeric value to a PLC. A pushbutton also includes a two state indicator whose message can be controlled by a PLC output.
Numeric input, either from pushbuttons or a pop-up keypad
List Selector, used to select one, from several, control options (e.g. calibrate, test, run, shutdown)
Control functions such as select a different screen or accept an alarm.

Panelview screens are built using the Panelbuilder software running on a normal PC screens. Figure 6.24 shows the steps used to add a push-button to a screen.

Figure 6.24(a) shows a summary of the Panelview contents on the left-hand side. The Panelview screen Timed Cut has been selected followed by the pull down menu for objects and selection of Pushbuttons. Clicking on Momentary bring up a pushbutton which is placed and sized on the screen with the mouse as Figure 6.24(b).

The button must now be configured. Double clicking on the button brings up the window shown in Figure 6.24(c). The button has been selected to be a single bit, buttons can also be configured to write a value to the PLC.

A button has two links to the PLC; an input where its action will be performed and an output used to control the state (e.g. Running/Stopped) on the button display. The links are done by tags. These are text descriptions of the link. The new pushbutton is an alarm accept button, and the two links are Reset_PB (for the input) and Alarm_Present (for the output). These tags can be defined in advance (which is quickest when starting a project from scratch) or defined as each object is added. Selecting a tag, Reset_PB say, and clicking the Edit Tag button brings up the Tag form of Figure 6.24(d). Here the data type (bit) and the destination address (I:12/03) are entered. The Node Name is the PLC to which this Tag applies. For the simple Remote I/O there can be only one PLC, but Panelviews with Peer to Peer links such as Data Highway can exchange data with many PLCs. Node then defines the PLC to which the tag applies. The initial value defines the data sent to the PLC on Panelview power-up. When all data has been entered the OK button is clicked. The Alarm_Present output tag is entered in a similar way.

Next the states of the button display must be defined. Clicking on States on Figure 6.24(c) brings up the state window of Figure 6.24(e). The button should be invisible when there are no alarms, and appear when there are alarms. Foreground and background colours can be defined for the text and the object itself. State 0 (healthy) thus has all foreground and background colours set to black and no text. State 1 has the text Alarm Accept (with /*R*/ denoting a new line). Colours can be defined with pull-down menus as shown or by clicking with left or right mouse buttons on the colour pallets visible on Figure 6.24(b). Font size and aligment controls can also be seen on this figure.

Figure 6.25 shows a typical complete Panelview screen.

6.6.3   Pixel graphics; the CEGELEC Imagem

The Panelview built up a graphics screen with semigraphic characters placed on a screen with a 'pick and place' menu approach. The GEM-80 Imagem is a true pixel graphics system where the programmer has

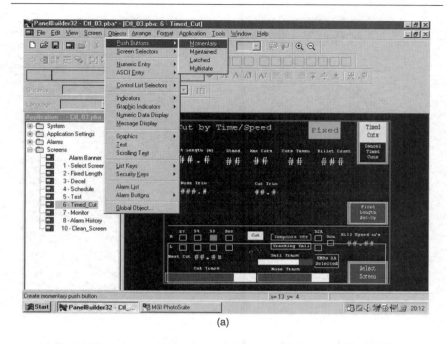

(a)

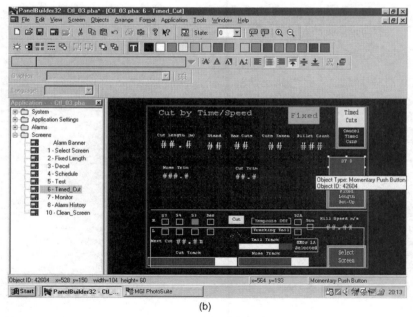

(b)

Figure 6.24 *Adding a pushbutton to a Panelview screen: (a) selection of Pushbutton from the Object pulldown menu; (b) pushbutton positioned on screen and sized to suit*

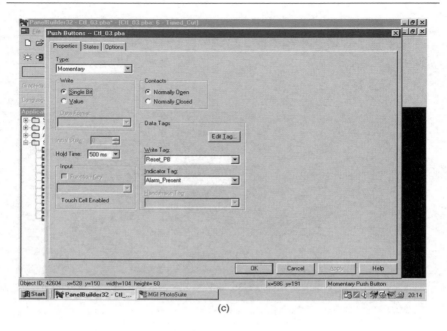

(c)

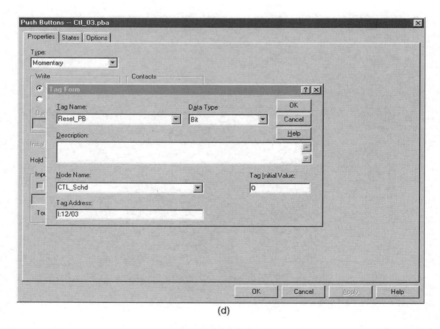

(d)

Figure 6.24 *(cont.) (c) pushbutton type and tag names defined, one tag for the input and one for the display; (d) definition of one tag linking name to a PLC address*

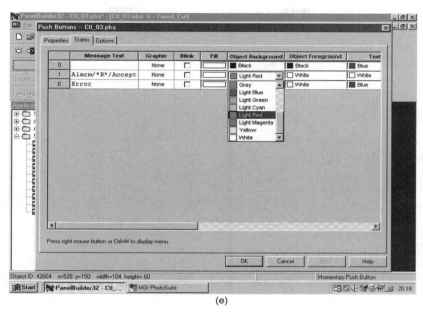

Figure 6.24 *(cont.) (e) text entry and colour selection for the display states*

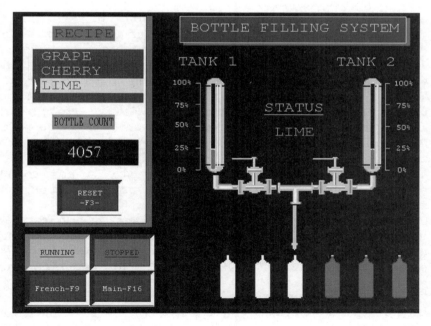

Figure 6.25 *Typical Panelview screen. Picture courtesy of Rockwell Automation*

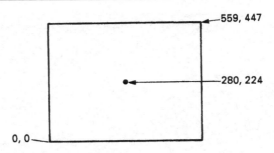

Figure 6.26 *Imagem display area*

access to every individual pixel on the screen. It should, perhaps, be noted at this point that Allen Bradley have a pixel graphics system (called Controlview) and Siemens the SCADIX graphics video processor. Most PLC manufacturers have both types.

The display area on an Imagem screen consists of over a quarter of a million pixels arranged 560 pixels wide and 448 pixels high. A point on the screen is defined by an X and Y coordinate arranged as shown in Figure 6.26 with point (0,0) arranged at the bottom left-hand side. (IBM-PC clones have an awkward system with 0,0 at top left.)

On any screen the programmer can use a palette of 16 colours. These are 16 genuine colours which can be chosen from a palette of 4096 colours and the tints can be changed by variables in the PLC program.

Pixel graphics are more powerful than block graphics, and this means that the programming is more involved to use the additional features. The Imagem uses a display language which has features common to the graphics command on a good quality desktop computer (move, draw, etc.) and mathematical functions for graphical construction involving trigonometry. The programming function CONSTRUCT is also provided, allowing the routine for drawing a valve, or a conveyor, or a hopper to be defined once and then called by a name (such as VALVE). Characters are based on the ASCII set (see Table 5.1) with 128 undefined characters. These can all be redefined with a character editor to produce any desired symbols. Characters can be drawn in any size (with different magnification in the X and Y direction).

In the rest of this section we will look at some of the features of the Imagem display language. In the space available this can be little more than an overview (the Imagem Programming Manual has over 200 pages) but should give an appreciation of the approach used with pixel graphics systems.

The display language is closely related to high level languages such as BASIC or Pascal. For example, to put a simple message on the screen you would write

```
FOREGROUND WHITE
BACKGROUND BLUE
WINDOW 0,0 559,447
MOVE 205,210
SIZE 2
'Have a Nice Day'
```

Here the colours are defined by the FOREGROUND and BACK-GROUND commands, the WINDOW command says the area this definition applies over (the whole screen), the MOVE command places the cursor at X=205, Y=210 (near the screen centre) and SIZE 2 gives double height text.

Data in the PLC program can be added to text. Assuming colours and size and position (MOVE) have been set up we can write

```
'Feed Rate ='
DECIMAL ^##.##,W[235]
'l/min'
NEWLINE
'Tank Level='
DECIMAL ^^#.#,W[236]
'metres'
```

Here the PLC program is accessed for the values to be displayed. The feed rate is held in the GEM location [235] and the tank level in location [236] (see Section 2.3.4 for a description of the GEM-80 data table). The DECIMAL command says how this is to be displayed, defining both the number of characters and the decimal point position. The caret ^ is a blank position and the hash # a number. The above instructions could produce:

Feed Rate=15.25 L/min
Tank Level=3.7 metres

Mathematical functions can also be included, for example:

```
'Total Water Flow='
DECIMAL, ^####,G[17]+10*G[23]+W[146]/2
'gpm'
```

We set the foreground and background colour earlier with default colours. The programmer can define a colour (up to a palette of 16 on any one screen) and even have colours varied by PLC variables. A colour has a name, and, in its simplest form, percentages 0–100% of the component colours red, green and blue. You could define, for example,

COLOUR PINK 80,10,20

or a colour controlled by plant variables

COLOUR PWRLEVEL G[10],G[11],G[12]

With the latter, a definition FOREGROUND PWRLEVEL will cause text (or graphics) to the next foreground definition to be set by the values in the variables G[10] to G[12].

Flashing characters have a larger definition with main colours (RGB) followed by on time (in tenths of a second) followed by inverse colours (RGB) and off time (again in tenths of a second), for example

COLOUR ALARM 100,0,0,4;50,0,0,2

which goes from bright red to dim red.

Symbols and mimics are drawn with lines, arcs and filled blocks. A line is drawn with the command DRAW X,Y which draws a line from the current cursor position to position X,Y. The instruction

```
MOVE 100,100
DRAW 200,200
```

would thus produce Figure 6.27(a) and

```
MOVE 100,100
DRAW 200,200
DRAW 300,100
DRAW 100,100
```

would give Figure 6.27(b).

These have defined positions in absolute screen positions. It is often useful to use relative positions, particularly when the graphics involve variables. Relative commands are RMOVE and RDRAW. Figure 6.27(b) could also be produced by

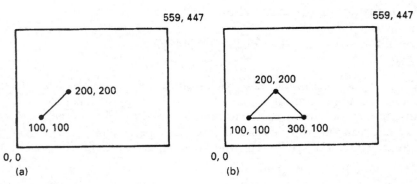

Figure 6.27 *The DRAW instruction: (a) simple line; (b) a triangle*

```
MOVE 100,100
RDRAW 100,100
RDRAW 100, -100
RDRAW -200,0
```

The line width can be set with a size command; SIZE 5 gives lines 5 pixels wide. The line type (dotted, dashes, etc.) can be defined with a suffix; for example

```
DRAW 157,203,3
```

gives a dotted line (denoted by suffix 3) to position 157,203.

The draw commands can be related to a value in a PLC data table location. For example, to draw a bar graph we would write

```
SIZE 7
MOVE 0,400
DRAW W[62],400
MOVE 20,20
DRAW 20,W[45]/62.5+15
```

This would draw two bar graphs; one horizontal directly linked to W[62] and a vertical one determined by a bit of maths on the contents of W[45].

A block of bar graphs can be drawn by using the FOR command. This operates like a FOR/NEXT loop in BASIC, or the FOR/BEGIN/ END construct in Pascal. The form is very similar to Pascal, in fact, with open brackets '(' being used for begin and close brackets ')' for end. For example

```
SIZE 7
FOR A=1 TO 10
(MOVE 20*A,20
DRAW 20*A,W[15+A])
```

draws ten vertical bar graphs for the values in W[16], W[17], etc. to W[25]. Here 'A' is an internal Imagem variable.

At this point you might be confused about how graphics controlled by plant variables can be changed without erasing the old values or causing annoying flicker. The principle is shown in Figure 6.28. The Imagem has two screens; let us call them A and B. When screen A is being displayed, screen B is updated in the background from the program. When this update is complete, the screens are switched, B being displayed and A updated. This toggling is invisible to the user, but gives very smooth updates. It allows an analog meter pointer to be simulated with

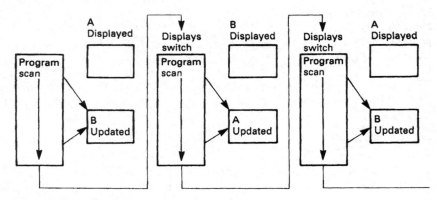

Figure 6.28 *Screen updating on Imagem*

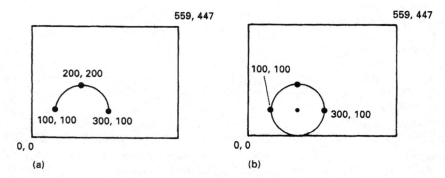

Figure 6.29 *The VIA instruction: (a) arc; (b) circle*

MOVE 200,200
DRAW 50*COS[90*W[127]/1500],50*SIN[90*W[127]/1500]

which draws a line 50 pixels long from 200,200 at an angle determined by the value in W[127] with 90 degrees corresponding to a value of 1500.
Arcs can be drawn with a VIA command

MOVE 100,100
DRAW 300,100 VIA 200,200

gives Figure 6.29(a), and

MOVE 100,100
DRAW 100,100 VIA 300,100

gives a circle as in Figure 6.29(b).
Predefined words are provided for a triangle

TRIANGLE XI,YI,X2,Y2,X3,Y3

and a rectangle

RECTANGLE X1,YI,X2,Y2

where the points define opposite corners. A suffix defines the line type (solid, dashed) as described for the DRAW command. The programmer can define a shape, a conveyor, say, and repeat it with

CALL CONVEYOR (parameter list, e.g. size, position, state)

Solid shapes can be produced by moving inside a CLOSED shape and using the command FILL.

Conditional tests are often needed to show the plant state; alarm conditions or running/stopped are common examples. These are provided by the IF/THEN/ELSE command which, in its simplest form, can change a colour:

```
MOVE 50,450
IF A[14].3 THEN
    FOREGROUND GREEN
    ('Running')
    ELSE (FOREGROUND RED
    'Stopped')
```

where the plant I/O signal A[14].3 changes the message, and colour, on the screen.

So far we have discussed just one screen. Obviously, to be useful, a graphics system must be capable of displaying several screens, and having some method of changing between screens in a controlled manner. The screens constructed above (the GEM calls them Formats') are edited and stored in a memory card which is accessible to the GEM-80 and one (or more) Imagem processors. An Imagem can display up to four formats (or screens) at any time (the effect is similar to windows on an IBM-PC clone). The controlling GEM-80 has an L table in its data table in which each Imagem has four locations, the first, not surprisingly, having L0, Ll, L2, L3. Into these are written the format numbers that are to be displayed. For simple, non-overlapping screens only one of the four will be used, and the others will contain zero. Figure 6.30 shows typical Imagem screens.

6.6.4 The Siemens Simatic HMI family

Siemens provide a vast range of operator terminals from simple text only message displays to complete computer based Scada systems.

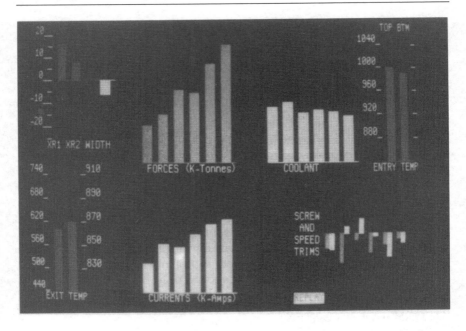

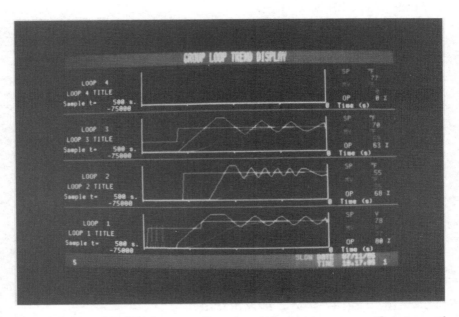

Figure 6.30 *GEM-80 Imagem Trend and Bargraph screens (Courtesy of CEGELEC)*

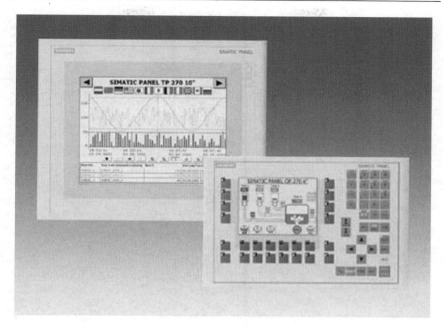

Figure 6.31 *Some members of the Siemens HMI family*

Typical members of the family are shown in Figure 6.31. Based on Windows ME they communicate with the controlling PLCs via Profibus DP (see Section 5.5.7) and other common serial communication standards.

The terminals are configured using the Siemens ProTool software shown on Figure 6.32. Here a trend chart is being added to a touchscreen TP170B terminal.

Screen based operator interfaces are becoming increasingly common, and the Siemens C7 device deserves particular mention. Rather than have a separate PLC and operator terminal this combines the two in one device. The C7 is the plant display screen *AND* the plant PLC. It uses the Siemens S7 standard PLC software for control and the ProTool software for screen configuration.

6.6.5 *Practical considerations*

One major advantage of graphical displays is that mistakes can be rectified without leaving scars, albeit at some cost. None the less it is far better not to make mistakes in the first place. Perhaps the most important consideration is to realize that the system is being designed for an operator

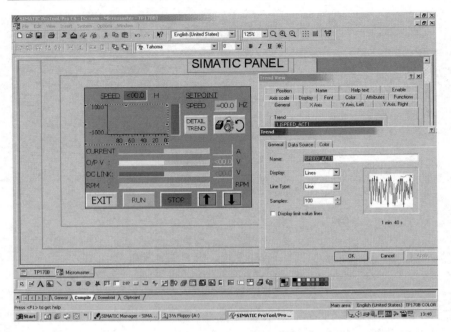

Figure 6.32   *Adding and configuring a trend chart on a TP170B using ProTool software*

who will be sitting in front of it for about eight hours a day, and not for the managing director and visitors. By all means include impressive 'visitor's screens', but remember the poor operator. Above all, remember that you are *not* designing a video game.

A typical mistake is overuse of flashing displays (and flashing lights on desks). A video display (or a desk) in normal operating conditions should not flash or blink. Flashing should only be used to attract an operator's attention (an unaccepted alarm, for example) and should go steady when the operator acts. On a screen, a flashing small box alongside fixed text is much more friendly and easier to assimilate than flashing text. Text flashing in varying intensity (bright to dim) is better than text on/off and avoid text and background switching colours, which is almost impossible to read.

Another common problem is screens which SHOUT AT YOU IN UPPER CASE all the time. Only Death in the Discworld novels SPEAKS IN UPPER CASE. Text in lowercase is much easier to read. The use of Capital Letters for the first Letter of each Noun or Verb also makes interpretation easier. Look how the text on road signs is displayed. All modern screens can support upper case and lower case. Careful use of initial capital letters can be used to emphasize text and draw the operator's attention. Think about layout.

Screens should be uncluttered and consistent. The 25–35% usage rule is a good starting point as it allows an operator to quickly scan a screen for relevant information. Consistency ensures that similar operations are performed in similar ways, with colours having the same meaning on different screens. Pumps should not, for example, be run with separate start/stop pushbuttons on one screen, and with push on/push off single buttons on another. If an 'End of Travel' limit switch is yellow on one screen, it should be yellow on all (and not red, green or blue). Consistency problems normally arise where more than one person has been involved, and can generally be overcome by laying down standards at the start of a project.

Bright colours (yellow/white) tire the eye, and should not be used in large areas (and should be avoided for background colours). Overuse will cause the operator to turn the brilliance down, possibly losing information in dark colours as a result. Grey is a much more restful background colour. Blue characters on a black background are particularly vulnerable to vanishing if the contrast or brilliance is turned down.

Good colour combinations are black on green, black on yellow, black on red, red on white, blue on white, green on white, red on black, green on black and white on blue (the latter is very good for large areas of text and is often used for word processors).

Colour combinations to avoid are yellow/green and yellow/white (which merge together) and blue on black for fine detail (the visibility is very dependent on the setting of the contrast and brilliance control). Cyan/blue is also poor to the point of vanishing into an unreadable bluish block.

With touchscreens, a useful standard is colour on black for an unactivated pushbutton, and black on the colour for an activated state. With start and stop buttons, for example, in the stopped state the start button would show 'Start' in green on black, and the stop button show 'Stopped' in black on red. When the start button is pressed, it changes to 'Running' in black on green, whilst the stop button changes to 'Stop' in red on black. Note the text changes from allowed action to state.

The environment around a display needs to be carefully considered. Most screens are mounted angled up, and are prone to annoying reflections from overhead lights and windows. Bright lighting (and above all direct sunlight) can make a display impossible to read.

Displays are also adversely affected by magnetic fields. Close proximity to electric motors, transformers or high-current cables will cause a picture to wobble and the colours to change. The effect can be overcome by screening the monitor with a mu-metal cage (normal steel or iron does not work).

The size and weight of the monitors are often overlooked, making them difficult to mount neatly, and even more difficult to change. Access should be made as easy as possible; trying to hold a 25-kg display in place with one hand whilst undoing interminably long mounting screws is not much fun.

Displays fail, and the implication of this needs to be considered in the design. If all the plant control is performed by screens, what will happen during the ten or so minutes it will take to locate a spare and change the faulty unit? Often, dual displays (main and standby) are used to overcome this problem.

6.6.6 Data entry

The operator will obviously need to input data and initiate actions. Keyboards are one approach, but many people are nervous of them (home computers help here) and the cable connecting the keyboard always seems prone to damage. In dirty environments keys can become blocked with dirt and membrane keypads with tactile (feel) feedback should be used.

Another useful approach is softkeys. Here a set of buttons (often 10) is positioned on the keyboard below a set of (software-driven) blocks on the screen. The pushbutton can thus change their meaning as the screen changes, as shown in Figure 6.31.

If the operator has to access points anywhere on the screen, a tracker ball is a useful device. Rather like an upside-down mouse it controls the movement of a cursor on the screen. All normal actions can be performed with three buttons on the trackerball and a numerical keypad. Trackerballs work surprisingly well in dirty environments as they are open underneath and dirt seems to fall straight through. Mice perform a similar function but are vulnerable to damage and dirt and are hence more suited to an office environment.

Touchscreens have already been mentioned briefly. Combining a display area with operator controls they provide a very compact interface, but their use should be tempered with care. There is absolutely no tactile feedback for the operator to sense a button, so their use is not recommended for an application where the operator has to look at the plant (and not the screen) when operating controls.

It is also easy to operate buttons by mistake. A similar effect can occur when a touchscreen is cleaned; a blank screen should always be provided for this purpose. The continued touching of the screen leads to a build-up of greasy fingermarks, accentuating this problem.

6.7 Message displays

Where a simple text message is to be displayed, possibly with embedded data, message displays driven by a simple serial link carrying ASCII coded characters can be used. The controlling PLC simply stores an ASCII string in its memory and outputs it, with added variable data, via a device such as the ASCII module described in Section 5.2.8. An alternative approach is to store precoded messages in the display itself, with the PLC having a few parallel lines allowing it to say 'Display message 23'.

6.8 SCADA packages

PCs have much better graphics capability than the average PLC, so it is not surprising that PCs are increasingly being used as the link between humans and PLC systems. These are usually called SCADA systems (for Supervisory Control and Data Acquisition). As the name implies these act as a higher level supervisor, and are commonly used for determining plant set-ups and displaying plant status on high quality screens. They also provide storage for several days performance records allowing problems to be investigated after the event.

Other common features are trending (producing time based historical graphs of plant data) and alarm annunciation. The use of the latter feature needs some care. The ease of adding alarms into a SCADA system can lead to a large number of alarms which can swamp the operator and be ignored. An alarm overkill is thought to have been a major contributory factor in the explosion at Milford Haven oil refinery in 1994 when the operators were exposed for several hours to alarm messages at the rate of over thirty per minute. SCADA systems are vulnerable to this problem, but any alarm system should have priority groups and some form of first-up system which blocks consequential alarms. The topic of alarm annunciation is discussed further in Section 6.4.

The security of the system should also be considered. If a SCADA system fails, what controls will be needed during the thirty or so minutes that it will take to change for a spare, can the plant run blind for this time, or should there always be a second, standby system?

Figure 6.33 shows a typical high quality SCADA display built on the popular Citect package. This consists of many objects which are linked to data inside a PLC. The operation, like the Panelview described in Section 6.6.2, is built around 'Tags'. These can be considered as a database of plant information which is built by the PLC. Definition and construction of the Tag database is the main part of the work in building a SCADA system; the construction of the screens is pure undiluted fun!

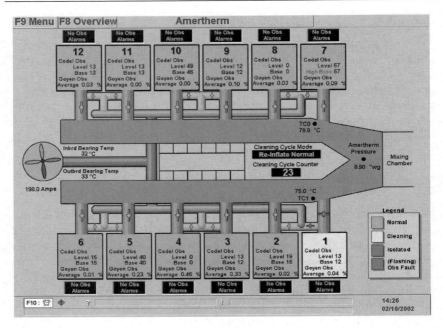

Figure 6.33 *Typical SCADA display built using Citect. Every object on the screen can be linked to PLC data*

Figure 6.34 shows two typical tags. The first is a digital signal saying a pump is running. The Field *I/O Device Name* defines the PLC from which this signal can be obtained. Elsewhere the communication method (e.g. Ethernet) for obtaining data from this PLC has been defined. The address, bit 8 of word N70:0, defines from where the data can be read. If this is true, Pump 1 contactor is energized, and this information can be used on a screen to, say, turn the colour of a graphic symbol for a pump to green.

The second tag, on Figure 6.34b, is numeric and gives a temperature. This is obtained from the PLC known as *Env_C_Pulpit*. Scaling can be performed on numeric values, in this case 0–1000 corresponds to a temperature of 0–100.0 °C. Engineering units (e.g. psi, °C) can be permanently attached to the value and a fixed format (e.g. ##.#) can be defined but they have not been used in this example.

Once the tags have been defined they can be used on the display. Objects are selected in a way similar to most paint packages. Each object has a vast array of parameters which can be linked to the tags. Figure 6.35 shows a very simple example for a text object. Here there are two limit switches saying a damper is in the open or closed

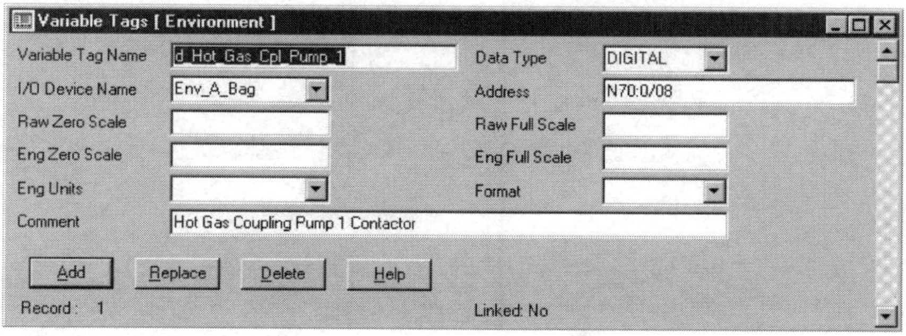

(a)

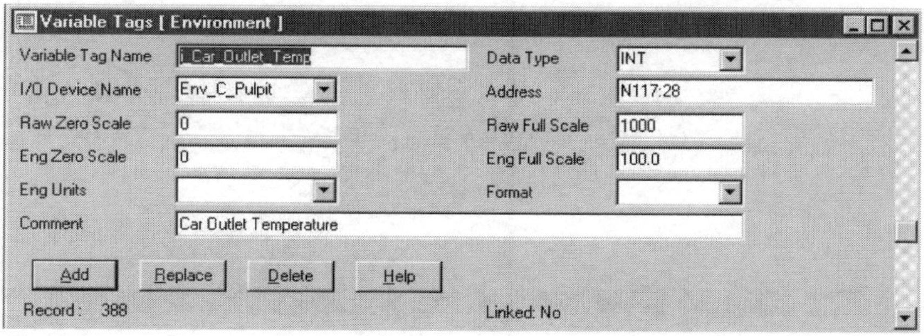

(b)

Figure 6.34 *Tag definition on a SCADA system. A consistent method of defining tags must be used. (a) Digital (On/Off) tag. Note the tag name starts d for digital; (b) Integer tag, again note the tag name starts i for integer*

position. Figure 6.35a shows the text associated with these limits. If only one is present the text gives the state. If neither is present, during movement for example, ???? is displayed. If both are present there is a fault.

Figure 6.35b sets the colour of the text for each of the four possible states. Legitimate states (open/closed) are Green, the transitional state is Yellow and the fault state is Red. Note, though, the other tabs around the object for other functions that could be applied. The range of possibilities is vast.

Trending is one of the strengths of SCADA systems. Any tagged variable can be trended, and custom displays can be built by operators. Range in the Y direction and time-base in the X direction can be

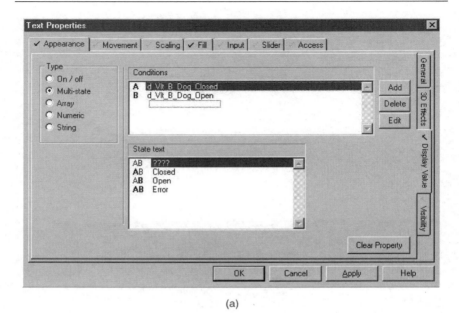

(a)

(b)

Figure 6.35   *Definition of properties for an object: (a) text definition for a simple text object; (b) colour definition for a text object*

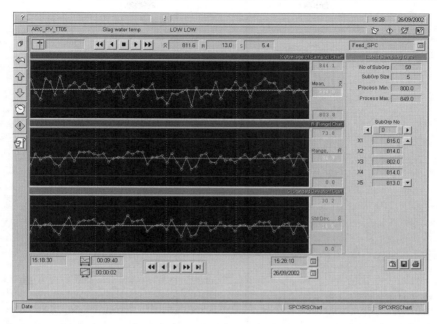

Figure 6.36 *Trending on a Scada system, in this case for a statistical process control (SPC) system showing the variation of crucial plant signals. Figure courtesy of Citect*

selected and changed, to aid fault finding for example. Figure 6.36 shows a typical trend display used, in this example, for statistical process control (SPC). Note the scrolling, zooming and scaling controls on the screen.

7 Industrial control with conventional computers

7.1 Introduction

The vast majority of this book has been concerned with computer control based on programmable controllers. In this chapter we will look at how conventional computers can fulfil a similar role. The idea of a control hierarchy was introduced in Section 5.3.5, and a typical layout was shown in Figure 5.28(a). Here the top level 3 is the company mainframe(s) used for accounting, reports and management. Level 2 is based on powerful minicomputers such as the DEC-VAX and 11/73s. These are concerned with scheduling, report generation based on plant events and data collection. The bottom level 1 is directly connected to the plant and performs real time control. PLCs operate at this level, and we will consider how general purpose computers can operate at this level, often in conjunction with PLCs.

We have seen how powerful PLCs are, so an obvious question is why use a computer at all. The main advantage is brute mathematical computing power, high speed and ease of connection to printers, keyboards and the like. Usually the programs require specialist knowledge to change. This can be an advantage or a disadvantage depending on the application. A PLC program is easy to understand and easy to modify. It can be quite difficult for engineering management to control and keep track of program changes.

All PLCs have some form of access control via keys and password, but these are actually little protection. Access has to be provided for maintenance staff, who must be in possession of keys or password. Inevitably there will be 'midnight programmers'. A computer with a program written in 'C' which is compiled and stored in ROM is as secure as a bank vault. If an application has few real I/O, needs a lot of mathematical operations, is unlikely to change, or has a need for several printers or

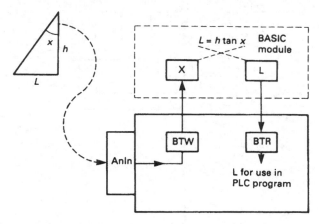

Figure 7.1 *The Allen Bradley BASIC module*

graphics monitors, or security is important, a conventional computer may be a sensible choice.

To some extent these features can be provided by specialist modules which can be added to PLCs. Allen Bradley, for example, have a BASIC module (1771-DB) which can be added to a PLC-5 system. This is a small computer which can be programmed in BASIC and can communicate with peripherals (graphics terminals, keyboards, printers, etc.) via RS232 ports, and with data in the PLC via block transfer reads and writes (described in Section 4.4.5). Figure 7.1 shows an example where a mathematical operation is passed to the BASIC module (which acts rather like a maths co-processor). Another common application is to produce printed reports in the absence of a higher level computer.

In general, computers at this level fall into two categories: bus-based systems and industrialized clones of the ubiquitous IBM PC family. The rest of this chapter briefly describes industrial computers belonging to these classes. More detailed information is given in *Bus Based Industrial Process Control*, and *PC Based Instrumentation and Control*. Both books are by M. Tooley and are published by Butterworth-Heinemann.

7.2 Bus-based machines

7.2.1 Introduction

The architecture of any computer (be it PLC, personal computer, mini-computer, games machine or company mainframe) can be represented by Figure 7.2 and consists of a central processor (CPU), memory store and input/output (I/O) linked to the outside world. These are linked by

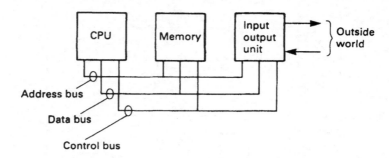

Figure 7.2   *A computer architecture*

a bus system (for busbar or omnibus, depending on what source you are first exposed to) which has three components. The data bus carries data between the various elements: I/O to store, store to CPU and so on. The address bus carries the address of the store or I/O port concerned with the data movement, e.g. 'bring data from I/O port 17 to register C' or 'store the contents of register D in store location address hex E147'. The final bus is the control bus. This carries timing and direction signals.

This structure allows the idea of an expandable DIY computer to be implemented. The bus is laid out on a printed circuit backplane with established connections for the data, address, and control signals. The designer can then plug in CPU, memory, video and I/O cards to build the computer needed to perform the required task. There are several bus standards, and the commonest are described briefly in Section 7.2.3. The IBM PC clone has this form as well, but the designer has less choice in the selection of the CPU.

There are actually two uses of the term 'Bus-based machine'. In the second form a complete master computer is linked to several external devices via a ribbon cable. Data can be read from, or written to, these external instruments. We will first look at the common GPIB (IEEE-488) bus which is of this second type.

7.2.2   *IEEE-488 parallel interface bus*

This system was originally developed by Hewlett Packard to link HP computers to HP instruments. In its original form it was known as HP-IB (Hewlett Packard Instrumentation Bus). In 1975, the standard was formulated by the American Institute of Electrical and Electronic Engineers as standard IEEE-488 (also popularly known as GP-IB for General Purpose Interface Bus). This allows the linking of up to 15 devices and a computer with a total transmission length of 20 m.

The IEEE-488 bus can support three types of device: listeners, talkers and controllers. Listeners accept data from the bus; typical examples are a printer or a display. Talkers place data, on request, onto the bus; a measuring instrument is a typical talker. A controller assigns the role of any other devices on the bus but only one controller can be active at any one time. The designations listener, talker and controller are attributes of a unit (rather than a description of a unit's function) and many devices can fill more than one role. A computer, for example, can act as all three.

Signals on the bus can be grouped into a bidirectional data bus (which serves the three roles of data transfer, address selection and control selection), transfer control, interface management and grounds/ shields as summarized in Table 7.1.

Signalling is done at TTL levels, with 0 V representing '1' and 3.5 V '0' (the inverse of normal TTL signals). Open collectors are used to

Table 7.1 Signals on the IEEE-488 bus

Group	Designation	Description	Pin
Data bus	DIO 1	Data Input/Output 1	1
	2	2	2
	3	3	3
	4	4	4
	5	5	13
	6	6	14
	7	7	15
	8	8	16
Transfer control	DAV	Data valid	6
	NRFD	Not ready for data	7
	NDAC	Not data accepted	8
Interface management	IFC	Interface clear	9
	SRQ	Service request	10
	ATN	Attention	11
	REN	Remote enable	17
	EOI	End or identify	5
Grounds/shield	Shield	12	
	DAV ground	18	
	NRFD ground	19	
	NDAC ground	20	
	IFC ground	21	
	SRQ ground	22	
	ATN ground	23	
	LOGIC ground	24	

Table 7.2   Control selection with ATN line. Bit 7 is not used

Function	Bit							
	7	6	5	4	3	2	1	0
Bus Command CCCCC	X	0	0	C	C	C	C	C
Enable Listen Address LLLLL	X	0	1	L	L	L	L	L
Enable Talk Address TTTTT	X	1	0	T	T	T	T	T
Enable Secondary Address SSSSS	X	1	1	S	S	S	S	S

allow the bidirectional data bus and bidirectional control signals (such as NDAC and NFRD) to operate.

The data bus is used for several purposes. It can obviously carry data to one (or more) listeners or data from talkers. It can be used as an address bus to enable or disable one (or more) devices. Up to 15 primary device addresses and 16 secondary addresses can be supported. Normally a secondary address controls an auxiliary function within a primary device. For an analog input device, for example, the channel would be selected with a secondary address, and the input value read with the primary address. Address 31 has a special function, being used to disable all active listeners.

The data bus action is determined by the active controller which uses the ATN line to signal whether the data bus is carrying data or control (address) information. With the ATN line taken low, any active talker is disabled and the new control mode selected as in Table 7.2.

The major attraction of the IEEE-488 bus is its ease of use, with the operation being totally transparent to a programmer working in a high level language. For example, the instruction

OUTPUT 702, Setpoint

sends the data in the variable 'Setpoint' in the computer (address 7 represented as 700 and acting as a talker) to the listener with address 02. The actual bus operation corresponding to this instrument has four steps:

1   Deselect all listeners.
2   Select talker (7).
3   Select listener (02).
4   Perform data transfer.

IEEE-488 interface cards are available for many devices, both instruments and computers such as IBM PC clones.

7.2.3 Backplane bus systems

Backplane bus systems are built in the form of Figure 7.3. A backplane provides the data, address and control bus signals, and various cards can be plugged into the bus to configure the system as required.

The advantages of a bus system are obvious: standardization, use of off-the-shelf cards, ease of expansion and a DIY bolt it together yourself approach. Unfortunately each and every microprocessor manufacturer and many equipment manufacturers devised their own standards, with different-sized data words (8 bit, 16 bit or 32 bit), different address ranges and, of course, different edge connectors and pin layouts. Fortunately some common standards do seem to be emerging, notably the VME and STE bus standards.

VME bus is a system designed originally for 16-bit machines and based on an earlier 16-bit bus system known as Versabus. A 24-bit address bus gives a large address range. The introduction of 32-bit microprocessors such as Intel's 80386 or Motorola's 68020 led to the upgrade of the VME bus to handle 32-bit data with a second connector. The bus thus exists in 16-bit (single 96-way DIN 41612 connector) or 32-bit (two 96-way connector) forms. In 32-bit form, the address bus was also extended to 32 bit. VME bus cards are generally compatible with either form.

With a high-speed clock (24 MHz data transfer rate), 32-bit data bus and an enormous address range, the VME bus is arguably the most powerful industrial control system available. If speed of performance is

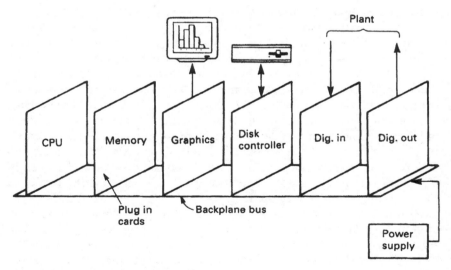

Figure 7.3 *Bus-based computer*

critical it cannot be beaten. It is, however, sophisticated and expensive and, as one user said, 'You don't go shopping in a Formula One racing car, you walk, go by bike or take the family estate car'.

In this latter category comes the STE bus (and the PLCs discussed in the rest of this book). The STE bus is an 8-bit data bus system with origins in the earlier STD bus. It has 20 address lines (allowing over 1 Mbyte of memory space) and 4 kbytes of addressable I/O, and has been formalized under the IEEE-1000 standard. This ensures compatibility between different manufacturers, and has led to it becoming a general purpose accepted standard where the higher performance of the VME bus is not required.

It has many attractive features. The cards are based on the compact Eurocard size (100×160 mm) and connections between the cards and the backplane are made by a robust two-part connector (DIN 41612) which is resistant to vibration and shock loads.

STE bus has been designed around an interface between a master processor and a wide range of I/O boards. The microprocessor type is not defined, and any microprocessor can be used providing the interface between the CPU board and the backplane meets the defined standards. The bus can support up to three master CPU boards although, of course, only one of these can be active at any one time. A well-defined procedure is laid down for the selection of control of the bus when contention occurs between the different masters.

7.2.4   IBM PC clones

Up to the early 1980s, the world of personal desktop computers (PCs) was very varied, with a wide range of different machines and no common standard between them. This was matched by a plethora of operating systems, with usually each manufacturer devising its own.

In 1981 the major mainframe manufacturer IBM entered the PC market. The effect of this was dramatic. IBM dominates the commercial computer market, and its timing for entering the PC market was, through design or luck, immaculate. Personal computers were falling in price and had reached a level where their widespread purchase could be justified by most companies. Although the specification for the PC was not particularly noteworthy (and the graphics ability of the early machines was poor), IBM's reputation ensured that the IBM family of PCs rapidly dominated the market.

IBM chose the software company Microsoft to provide the operating system, known originally as PCDOS. This operating system had a loose relationship to an earlier Z80-based operating system called CP/M. To its credit, IBM was very open about the hardware and software of the PC and designed a bus system which allowed easy expansion. A vast market of add-on cards appeared, along with cheap IBM clone computers. For these Microsoft provide an operating system identical to PCDOS (for all practical purposes) known as MSDOS.

Since its introduction in 1981 the IBM PC family has undergone a steady development. The original machine, with floppy disk storage and known simply as the IBM PC, was based on the Intel 8088 micro-processor, a 16-bit version of the ubiquitous Intel 8080 (and the Zilog Z80) family. This was followed rapidly in 1982 by the IBM-XT, again an 8088-based machine with hard disk for storage.

The next step occurred in 1984 with the introduction of the IBM-AT. This was based on the more powerful Intel 80286 micro which could handle a large amount of memory (16 Mbyte compared with the 1 Mbyte of the 8088 and 64 kbyte of earlier micros such as the Z80). Unfortunately MSDOS (and PCDOS) were designed around 1-Mbyte memories and could not directly use all the memory capability of the 286. The 80286 could also support multitasking where the processor can operate more than one task at a time.

The PC and XT had been constructed with well-documented bus systems allowing the addition of cards such as modems, interface cards, etc. The AT introduced a bus with additional capabilities (which was still compatible with the earlier standard). We will describe these bus systems shortly.

In 1987 IBM introduced a new family, the PS/2 (for Personal System 2) computers with a variety of number suffixes (PS/2-30, PS/2-50, etc.). These are based on a range of Intel processors from the 286 (in the PS/2-30, PS/2-50, PS/2-60), to the 32-bit 80386 (used in the PS/2-55, PS/2-70 and PS/2-80) and 80486 (used in the PS/2-486). These machines also introduced improved graphics facilities and a totally new bus system called MCA (for Micro Channel Architecture).

This new bus system was incompatible with the earlier bus standards and has resulted in two divergent systems. The early bus (in the original PC and XT) was based on a 62-way edge connector with an 8-bit data bus and a 19-bit address bus plus control, timing and power supply lines (+12 V, +5 V, 0 V and −12 V). The 8-bit data bus was a restriction and IBM increased the data bus to 16 bits in the IBM-AT bus with the addition of an extra 36-pin connector. Both this (and the 62-way connector) use printed circuit board edge connectors which are far less robust than the two-part connectors used on STE bus. The two common standards, usually known as the PC bus and AT bus, appear as Figure 7.4 along with smaller half-card formats. These standards are also known as Industry Standard Architecture (or ISA, which has nothing to do with the Instrument Society of America). Although IBM had originated the ISA bus, it never actively pursued legal action against clone and interface board

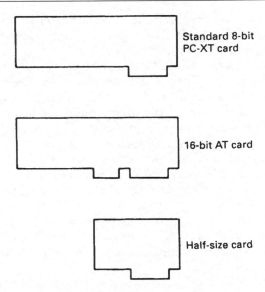

Figure 7.4   *Expansion cards for IBM PC family*

manufacturers, and a flourishing industry in machines and add-ons developed.

IBM's approach to the PS/2 MCA bus has, however, been markedly different. The MCA is superior to, but incompatible with, the ISA bus, introducing features such as automatic configuration, higher speed and more tolerance to electrical interference. IBM has protected its rights to the MCA with strict licensing fees and a far less open approach.

Many of the clone manufacturers chose to co-operate against IBM and jointly developed an improved version of ISA called EISA (for Extended Industry Standard Architecture) as an alternative. There are thus four different bus systems for the IBM family: the PC bus, the AT bus, the EISA bus and the MCA. Future developments are not clear at the time of writing, but the ISA bus will be available for the foreseeable future.

Although the original PC had poor graphics, these have improved through the XT, AT and PS/2, with common standards being shown in Table 7.4. The poor graphics capability of the early IBM PC led to the

Table 7.3   Intel microprocessors used in PC-compatibles and clones

	8086	8088	80286	80386	80486	Pentium
Data bus (bits)	16	8	16	32	32	32
Max. clock rate (MHz)	5	8	10	16	66	>200

Table 7.4 Various graphics adaptors

Mode	Colours	Horiz. × Vert.
CGA	4	320×200
	2	640×200
Hercules	2	720×348
EGA	16	640×200
	16	640×350
VGA	16	640×200
	16	640×350
	16	640×480
IBM 8514	256	1024×768

development of graphics cards by external manufacturers, such as the monochrome Hercules Graphics Adaptor. These have largely been superseded by the EGA and VGA standards. These latter standards are useful for industrial graphics terminals.

Industrial PCs are based on clones and clone adaptor boards, usually to the AT bus standard. Using serial communications they often act as programming terminals or operator stations with PLCs.

7.3 Programming for real time control

The use of conventional programming languages was briefly discussed in Section 1.3.3. Languages such as BASIC, FORTRAN, Pascal and C were designed for general purpose, or scientific, computing and do not normally provide functions for real time control. There are exceptions, however, with real time variations on the standard language. MACBASIC, for example, is a version of BASIC, with instructions such as AIN(M,N) which gets an analog input from channel N on card M. Most of the single board and bus board computers described earlier operate with non-standard additional instructions to BASIC or C. The Pyramid Integrator, from Allen Bradley, for example, is a rack containing a 5/250 PLC (the top of the PLC-5 family) and a DEC-Vax computer. These communicate via the backplane. The PLC-5/Vax link is designed to be controlled via a program written in C, with additional C instructions (called the Data Table Library) available to access the PLC data table.

The programmer has to ensure that the computer program responds to plant actions and operator inputs in a reasonable time. One way of achieving this is simply to write a version of the PLC program scan of Figure 7.5 which would have the form

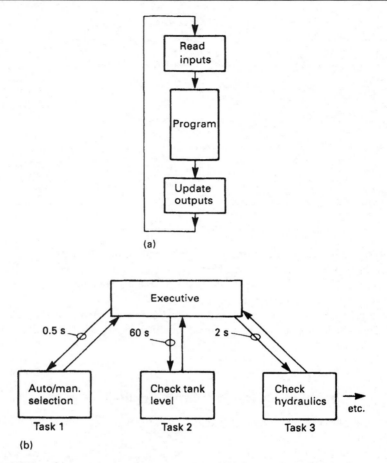

Figure 7.5　*Comparison of computer and PLC operation: (a) PLC scan; (b) computer tasks*

```
Begin
  Repeat
    Read Plant Inputs
    Work out Required Actions
    Write Plant Outputs
  Until Hellfreezesover
End. {of program}
```

This is possibly satisfactory for small schemes, but could be wasteful of computing time in large schemes because manual operations would expect a response in under 0.5 s, but the level of water in, say, a large storage tank would only need to be examined perhaps once a minute. Combined in a single scan, activities with widely differing speed requirements would be difficult to manage.

An alternative is to have the required action broken down into a series of tasks which are controlled by a common executive as shown in Figure 7.5(b). The executive can call tasks at different intervals. Task 1, for example, is an automanual changeover run at 0.5-s intervals, task 2 is checking the level in a water tank at 60-s intervals, task 3 is checking the oil level, pressure and filter state in a hydraulic system every 2 s, and so on. This helps to streamline the process by having no wasted time, and aids programming, as each task is totally separate from every other task and can be written independently.

Specialist real time control languages are available, such as ICI's RTL (for Real Time Language) the US defense language ADA and the CEGB's CUTLASS (designed initially for power station control).

CUTLASS is a compiled language, originally written for DEC minicomputers, which follows the idea of Figure 7.5(b). A control scheme is broken down into tasks which are activated at preset time intervals. A task program starts with a definition giving its name, priority (in the event of a clash, tasks with highest priority are run first) and its run rate, for example

TASK AUTOSLEW PRIORITY = 236 RUN EVERY 600 ms

Next comes the definition of variables. These can be *global* (for the whole program and usable in any task) or *local* to the task. CUTLASS supports the usual forms of reals, integers and booleans (the latter being called *logic*) but introduces the concept of good and bad data. Any data coming from the outside world have the possibility of being erroneous due to plant failures. In CUTLASS, data can have a value or the state *bad*. Reals, for example, can have a numeric value or bad. Logic (Boolean) can be true, false, or bad. This state is carried through operations; for example

Aver := (temp1 + temp2)/2

will yield the average temperature if both temp1 and temp2 are good values, but bad if either temperature reading is faulty. Some operations can produce good outputs with some bad data. A majority vote, for example, of two out of three, will give a good output with one bad value.

A digital input command has the form

DIGIN CARD n m, variable

where n is the card number, m the channel number and variable the name of the variable in the program to which the value is assigned, for example

DIGIN CARD 36 12, StartPB

Digital outputs have the form

DIGOUT variable CARD n m, action 1, action 2, action 3

Here variables n and m have the same meaning, and action 1 is performed if the variable is false, action 2 if it is true and action 3 if it is bad. For example

DIGOUT RUNLAMP CARD 23 7, CLEAR, SET, FLASH

To show the principle, the small code segment below is part of a task controlling auto/manual selection. Variables (whose meaning is obvious from their names) have been previously declared. AutoPermit is a global variable coming in from outside this program segment:

```
DIGIN CARD 63, 15 AutoSW
AutoReq :=AutoSW AND AutoPermit
AUTOMAN AutoReq {A built in function putting mode to Auto}
IF AutoReq=TRUE THEN
   AutoLamp=TRUE
   ELSE
   IF AutoPermit=TRUE THEN
     AutoLamp :=FALSE
     ELSE
     AutoLamp :=BAD
     ENDIF {Inner IF}
ENDIF {Outer IF}
DIGOUT AutoLamp CARD 14 7 CLEAR, SET, FLASH
```

Analog inputs are read with a Multiplexed Analog Input (MXANIN) instruction with the form

MXANIN CARD m n, variable1 p, variable2 etc.

where m is the card number, and n, p etc. are the channels. For example

MXANIN CARD 47 1 Setpoint 2, Actual Value

CUTLASS has many built-in control functions such as filters, rate of change, limiters and controllers. We can use the inputs from above as

Error := (Setpoint − Actual Value)
Actuator := PID (Error, Gain. Ti, Td, Tf)

where PID is a three-term control function, and Gain, Ti and Td are variables holding the settings for the controller and Tf is the high-frequency

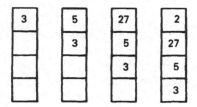

Figure 7.6 *The Forth stack*

roll-off filter. The value in the variable Actuator can be written to the outside world with an ANOUT instruction.

Forth is a language also designed for real time control. Most languages come from academic and research backgrounds. Forth was designed by an astronomer for the control of a telescope at Kitts Peak in the USA. It is an unconventional language in many respects, but once its peculiarities are learned it is ideally suited for industrial use.

Forth uses the idea of a pushdown stack, which can be considered similar to the spring-loaded piles of plates seen occasionally in cafeterias. As a plate is added, the pile moves down. Numbers in Forth are treated in a similar way; Figure 7.6 shows the numbers 3, 5, 27, 2 being placed in the stack.

Most operations are concerned with numbers at the top of the stack. Polish notation is used, with the arithmetic symbol or operation following the data. The addition $273 + 28$ is written

273 28+

and behaves as in Figure 7.7(a). The more involved expression $(412 + 27 - 16) \times 3$ is written

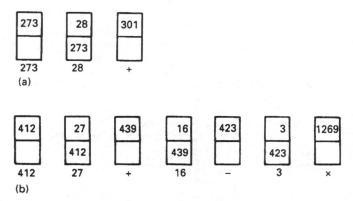

Figure 7.7 *Arithmetic and the Forth stack: (a) simple arithmetic;*
(b) evaluation of $(412 + 27 - 16) \times 3$

412 27+16−3*

and performed as in Figure 7.7(b). In each case we are working with the top pair of digits.

In Forth, the programmer extends the language by defining a series of instructions and giving them a name. For example, we will write a series of instructions to convert a temperature in Fahrenheit to Centigrade. It assumes that the temperature in °F is in the top of the stack and leaves the corresponding temperature in °C on the stack. The definition of a new instruction called FTOC goes

```
: FTOC  {: means this is a definition}
32      {goes to top of stack, pushing ×F down}
−       {subtract top of stack from next down}
5*
9/      {×C now on stack}
;       {; means end of definition}
```

We can now write

68 FTOC

and 20 will be left on the stack.

Suppose we want to control the batch process of Figure 7.8, where two chemicals are added to a vat, mixed, heated to some preset temperature, mixed for some time again, and then drained ready for a new batch. We could define a new word BATCH

```
: BATCH
  ADD1 ADD2 MIX1 HEAT MIX2 DRAIN;
```

These are all new words, with new definitions, for example

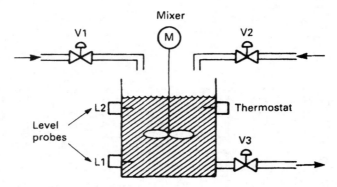

Figure 7.8   *Forth control of batch process*

```
: ADD1
  OPENV1
  BEGIN TESTL1 UNTIL {This is a Forth loop}
  SHUTV1;
```

and

```
: MIX1
  MOTOR ON
  BEGIN TIME1 UP UNTIL
  MOTOROFF
```

Again, new words are introduced (OPENV1, TESTL1) which are again defined until eventually the 'built-in' Forth words can be used. OPENV1 uses standard words:

```
: OPENV1
    1             {state bit 1 = 'ON'}
    3             {channel number}
    5             {card number}
  ; DIGOUT        {standard word in TCS Forth}
  ;
```

and TESTL1 is simply

```
: TESTL1
    4             {channel number}
    2             {card number}
    DIGIN         {leaves state 1 or 0 of digital input 4 of card 2 on the stack}
  ;
```

Analog inputs and outputs are handled in a similar manner.

When all the user-defined words have been broken down to original Forth words, the sequence is run with the one word BATCH.

Forth programs are perfect examples of top-down programming, where a requirement is split into smaller tasks, which are split into subtasks and so on until units of small size and minimal complexity are created.

Where speed or minimal memory is of absolute importance, the programmer has little choice but to work in machine code. Normally programs are written in assembly code and turned into machine code by an assembler provided by the suppliers of the target computer system. The resulting program will be compact and fast, but can be difficult to change or maintain if the documentation is not good. The ability to monitor the running program (standard in all PLCs) will not be available unless fault-finding procedures have been written in as part of the specification.

7.4 Soft PLCs

Traditionally PLCs have had their own operating systems, written by the supplier and held in the PLC in a Read Only Memory (ROM). This gives the PLC a great deal of security; the software is well proven and cannot be modified, accidentally or intentionally, by the end user. It also provides almost total immunity to the malign influences of computer viruses. The author is almost tempted to say complete immunity, never having heard of a PLC being infected by a virus.

Problems arise, though, when the PLC has to link to computer systems for SCADA systems or to access database or spreadsheet files. This is usually done via serial communications; point to point RS232 at the simplest level, networks such as Ethernet where there is more than one PLC or computer. Allen Bradley did provide a solution for this called the Pyramid Integrator which contained a PLC5 processor and a VAX computer in a common chassis, but it was expensive and has been discontinued.

PLC manufacturers now sell the PLC operating system, and a common solution is to run this on a normal PC. Cards fitted in the PC communicate with the normal PLC plant I/O. This allows easy exchange of data between the 'PLC' and the rest of the computer, but does bring some problems which should be seriously considered. First is robustness.

Computers fall over from time to time, usually waiting until the most inconvenient moment as anyone who has used a word-processor will testify. Usually crashes are caused by clashes between software installed on the machine. The system can be made more robust by strictly controlling the software running on the computer and removing all unnecessary program files (e.g. the games and accessories that are normally installed).

The second problem is the power-up time. After a power failure a PLC will normally be working again within a second. A computer, however, can take tens of seconds to come alive and this may be crucial. Some protection can be given by battery backed Uninterruptable Power Supplies (UPS), but in the author's experience these seem to cause as much trouble as they prevent.

A desktop PC is vulnerable to theft. Even if someone cannot walk out with the PC tucked under their arm, the internal motherboard and memory cards can be stolen and are easily hidden. Even worse, 'dongles' are often used to prevent copying of software and a thief, in a hurry, may take the dongles. If a PC based system is left unattended the security of the system should be seriously considered.

Finally, of course, there are computer viruses. These are best controlled by restricting access to the computer and running up to date virus detection software. Under no circumstances allow people to use the computer to view their digital camera photos or print out a word processor file from home.

8 Practical aspects

8.1 Introduction

Programmable controllers are, simply, tools that enable a plant to function reliably, economically and safely. This chapter considers some of the factors that must be included in the design of a control system to meet these criteria.

8.2 Safety

8.2.1 Introduction

Most industrial plant has the capacity to maim or kill. It is therefore the responsibility of all people, both employers and employees, to ensure that no harm comes to any person as a result of activities on an industrial site.

Not surprisingly, this moral duty is also backed up by legislation. It is interesting that most safety legislation is reactive, i.e. responding to incidents which have occurred and trying to prevent them happening again. Most safety legislation has a common theme. Employers and employees are deemed to have a *duty of care* to ensure the health, safety and welfare of the employees, visitors and the public. Failure in this duty of care is called *negligence*. Legislation defines required actions at three levels.

- *Shall* or *must* are absolute duties which have to be obeyed without regard to cost. If the duty is not feasible the related activity must not take place.
- *If practicable* means the duty must be obeyed if feasible. Cost is not a consideration. If an individual deems the duty not to be feasible, proof of this assertion will be required if an incident occurs.
- *Reasonably practicable* is the trickiest as it requires a balance of risk against cost. In the event of an incident an individual will be required to justify the actions taken.

Safety legislation differs from country to country, although harmonization is under way in Europe. This section describes safety from a British viewpoint, although the general principles apply throughout the European community and are applicable in principle throughout the world. The descriptions are, of course, a personal view and should only be taken as a guide. The reader is advised to study the original legislation before taking any safety related decisions.

The Health and Safety at Work Act 1974 (HASWA) lays down the main safety provisions in the United Kingdom. It is wide ranging and covers everyone involved with work (both employers and employees) or affected by it. In the USA the Occupational Safety and Health Act (OSHA) affords similar protection.

HASWA defines and builds on general duties to avoid all possible hazards, and its main requirement is described in section 2(l) of the act:

It shall be the duty of every employer to ensure, so far as is reasonably practicable, the health, safety and welfare at work for his employees.

This duty is extended in later sections to visitors, customers, the general public and (upheld in the courts) even trespassers. The onus of proof of 'reasonably practicable' lies with the employer in the event of an incident.

8.2.2   Risk assessment

It is all but impossible to design a system which is totally and absolutely fail-safe. Modern safety legislation (such as the 'six pack' listed in Section 8.2.6) recognizes the need to balance the cost and complexity of the safety system against the likelihood and severity of injury. The procedure, known as *risk assessment*, uses common terms with specific definitions:

- *Hazard*   The potential to cause harm.
- *Risk*   A function of the likelihood of the hazard occurring and the severity.
- *Danger*   The risk of injury.

Risk assessment is a legal requirement under most modern legislation, and is covered in detail in standard prEN1050 'Principles of Risk Assessment'.

The first stage is identification of the hazards on the machine or process. This can be done by inspections, audits, study of incidents (near misses) and, for new plant, by investigation at the design stage. Examples of hazards are: impact/crush, snag points leading to entanglement, drawing in, cutting from moving edges, stabbing, shearing (leading to amputation), electrical hazards, temperature hazards (hot and cold), contact with dangerous material and so on. Failure modes should also be considered, using standard methods such as HAZOPS (Hazard and Operability Study, with key words 'too much of' and 'too little of'), FMEA (Failure Modes and Effects Analysis) and Fault Tree Analysis.

With the hazards documented the next stage is to assess the risk for each. There is no real definitive method for doing this, as each plant has different levels of operator competence and maintenance standards. A risk assessment, however, needs to be performed and the results and conclusions documented. In the event of an accident, the authorities will ask to see the risk assessment. There are many methods of risk assessment, some quantitative assigning points, and some using broad qualitative judgements.

Whichever method is used there are several factors that need to be considered. The first is the severity of the possible injury. Many sources suggest the following four classifications:

- *Fatality* One or more deaths.
- *Major* Non-reversible injury, e.g. amputation, loss of sight, disability.
- *Serious* Reversible but requiring medical attention, e.g. burn, broken joint.
- *Minor* Small cut, bruise, etc.

The next step is to consider how often people are exposed to the risk. Suggestions here are:

- *Frequent* Several times per day or shift.
- *Occasional* Once per day or shift.
- *Seldom* Less than once per week.

Linked to this is how long the exposure lasts. Is the person exposed to danger for a few seconds per event or (as with major maintenance work) several hours? There may also be a need to consider the number of people who may be at risk; often a factor in petrochemical plants.

Where the speed of a machine or process is slow, the exposed person can easily move out of danger in time. There is less risk here than with a silent high-speed machine which can operate before the person can move. From studying the machine operation, the probability of injury in the event of failure of the safety system can be assessed as:

Certain, Probable, Possible, Unlikely

From this study, the risk of each activity is classified. This classification will depend on the application. Some sources suggest applying a points scoring scheme to each of the factors above, then using the total score to determine *high, medium* and *low* risks. Maximum possible loss (MPL), for example, uses a 50 point scale ranging from 1 for a minor scratch to 50

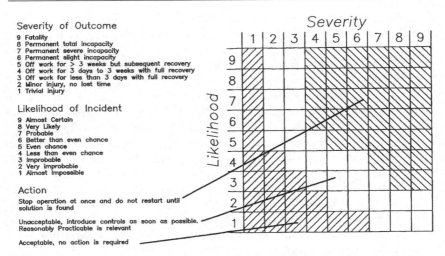

Severity of Outcome

9 Fatality
8 Permanent total incapacity
7 Permanent severe incapacity
6 Permanent slight incapacity
5 Off work for > 3 weeks but subsequent recovery
4 Off work for 3 days to 3 weeks with full recovery
3 Off work for less than 3 days with full recovery
2 Minor injury, no lost time
1 Trivial injury

Likelihood of Incident

9 Almost Certain
8 Very Likely
7 Probable
6 Better than even chance
5 Even chance
4 Less than even chance
3 Improbable
2 Very improbable
1 Almost Impossible

Action

Stop operation at once and do not restart until solution is found

Unacceptable, introduce controls as soon as possible. Reasonably Practicable is relevant

Acceptable, no action is required

Figure 8.1　*A simple risk assessment chart. This is only a guide, and should not be adopted without further study and only then if the conclusions can be justified for a specific hazard*

for a multi-fatality. This is combined with the frequency of the hazardous activity (F) and the probability of injury (again on a 1–50 scale) in the formula:

$$\text{risk rating (RR)} = F \times (\text{MPL} + P)$$

The course of action is then based on the risk rating.

An alternative and simpler (but less detailed approach) uses a table as in Figure 8.1 from which the required action can be quickly read.

There is, however, no single definitive method, but the procedure used must suit the application and be documented. The study and reduction of risks is the important aim of the activity.

The final stage is to devise methods of reducing the residual risk to an acceptable level. These methods will include removal of risk by good design (e.g. removal of trap points), reduction of the risk at source (e.g. lowest possible speed and pressures, less hazardous material), containment by guarding, reducing exposure times, provision of personal protective equipment and establishing written safe working procedures which must be followed. The latter implies competent employees and training programs.

8.2.3　PLCs, computers and safety

A PLC can introduce potentially dangerous situations in different ways. The first (and probably commonest) route is via logical errors in the program. These can be the result of oversight, or misunderstanding, on behalf of the original designer who did not appreciate that *this* set of actions could be dangerous, or by later modifications by people who deliberately (or accidentally) removed some protection to overcome a failure in the middle of the night. *Midnight programming* is particularly worrying as usually the only person who knows it has been done is the offending person, and the danger may not be apparent until a considerable time (days, weeks, months, years) passes and the hazardous condition occurs.

The second possible cause is failure of the input and output modules; in particular the components connected directly to the plant which will be exposed to high-voltage interference (and possibly direct-connected high voltages in the not unlikely event of cable damage). Output modules can also suffer high currents in the (again not unlikely) event of a short circuit.

Typical output devices are triacs, thyristors or transistors. The failure mode of these cannot be predicted; all can fail short circuit or open circuit. In these conditions the PLC would be unable to control the outputs. Similarly an input signal card can fail in either the 'On' or 'Off' state, leaving the PLC misinterpreting a possibly important signal.

The next failure mode is the PLC itself. This can be further divided into hardware, software and environmental failures. A hardware failure is concerned with the machine itself; its power supply, its processor, the memory (which contains the supplier's software with the 'personality' of the PLC, the user's program, and the data storage). Some of these failures will have predictable effects; a power supply failing will cause all outputs to de-energize, and the PLC supplier will have included checks on the memory in his design (using techniques similar to the CRC discussed in Section 5.2.7). Environmental effects arise from peculiarities in the installation such as dust, temperature (and rapid temperature changes) and vibration.

The final cause is electrical interference (usually called noise). Internally almost all PLCs work with 5 volt signals, but are surrounded externally by high-voltage, high-current devices. Noise can cause input signals to be misread by the PLC, and in extreme cases can corrupt the PLC's internal memory. PLCs generally have internal protection against memory corruption and noise on remote I/O serial lines (again using CRC and similar ideas) so the usual effect of noise is to cause a PLC to stop (and outputs to de-energize). This cannot, however, be relied upon.

There is no such thing as an absolutely safe process; it is always possible to identify some means of failure which could result in an unsafe condition. In a well-designed system these failure modes are exceedingly unlikely.

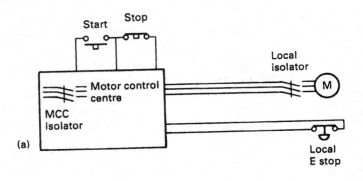

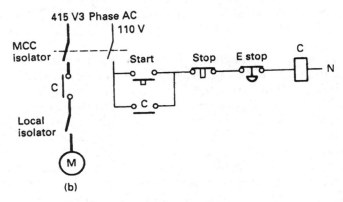

Figure 8.2   *Conventional motor starter: (a) physical arrangement; (b) circuit diagram*

Figure 8.2 shows a normal motor starter circuit built without a PLC. We will deal with emergency stop circuits in the following section, but for the time being the safety precautions here are

(a)   Isolation switch at the MCC removes the supply.
(b)   Local isolation switch by the motor. This, and (a), are for protection during maintenance work on the motor or its load.
(c)   Normally closed contact on the stop and emergency stop buttons. A broken wire will act as if a stop button has been pressed, as will loss of the supply.
(d)   If the emergency stop is pressed and released, the motor will not restart.
(e)   Isolation, stop and emergency stop have priority over start.

It is possible, though, to identify dangerous failure modes. The button head of the emergency stop button could unscrew and fall off, or the contacts of the contactor could weld made (albeit two welding together

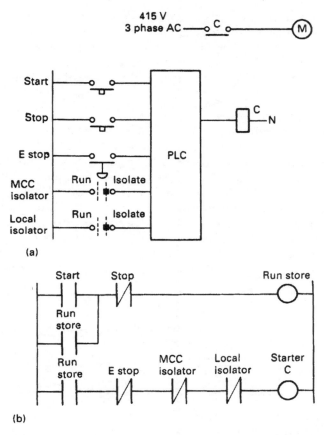

Figure 8.3 *Unsafe PLC-based motor starter: (a) physical arrangement; (b) PLC program*

are needed to cause an unsafe condition), but these failure modes are exceedingly rare, and, without discussing further details of the emergency stop function, Figure 8.2 would be generally accepted as safe.

In Figure 8.3 the same *function* has been provided by an *unsafe* PLC system. To save costs the MCC and local isolators have been replaced by simple switches which *make* to say 'isolate'. Similarly normally open contacts have been used for stop and emergency stop. This is controlled by the *unsafe* program of Figure 8.3(b).

It is important to realize that to the casual user, Figures 8.2 and 8.3 behave in an identical manner. The differences (and dangers) come in fault, or unusual, conditions. In particular

(a)   A person using a programming terminal can force inputs or outputs and override the isolation. Although it is unlikely that anyone

would do this deliberately, it is easy to confuse similar addresses and swap digits (forcing 0:23/01 instead of 0:32/01, for example).
(b)   A loss of the input control supply during running will mean the motor cannot be stopped by any means.
(c)   If the emergency stop is pressed and released, the motor will restart.

None of these are apparent to the user until they are needed in an emergency.

A prime rule, therefore, for using PLCs and computers is 'The system should be at least as safe as a conventional system.'

Figure 8.4 is a revised PLC version of Figure 8.2. The isolators have been reinstated with auxiliary contacts as PLC inputs, and normally closed contacts used for the stop and emergency stop buttons. An auxiliary contact has been added to the starter, and this is used to latch the PLC program of Figure 8.4(b). The emergency stop is hardwired into the output and is independent of the PLC, and on release the motor will not restart (because the latching auxiliary contact in the program will have been lost). On loss of control supply the program will think the

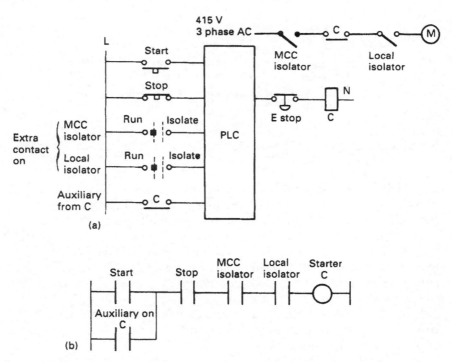

Figure 8.4   *Safe PLC motor starter: (a) physical arrangement; (b) PLC program*

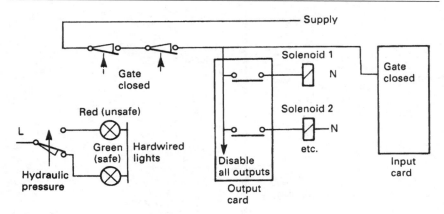

Figure 8.5 *Gate-controlled safety access*

stop button has been pressed, and the motor will stop. Figure 8.4 thus behaves in failure as Figure 8.2, and meets the rule above.

Figure 8.5 shows a similar idea used to disable a hydraulic system when the operator opens a gate giving access to a machine. The gate removes power to the PLC output card driving the solenoids which will all de-energize regardless of what the PLC is doing. A separate input to the PLC also software disables outputs. One of the solenoids is the loading valve which, when de-energized, causes the manifold pressure to fall to near zero. This pressure is monitored by hardwired traffic lights.

Although these examples are simple, the necessary analysis and considerations are identical in more complex systems.

Complex electronic systems *can* bring increased safety. In Figure 8.6 a thyristor drive is controlling the speed of a large DC motor. The arrangement is typical; switched isolator for maintenance and an upstream AC contactor. It is required to add an emergency stop to this drive. Using this to hardwire the AC contactors will obviously stop the drive, but the inertia of the motor and the load will keep it rotating for several seconds. A thyristor drive, however, can stop the load in less than one second by regenerative braking the motor, but this requires the drive to be alive and functional.

The operation of the emergency stop implies a dangerous condition in which the fastest possible stop is required. It is almost certain that at this time the drive controls are functional and there are no 'latent' faults. Faults with the speed control system would have been noticed by the operator, for example. From a risk assessment, the author would argue that if a guarding system was not practicable, the emergency stop should operate in two ways. First, it initiates an electronic regenerative crash stop via the control system which should stop the drive in less

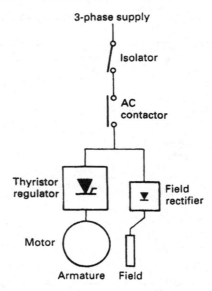

Figure 8.6 *A DC drive requiring an E stop*

than one second (albeit at great strain to the motor and mechanics). The emergency stop also releases a delay drop-out hardwire relay set for 1.5 seconds which releases the AC contactor. This gives the safest possible reaction to the pressing of the emergency stop button.

Safety considerations do not therefore explicitly require relay-based, non-electronic hardware, but the designer must be prepared to justify the design decisions and the methods used. Where complex control systems are to be used, a common method is to duplicate sensors, control systems and actuators. This is known as redundancy.

A typical application is a boiler with feed water being held in a drum. Deviations in water level are dangerous; too low and the boiler will overheat, possibly to the point of melting the boiler tubes; too high and water can be carried over to the downstream turbine with risk of catastrophic blade failure. High- and low-level sensors are provided and each are duplicated. The control system reacts to *any* fault signal, so two sensors have to fail for a dangerous condition to arise. If the probability of a sensor failure in time T is p (where $0 < p < 1$) the probability of both failing is p^2. In a typical case, p will be of the order of 10^{-4} giving p^2 a probability of 10^{-8}.

There are two disadvantages. The first is that a sensor can fail into a permanently safe signal state, and this failure will be 'latent', i.e. hidden from the user with the plant running on one sensor. The second problem is that the plant reliability will go down, since the number of sensors

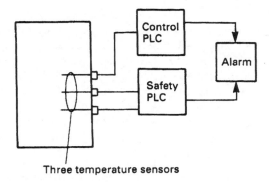

Three temperature sensors

Figure 8.7 *Duplication of sensor and control/protection systems*

goes up and any sensor failure can result in a shutdown. Both of these effects can be reduced by using 'majority voting' circuits, taking the vote of two out of three or three out of five signals.

So far we have duplicated the sensors. To give true redundancy, it is sensible to provide duplication in the control system as well to protect against hardware and software failures in the system itself. Figure 8.7 shows three temperature sensors (for redundancy) connected into two separate and independent PLC systems, one concerned with control and safety and one purely concerned with control. A simultaneous failure of both is required to give a hazard.

Redundancy can be defeated by 'Common mode' failures. These are failures which affect all the parallel paths simultaneously. Power supplies, electrical interference on cables following the same route and identical components from the same batch from the same supplier are all prone to common mode failure. For true protection, diverse redundancy must be used, with differences in components, routes and implementation to reduce the possibility of simultaneous failure.

Examples such as the duplicate control scheme of Figure 8.7 are also vulnerable to a form of common mode failure called a 'systematic failure'. Suppose the temperature sensors are compared, inside the program, with an alarm temperature. Suppose both are identical systems, running the same program containing a bug which inadvertently (but rarely, so it does not show up during simple testing) changes the setting for the alarm temperature (from, say, $60\,°C$ to $32\,053\,°C$). Such an effect could easily occur by a mistype in a *move* instruction in a totally unrelated part of the program. This error will affect *both* control systems, and totally remove the redundancy.

If reliance is being made on redundant control systems, they should be totally different; different machines with different I/O and different programs written by different people with the machines installed running on different power supplies with different types of sensors connected by different cable routes. This is what true redundancy means.

The Health and Safety Executive (HSE) became concerned about the safety of direct plant control with computers, and produced an occasional paper OP2 'Microprocessors in Industry' in 1981. This was followed in 1987 by two booklets *Programmable Electronics Systems in Safety Related Applications*, book 1, an *Introductory Guide* and book 2, *General Technical Guidelines*. Book 1, like the earlier 1981 publication, is a general discussion of the topic, with book 2 going through the necessary design stages. They suggest a five stage process:

(i)   Perform a hazard analysis of the plant or process (key phrases like 'too much of', 'too little of', 'over', 'under' are useful here).
(ii)   From this, identify which parts of the control system are concerned with safety and which are concerned purely with efficient production. The latter can be ignored for the rest of the analysis.
(iii) Determine the required safety level (based on accepted attainable standards or published material).
(iv)   Design safety systems to meet or exceed these standards. The HSE stress the importance of 'quality' in the design; quality of components, quality of the suppliers and so on.
(v)   Assess the achieved level (by using predicted probability of failure for individual parts of the design). Revise the design if the required level has not been achieved.

Testing is a crucial part of safety, and it can be difficult to complete a sensible test routine with the unavoidable pressures to get a plant into production. Once the control room lights are on, there is a 'gung-ho, let's be away' attitude. This can be hard to resist, particularly if the project is late, but it can be lethal. The only way to avoid this trap is to have a pre-agreed safety checklist (written in the cold light of day well before testing starts) which can be ticked off item by item. The engineer then has firm grounds for not releasing the plant until all items are cleared.

Maintenance is, perhaps, the most dangerous time. Chernobyl, Flixborough, Three Mile Island, Bhopal, Piper Alpha and the Charing Cross rail accident all had seeds in ill-advised maintenance activity. The author has seen faulty protection systems bypassed 'to get the plant away', with the bypasses still in place weeks later. The ease of programming of PLCs makes them very vulnerable (and it is for this reason that programs held in ROM are preferred for safety applications).

Plant must be put into a safe condition before people work on it. The need for electrical isolation and a Permit to Work system is generally appreciated (and the isolators in Figure 8.2 are provided for this purpose) but the danger from pneumatic or hydraulic actuators is often overlooked.

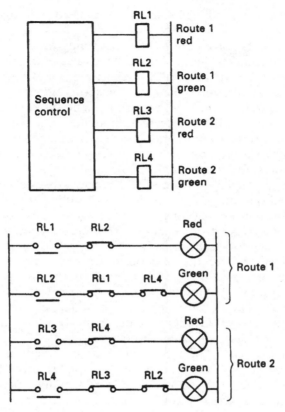

Figure 8.8   *Safety system with hardwire relays*

If a PLC stops or is powered down (a not unlikely event on maintenance periods) all solenoids will de-energize and plant may move. Isolation procedures must therefore include *all* actuators and not just the obvious electric motors.

Using relays it is possible to design circuits which are *almost* fail-safe using the principle shown for simple traffic lights in Figure 8.8. (No circuit can ever be totally fail-safe, of course.) Here the cross-coupling of the relays, along with the use of techniques such as spring-loaded terminals to prevent cores coming loose under vibration, ensures that the system can fail with no lights, or locked onto one route, or showing both red, but it cannot fail showing both green. To achieve this, high-quality relays are used, constructed in such a manner that an internal mechanical collapse cannot lead to normally open and normally closed contacts being made together. With a very high degree of safety, the idea of

Figure 8.8 is widely used in lifts, traffic lights, burner control and railway signalling.

The safety levels of Figure 8.8 are becoming achievable with some PLCs. Siemens market the 115F PLC which has been approved by the German TUV Bayern (Technical Inspectorate of Bavaria) for use in safety critical applications such as transport systems, underground railways, road traffic control and elevators. The system is based on two 115 PLCs and is a model of diverse redundancy. The two machines run diverse system software and check each other's actions. There is still a responsibility on the user to ensure that no systematic faults exist in the application software.

Inputs are handled as in Figure 8.9. Diverse (separate) sensors are fed from a pulsed output. A signal is dealt with only if the two processors agree. Obviously the choice of sense of the signal for safety is important. For an over-travel limit, for example, the sensors should be made for healthy and open for a fault.

Actuators use two outputs (of opposite sense) and two inputs to check the operation as Figure 8.10. Each subunit checks the operation of the other by brief pulsing of the outputs allowing the circuit to detect cable damage, faulty output modules and open circuit actuators. If, for example, output B fails On, both inputs A and B will go high in the Off state (but the actuator will safely de-energize).

The operation of Figures 8.9 and 8.10 is straightforward, but it should not be taken as an immediately acceptable way of providing a fail-safe PLC. The 115F is truly diverse redundant, even the internal integrated circuits are selected from different batches and different manufacturers, and it contains well-tested diverse self-checking internal

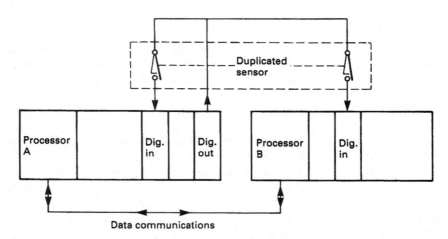

Figure 8.9 *Safety-critical input with the Siemens 115F PLC*

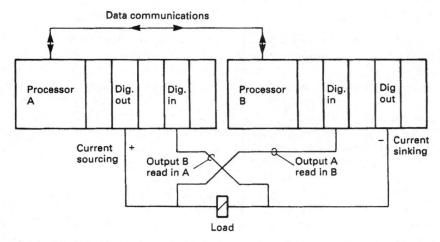

Figure 8.10 *Safety-critical output with the Siemens 115F PLC*

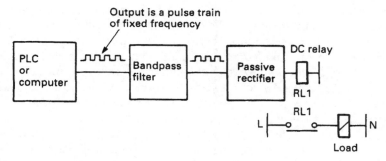

Figure 8.11 *Dynamic 'failsafe' circuit*

software. A DIY system would not have these features, and could be prone to common mode or systematic failures.

Figure 8.11 shows a dynamic fail-safe circuit sometimes used in applications such as gas burner control. Here a valid output signal is a square wave pulse train of a specific frequency (obtained by turning an output on and off rapidly). This oscillating signal passes through a narrow bandpass filter and, when rectified, energizes the actuator. Failure of the CPU would lead to failure of the pulse train (or a shift in frequency of the pulse train which would then be rejected by the filter). Failure of the output would give a DC signal which would again be rejected by the filter. Failure of any component in the filter will cause a shift in the bandpass frequency and a failure to respond to normal outputs. The principle of Figure 8.11 is often used as a single 'watchdog' output which can be used as an interlock to say the PLC is healthy.

8.2.4 Emergency stops

Most plants have moving parts with the ability to cause harm. There is therefore a need to provide some method of stopping the operation in the shortest possible time when some form of danger is seen by the operators. Usually the initiation is provided by emergency stop push-buttons at strategic points around the plant. These must be red, mushroom headed buttons surrounded by a yellow surface. They must latch and need some form of manual action (key, twist or pull) to release. Even when the button is released the plant must not start again without some separate restart operation. Conveyors and similar items use pull wire emergency stops which have to be physically reset.

Until fairly recently, an emergency stop circuit would operate as shown in Figure 8.12(a), and indeed the earlier circuit of Figure 8.2 implemented the emergency stop in this way. Operation of the button breaks the control circuit to the contactor, causing it to de-energize and remove power from the machine. This circuit, however, has several failure modes which may be dangerous. In particular, the contactor contacts may weld made or the opening spring in the contactor may break. In these circumstances the emergency stop will have no effect.

Figure 8.12(b) uses redundancy to give improved safety. With redundancy, care must be taken to ensure that the failure of one element cannot lead to continued, apparently normal, operation on the remaining elements with reduced safety. In this circuit both contactors must fail for the emergency stop to be inoperative. The two normally closed contacts in the left-hand leg give some protection against welding or sticking of a contactor. A firm fault with one contactor hard welded will cause its normally closed contact to open, and the circuit will fail to start. The

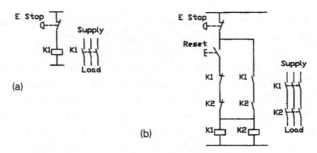

Figure 8.12 *(a) A simple emergency stop circuit. This circuit has several potentially dangerous failure modes. (b) An improved emergency stop circuit. For this circuit to work, however, the two contactors must have overlapping normally open and normally closed contacts and the circuit can fail in an unsafe condition*

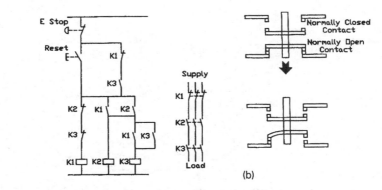

Figure 8.13 *(a) Emergency stop circuit built with non-overlapping contacts. This circuit has to be reset after the stop button has been pressed and is still safe with a single failure. (b) The principle of a non-overlapping positively driven contact set*

circuit, though, is not ideal. For it to operate at all there must be an overlap between the normally open and normally closed contacts (i.e. a short region where both are made together). It is thus feasible for a contactor to fail with both normally open and normally closed contacts made. In addition overlapping contacts must be spring loaded in some way, which introduces additional hazardous failure modes. Although much better than the circuit in Figure 8.12(a) it still has hazards.

Safety can be improved further by using three contactors as in Figure 8.13(a) and positively driven contacts. These ideas have been traditionally used in traffic lights and railway signalling to ensure, for example, that two routes cannot be given a green 'Go!' signal at the same time. Positively driven contacts are constructed in such a way that both normally open and normally closed contacts are moved by the same mechanism and cannot both make at the same time even in the event of failure. The principle is shown in Figure 8.13(b).

In Figure 8.13(a), three contactors are used in series. One, K1, uses normally closed contacts and the others normally open contacts. For normal operation, therefore, K1 must be de-energized and K2/K3 energized. When first powered up, or when the emergency stop has been pressed and released, all contactors are de-energized. When the reset button is pressed, K1 will first energize provided K2 and K3 have not stuck (positively driven contacts remember). Contacts from K1 then bring in K2 and K3 which de-energize K1 but hold themselves in via their own contacts.

A single failure of any contactor will cause the circuit to fail or prevent it from starting. It is still, however, vulnerable to *simultaneous*

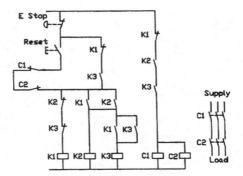

Figure 8.14 *Emergency stop circuit of Figure 8.13(a) modified for use with separate large contactors*

welding of K2 and K3, but the probability of this is usually thought to be acceptably low.

Figure 8.13(a) is acceptable for small loads of a few kilowatts, but is a bit impractical (and expensive) for large loads. In Figure 8.14 the safety circuit is constructed from low-power positively driven relays, which then control two redundant contactors C1 and C2. The (positively driven) auxiliary contacts of these contactors are connected in series with the reset pushbutton. Failure of a contactor in an unsafe mode thus prevents a restart and, like Figure 8.13(a), the circuit can only be started with both contactors in a healthy condition. Also like Figure 8.13(a) there is a residual risk of both contactors failing in a made state whilst running.

The arrangement of Figures 8.13(a) and 8.14 is commonly used and is available as a pre-made safety relay from many control gear manufacturers.

Although the residual risk in these circuits is very low, it may not be acceptable in some circumstances. In particular, the circuit relies on a single contact in the emergency stop itself, and it is possible for a cable fault to bridge out the pushbutton contact (a severed cable will, of course, cause the safety relay to stop the plant). If a risk assessment of the application demands a lower residual risk (often when flexible cable is used to the stop button), two contacts may be used on each pushbutton, with one switching the supply voltage to the safety relay and one the return as in Figure 8.15. If a four core cable is used to link the button, there is a very high probability that any cable fault will either de-energize the safety relay or cause the circuit protection fuse or breaker to open.

Emergency stop circuits need regular maintenance and testing. The author has seen buttons in a dusty and humid atmosphere build up an

Figure 8.15 *Emergency stop pushbutton with dual action to guard against contact failure or cable damage. This will be used with circuits similar to Figures 8.13 and 8.14 and disconnect both the supply and return when pressed*

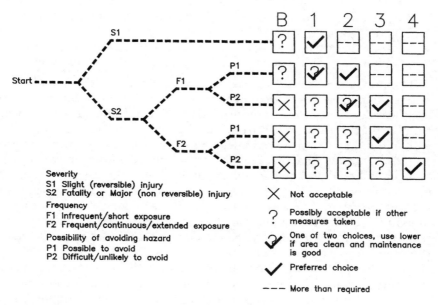

Figure 8.16 *Emergency stop selection chart*

almost concrete ring under the mushroom head which prevented the button operating even when hit with a hammer! Regular inspection and testing will prevent similar problems. Remember that inspection and maintenance of safety equipment is a legal requirement under most safety legislation.

Section 8.2.2 introduced the idea of Risk Assessment. European Standard prEN954-1 gives a risk assessment chart (shown in Figure 8.16) for fail-safe control equipment (which includes emergency stops and movable guards). This results in the following categories:

- *Category B* Minimum standards to meet operational requirements of the plant with factors such as consideration of humidity, temperature and vibration.

- *Category 1*   As 'B' but safety systems must use 'well-tried' principles and components. Sole reliance on electronic or programmable systems at this level is not permitted.
- *Category 2*   As 'B' but the machine must be inhibited from starting if a safety system fault is detected on starting. Single channel actuation (i.e. emergency stop buttons or gate limit switches) is permitted providing there is a well-defined regular manual testing procedure.
- *Category 3*   As 'B' but any fault in the safety system must not lead to the loss of the safety function and, where possible, the fault should be identified. This implies redundancy and dual channel switches as Figure 8.15.
- *Category 4*   As 'B' any single fault must be detected, and any three simultaneous faults shall not lead to the loss of the safety function.

The previous section discussed the safety aspects of PLCs and computers. In all bar the most complex system it is not financially viable to use the redundancy techniques needed to achieve adequate safety levels from a purely software/electronic emergency stop system. The best system is to have a hardwired safety system which acts in series between the PLC outputs and the actuators (contactors, valves or whatever). A contact should also be taken as an input to the PLC to say the safety system has operated (in practice, of course, an input is given which says the safety system has *not* operated so a fault or loss of supply causes motion to cease). This input will, via the program, cause the output to turn off so the system needs some manual action beyond the removal of the emergency signal to restart.

Where several devices are to be turned off together in an emergency (e.g. several solenoid valves), the emergency stop contacts can remove the supply to the relevant output cards as shown earlier in Figure 8.5.

One final comment is that an effective guarding system can reduce the requirements of the emergency stop system by reducing the exposure (i.e. the system will be inoperative whenever people are at risk). This is discussed in the following section.

8.2.5   Guarding

One very effective method of reducing risk is to restrict access to dangerous parts of a machine or process by fixed or movable guards. Fixed guards are simple to design, but movable guards, often required where access is required for production or maintenance, need careful design.

Movable guards must provide two safety functions. Firstly, they must ensure the machine cannot operate when the guards are open. Secondly, if the machine or plant has an extended stopping time, the

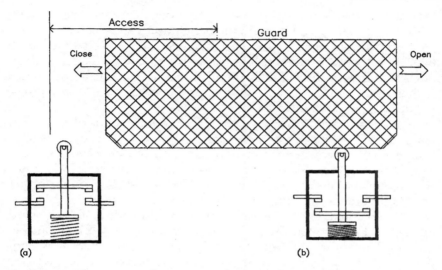

Figure 8.17 *Use of guarding limit switches. (a) Non-positive switch (guard makes, spring opens); (b) Positive switch (guard opens, spring makes)*

guards must be locked in some way until the machine has come to a standstill.

The intuitive, but incorrect, method of ensuring machine isolation when a guard is open is to use a limit switch which makes when the guard is closed as shown on Figure 8.17(a). This has two potentially dangerous shortcomings. In this arrangement (called negative or non-positive operation) the switch is closed by the guard and opened by a spring. Failure of this spring may result in the contacts still being made when the guard is open. In addition the switch is accessible when the guard is open and can easily be 'frigged' by the operators to bypass the safety feature.

In Figure 8.17(b) the guard itself pushes open the switch on opening, and a spring is used to make the switch when the guard recloses. Failure of the spring thus prevents the machine from operating. If the contacts weld, either the guard cannot be opened, or the opening action will break the weld or switch. With the switch being pushed down by the guard and raised for operation it is very difficult to bypass. This is called positive operation. A very low residual risk is obtained if the guard is equipped with both types of switch.

Photocell beams are another possible solution to guarding. These should be self-checking systems designed for safety applications, not simple photocell sensors. Care must be taken in the mounting to ensure the beam cannot be bypassed by going over or under, and the operator cannot pass totally through the beam and out the other side. Pressure mats may also be used but again the design needs some care.

In applications such as shears and presses, dual two-handed push-buttons offer good protection. The use of these builds on the relay circuits described above, and specialist dual control relays can be purchased. It should, though, be noted that dual pushbutton systems only protect the operator and not other people.

Maintenance and repair usually requires access to parts of the process barred to operators. This puts maintenance staff at higher risk from trapping, entanglement and crushing. Often the only way to achieve an acceptable residual risk is a full isolation and Permit to Work system with removal of energy at pre-determined points such as motor control centres, hydraulic accumulators, air lines, etc. The removal of energy should be tested, then the isolation points locked, isolation boards applied and the written permit issued. Only then can work commence. For smaller/shorter jobs with low risk, the risk assessment may show that a local removable safety key system may be acceptable. Note that a risk assessment should be done for each possible maintenance job.

8.2.6   Safety legislation

There is a vast amount of legislation covering health and safety, and a list is given below of those which are commonly encountered in industry. It is by no means complete, and a fuller description of this, and other, legislation is given in the third edition of the author's *Industrial Control Handbook*. An even more detailed study can be found in *Safety at Work* by John Ridley, both books published by Butterworth-Heinemann.

Health and Safety at Work Act 1974 (the prime UK legislation)

The following six regulations are based on EEC directives and are known collectively as 'the six pack':

    Management of Health and Safety at Work Regulations 1992
    Provision and Use of Work Equipment Regulations 1992 (PUWER)
    Manual Handling Regulations 1992
    Workplace Health, Safety and Welfare Regulations 1992
    Personal Protective Equipment Regulations 1992
    Display Screen Equipment Regulations 1992

Reporting of Injuries, Diseases and Dangerous Occurrences Regulations (RIDDOR) 1995
Construction (Design and Management) Regulations (CDM) 1994
Electricity at Work Regulations (1990)
Control of Substances Hazardous to Health (COSHH) 1989
Noise at Work Regulations 1989

Ionizing Radiation Regulations 1985
Safety Signs and Signals Regulations 1996
Highly Flammable Liquids and Liquefied Petroleum Gas Regulations
Fire Precautions Act 1971
Safety Representative and Safety Committee Regulations 1977
Health and Safety Consultation with Employees Regulations 1996
Health and Safety (First Aid) Regulations 1981
Pressure Systems and Transportable Gas Containers Regulations 1989.

8.2.7 IEC 61508

IEC61508 is the International Electrochemical Committee Standard on electrical, electronic and programmable (E/E/PE) safety systems. Like all modern safety concepts it is based on the idea of risk assessment and the implementation of measures to reduce the risk to an acceptable level. It is a complex document and this section can only give a brief introduction to the basic ideas. The reader is strongly advised to study the full standard or take professional advice before designing any system where safety will be an issue.

Because it is an international standard rather than a UK or European directive compliance is not mandatory, however the Health and Safety Executive's official view is '*IEC61508 will be used as a reference standard in determining whether a reasonably practical level of safety has been achieved when E/E/PE systems are used to carry out safety functions*'.

There are several terms in IEC61508 which carry specific meaning:

Hazard is the potential for causing harm to people or the environment.

Risk is a combination of the probability of the hazard and the severity of the result:

Risk = probability × consequence

You can reduce the risk by reducing the probability or the consequence. For example, with motor cars, imposing speed limits reduces both the probability of an accident and also the likely consequences. A cycle rider wearing a helmet reduces the concequences of an accident.

Equipment under Control (or EUC) is the plant under consideration. It consists of sensors, a logic control system and actuators.

Functional Safety relies on the correct operation of safety functions when required. Safety functions are the parts of the plant which provide safety such as flow sensors, pressure relief valves, safety gates, emergency stop buttons, pull wires, etc.

IEC 61508 defines the four levels of risk classification given in Table 8.1.

The level of accepted risks varies surprisingly from industry to industry, but IEC61508 suggests the risk classifications in Table 8.2 are typical.

Table 8.1   *Risk classification*

Risk class	Interpretation
I	Intolerable risk
II	Undesirable, tolerable only if risk reduction is impossible or costs are disproportionate to the benefits gained
III	Tolerable provided the costs of further improvements are disproportionate to the benefits gained
IV	Negligible risk. Acceptable

Table 8.2   Based on Table B1 in IEC61508-5

Frequency		Consequences			
		Catastrophic	Critical	Marginal	Negligible
Frequent	1 per year	I	I	I	II
Probable	1 per 5 years	I	I	II	III
Occasional	1 per 50 years	I	II	III	III
Remote	1 per 500 years	II	III	III	IV
Improbable	1 per 5000 years	III	III	IV	IV
Incredible	1 per 50 000 years	IV	IV	IV	IV

Catastrophic is more than one death.
Critical is one death or one or more serious injuries.
Marginal is one or more minor injuries.
Negligible is a trivial injury or plant damage and resultant loss of production.

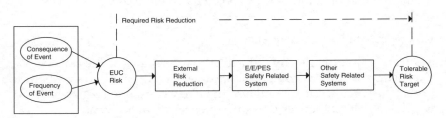

Figure 8.18   *Required risk reduction*

A hazardous plant can be represented by Figure 8.18. The probability and consequence of each hazard gives a risk. This can be reduced by a combination of external risk reduction facilities (e.g. enclosures), other systems (e.g. personal protective equipment) and in the centre the E/E/PE safety system itself. The combination of these three measures gives the necessary risk reduction to produce a tolerable risk.

Table 8.3 Low demand mode

SIL	Probability of failure
4	$>= 10^{-5}$ to 10^{-4}
3	$>= 10^{-4}$ to 10^{-3}
2	$>= 10^{-3}$ to 10^{-2}
1	$>= 10^{-2}$ to 10^{-1}

Table 8.4 High demand mode

SIL	Probability of dangerous failure per hour
4	$>= 10^{-9}$ to 10^{-8}
3	$>= 10^{-8}$ to 10^{-7}
2	$>= 10^{-7}$ to 10^{-6}
1	$>= 10^{-6}$ to 10^{-5}

Risk reduction is described in terms of Safety Integrity Levels or SILs. There are two groups of SIL. The first, called a Low Demand Mode of Operation will only be required to operate very rarely (if at all). A typical example is a car air bag which is a complex safety function designed to reduce the probability of injury to the occupants of a car in the event of an accident. Here the SIL is determined by the probability of failure to perform its design function on demand as shown on Table 8.3.

The second, called High Demand Mode or Continuous Mode is concerned with a safety function that monitors the EUC continuously. A typical example would be the drum water level system in a high pressure boiler. Here the SIL is the probability of dangerous failure per hour as defined in Table 8.4.

The study of Figure 8.18 will show the SIL level required. Note that a complete study is required with investigation of all failure modes of the safety functions. It is not just a case of obtaining the failure rate of a sensor and valve.

Suppose we have done a risk assessment and identified a risk frequency of once per year with a consequence of one injury. Table 8.2 gives this a risk classification of I. We need to reduce this from class I to class III which means we need to reduce the risk frequency to approximately 1 per 5000 years. This implies a *Risk Reduction Factor* (RRF) of 5000.

The plant designers, though, can provide a risk reduction factor of 15 by improved mechanical design. The demand on the E/E/PE safety system now becomes a RRF of 333 giving a required probability of failure on demand of 3×10^{-2}. From Table 8.3 this gives an aim integrity level of SIL-2 for our E/E/PE safety system.

The Heath and Safety Executive publish an excellent book called 'Out of Control', ISBN 0-7176-0847-6 which every PLC user should read. Part of this book analyses the major causes of control related accidents and comes up with the following worrying statistics:

44% caused by bad or incomplete specification
15% caused by design errors
 6% introduced during installation and commissioning
14% occur during operation and maintenance
21% caused by ill thought out modifications

In other words, the commonest cause of accidents were flaws in the original specification where the need for a safety function was not recognized or was badly described.

IEC61508 therefore lays down the safety life-cycle shown on Figure 8.19. Each stage produces output documentation which is the input for the following stages. This documentation should be available at all times. The sixteen stages are:

Stage 1 Concept. An understanding of the EUC and its environment is developed along with relevant legislation.

Stage 2. Overall Scope Definition. Define the boundaries of the EUC and its control system in all modes of operation (e.g. start up, normal operation, etc.). Specify the scope of the hazard and risk analysis.

Stage 3. Hazard and Risk Analysis. This should be done for all modes of operation and all reasonably foreseeable conditions including faults and operator errors.

Stage 4. Overall Safety Requirements. Develop the specification for the overall safety function requirements.

Stage 5. Safety Requirement Allocation. Determination of how a safety function is to be achieved and allocate a SIL to each safety function.

Stage 6. Overall Operation and Maintenance Planning. Developing a plan of for operating and maintaining the E/E/PE safety related systems so the safety system operates correctly.

Stage 7. Overall Safety Validation Planning. Developing a plan for validating the safety system.

Stage 8. Overall Installation and Commissioning. Developing a plan to ensure the safety system is correctly installed and operational.

Stage 9. E/E/PE Safety System Realization. Create the E/E/PE safety system that conforms to the safety requirements specification.

Stage 10. Realization of safety systems implemented in other technologies. Creation of non P/E/PE safety systems. This stage is not part of IEC61508.

IEC 61508 Safety Life-Cycle

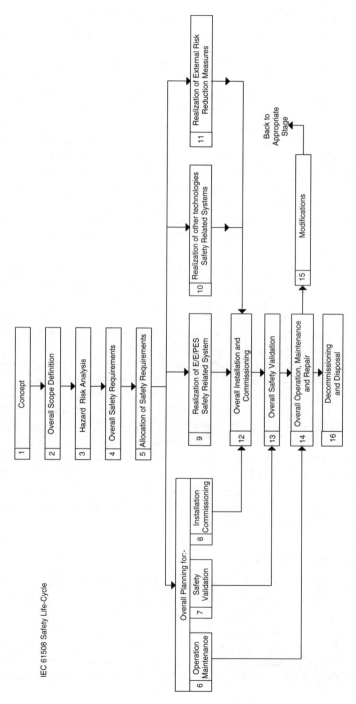

Figure 8.19 *The safety life-cycle*

Stage 11. Other Risk Reduction Factors. Again not part of IEC61508.

Stage 12. Overall Installation and Commissioning. Based on the plans from Stage 8.

Stage 13. Overall Safety Validation. Based on the plans from Stage 7.

Stage 14. Overall Operation, Maintenance and Repair. This covers the majority of the plant life. The safety systems should be operated, maintained and repaired in such a way that their integrity is ensured. This will be based on the plans from Stage 6.

Stage 15. Overall Modification and Retrofit. Plant modifications and changes are a very dangerous operation, see the earlier HSE study in Out of Control. Flixborough and Chernobyl originated because of poorly thought out maintenance work. If any maintenance work or plant changes have safety implications the safety life-cycle procedure should be repeated.

Stage 16. De-commissioning. Ensuring that the functional safety of the E/E/PE safety related system is appropriate for the final shut down and disposal of the EUC.

IEC61508 also imposes some system architectural constraints. The most important of these is that the safety system and the control system should be separate. This is normal in most PLC based systems where safety devices such as Emergency Stops or guards operate directly into the actuators. It can, though, be more difficult in a complex petro-chemical plant where a controlled shutdown sequence is required.

Another important architectural constraint is the failure mode of devices and the probability of 'safe failure' must be considered. For a cooling water modulating valve a safe failure is a failure to close. For a fuel control valve a safe failure is a failure to open. Tables in IEC61508 relate the required SIL and the probability of safe failure to the required level of redundancy. For example, with a required level of SIL-2 and a probability of safe failure of less than 0.6 for a safety related actuator, dual redundancy is mandatory even if the calculated SIL is adequate.

IEC61508 also emphasizes the need for quality in the design procedure and quality of the component parts. Needless to say the people concerned should be knowledgeable and competent.

8.3   Design criteria

In Chapter 3 the problems of defining what functions a PLC system is to perform were discussed. In this section we will consider similar points that have to be established for the PLC hardware.

The first (and possibly most important) of these is to establish the amount of I/O (how many digital inputs, how many digital outputs, etc.) and from this determine how many cards are needed. This can be surprisingly difficult, as it depends very much on the user's requirements.

Consider, for example, a simple hydraulic pump started and stopped from a main control desk. At the simplest level this requires one PLC input (the desk run switch) and one output (the pump contactor). With pushbuttons, desk lamps and useful diagnostics this could rise to eight inputs (start PB, stop PB, contactor auxiliary, trip healthy, emergency stop healthy, local isolator healthy, MCC on, hydraulic pressure switch and five outputs (the pump contactor plus indicators for running, stopped, tripped, fault, not available). The designer needs to know the point between these two extremes that the user expects. I much prefer to connect to everything that is available and then decide later whether it is to be used. Adding 'forgets' in later is messy and expensive. ('We've just decided it would be useful to know when a motor is tripped' says the user. 'There are 47 motors, no spare contacts on the overload relays in the MCC and no spare I/O in the PLC, and we want to start testing tomorrow.') Avoid problems like this.

The I/O is not simply the number of signals divided by the number of channels per card. It helps maintenance to group functions by card; Figure 8.20 shows a typical example. PLC cards have LEDs showing signal status and with sensible I/O allocation readily identifiable 'patterns' can be established which assist fault finding.

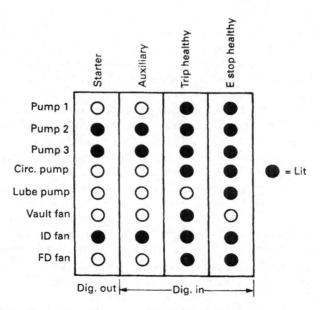

Figure 8.20 *Sensible allocation of I/O aids fault finding. It can readily be seen that the lube pump has tripped and the E stop is pressed on the vault fan*

This functional grouping also generates useful spare I/O. A PLC system should never be built without spare I/O (and never installed without the ability to expand). If the project is well documented and the user trusted implicitly, 10% installed spare I/O may be feasible. If there is no real written specification, 50% installed spare I/O is not unreasonable. Inputs and indicator lamps have a tendency to appear magically from out of nowhere, so be prepared.

Cabling is a major capital cost, and can be significantly reduced by the use of remote I/O which is connected via simple coaxial or twin-axial cable. The geography of the plant should be studied to establish where I/O can be clustered to minimize cabling. Desks, for example, can often be built with integral I/O cards only connecting to the outside world via the remote I/O cable, a power supply and a few hardwire signals such as emergency stop buttons. Such an approach allows useful offsite testing before installation.

8.4 Constructional notes

8.4.1 Power supplies

A PLC system obviously requires a power supply to operate. This will usually be a low voltage; 110 V AC is probably the commonest in industry. If at all possible, an entire system should be fed from one common supply. If separate supplies are used, any noise spikes on one supply can cause momentary loss of communication between different parts of the system and result in mysterious shutdowns. With a common supply, all component parts experience the same effects and are more tolerant of noise.

In each cubicle, a power supply distribution system similar to Figure 8.21 will be needed. It can be seen that this feeds several different areas, each requiring breakers, or fuses, for protection.

The PLC racks and processors obviously need a supply, and this should be clean and free of noise to prevent unexplained trips. Until recently, it was common practice to use constant-voltage transformers (CVTs) to give a smooth clean supply for the PLC power supplies. These act as a block to high-frequency noise on the supply. Unfortunately they also block high-frequency loads *from* the supply, and can give some very odd results with switched-mode power supplies. If the PLC is to work off a potentially noise-prone supply (on a 110-V supply derived direct from the bars of an MCC, for example) CVTs should be considered. CVTs have a very high inrush current, which results in higher rated upstream protection and cables than might be first thought. In-line filters are also useful, but these also are prone to odd effects with switched-mode power supplies.

In Figure 8.21 a single emergency stop relay has been used. This removes all power to output cards in slot 0 and disables one output in slot 1. If this latter arrangement is used, snubbers should be put across the load and/or the contact to reduce the inductive voltage spike as the contact opens. This voltage spike can be a major source of electrical interference and can even cause damage to the PLC output transistors or triacs.

It is good practice to have independent protection for each output card as shown. This limits the affected area in the event of any external fault, and aids fault finding by locating the problem to outputs connected to one card. Usually PLC output cards have their own internal protection, often at the level of one fuse per output (with a common fuse blown indicator). Often each output supply is fed back as an input to allow the PLC to check the supply state and give an alarm when a blown fuse or tripped breaker occurs.

With inputs there are two possible protection methods. In the first all inputs for one card are fed from the same breaker. This means that a card can be isolated at one point (but has the disadvantage that several different supplies may exist inside a desk or junction box). With the second approach, every point at one location is fed from a common supply. This has only one supply at a location (allowing isolation of several limit switches at one breaker) but means that a single card is fed from many different supplies. The author prefers the second scheme but this is only a personal opinion.

During commissioning, maintenance and fault finding, it is often useful to be able to shut down outputs or inputs whilst leaving the system running. Switches, breakers or fuses on Figure 8.21 allow this isolation to be performed.

A cubicle often contains non-PLC devices, 24-V power supplies, instruments and chart recorders. These also need individual protection.

Finally we have two often-overlooked essential supplies. Cubicle lighting and sockets for the programming terminal and tools such as a soldering iron can aid commissioning and fault finding. The author always includes a standard 15-A and 5-A 110-V socket in every cubicle.

It would obviously not be desirable to have the whole PLC system shut down by a simple fault like a stuck AC solenoid (which produces a high current) tripping the main breaker. A hierarchy such as that of Figure 8.21 needs discrimination between the various protection devices to ensure that a breaker trips or a fuse blows at only the lowest level. This is a complex subject, but as a rough rule the rating of the protection should be between five and ten times the rating of the protection at the next lowest level. Remember that the protection is for the *cable*, not the device it is connected to.

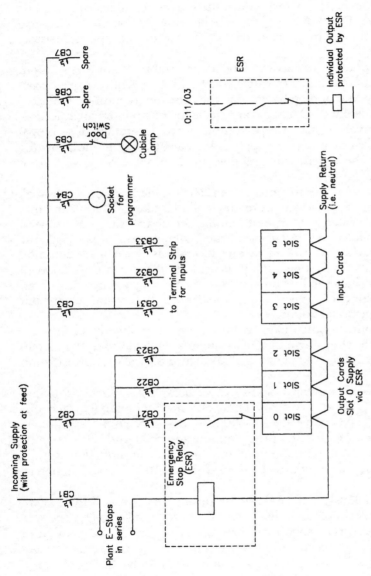

Figure 8.21 *Typical cubicle power distribution for the 'dirty' plant signals. The PLC processor will be fed from its own separate 'clean' supply. It helps system recovery after a power outage if all the processor and racks are fed from the same supply, albeit via individual protection. The cubicle and plant wiring should follow the appropriate electrical safety regulations (e.g. for safety earths)*

A supply hierarchy needs sensible labelling. The author uses a scheme with each level adding a one-digit suffix:

		L					
	L1	L2			L3		
L11	L12	L13	L21	L22	L31	L32	L33
	L131	L132...					

In this way the origin and route of any supply in the system can be quickly determined.

One common source of trouble is centre-tapped supplies (55/0/55 is often used). Although these reduce the voltage from any point to earth, they complicate fault finding and can bring increased danger if not properly installed. Protection in *each* leg (two pole breakers or two fuses), is needed, not just in the supply line. The author once nearly burnt out a cubicle which was connected to a 55/0/55 supply via a single pole breaker.

The relative merits of fuses and breakers are often discussed. Certainly DIN rail-mounted breakers simplify fault finding and maintenance, and a fault does not necessitate a trip to the stores to get a pocketful of fuses. If fuses are used, indicating holders should be used to allow a blown fuse to be quickly located. Standardization of fuse dimension in a particular area should be specified. There are few things more annoying than a cubicle equipped with different lengths of fuse in what appear to be identical holders.

Earthing is important for safety and reliable operation. There are many separate earths, typically:

(a) safety earths for cubicle, desks and junction box frames
(b) dirty earths for antisocial high current loads such as inductive relay and solenoid coils
(c) clean earths for low current signals

These should meet at one, and only one, common earth point to prevent earth loops (the connection of the screens on analog cables was discussed in Section 4.12). Where items such as PLC racks are mounted on the backplane of a cubicle, earthing via the mounting screws should not be assumed, and earth links should be added.

All supply wiring should be installed to local standards. In the United Kingdom these are the wiring regulations of the Institute of Electrical Engineers (IEE). This is currently at the 16th Edition.

8.4.2 Equipment protection

The designer must build the PLC and its associated equipment into a plant. To achieve this, cubicles, junction boxes and cabling are needed.

The cubicles serve to protect the PLC from the environment (notably dust and moisture), to deter unskilled tampering and separate dangerous voltages from production staff. The protection given by an enclosure is given by its IP number (for ingress protection). This is a two-digit number; the first digit refers to protection against solid objects, and the second to protection against liquids. The higher the number, the better the protection, as summarized in Table 8.5. Some IP numbers have commonly used names, but these have no official standing:

IP22   Drip-proof
IP54   Dust-proof
IP55   Weatherproof
IP57   Watertight

Most industrial applications require IP55, even if used indoors (but it should be remembered that IP55 is only IP55 with the door *closed*).

High ambient temperatures can often be a problem, and it is always wise to check what dissipation is expected inside a cubicle. Manufacturers give figures for their PLC equipment (these are generally low) but devices such as transformers (particularly CVTs) can generate a fair amount of heat.

For a standard cubicle, 5 W per m^2 of free cubicle surface will produce a temperature rise of 1 °C. For example, 400 W dissipation in a cubicle of 5 m^2 free surface area will give a temperature rise of about 16 °C inside the cubicle. The base and any sides in close proximity to walls should be ignored in calculating the free area.

If the elevation from this calculation above the expected highest ambient exceeds the manufacturer's maximum temperature specification (usually around 60 °C), cooling will be needed (or a larger cubicle). Recirculatory coolers as in Figure 8.22 allow around 100 W for a 1 °C rise per m^2 of free surface. In extreme cases, refrigeration can be used.

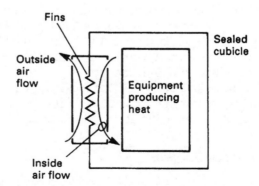

Figure 8.22   *Recirculatory cooler. Air flows are fan assisted*

Table 8.5 Ingress protection (IP) rating

Solid bodies			Liquids		
First number			Second number		
0	/	No protection	0	/	No protection
1	50 mm	Protection against large solid bodies. Hand cannot come into contact with live parts	1		Drops of condensed water falling on enclosure shall cause no harm
2	12 mm	Protection against medium solid bodies. Fingers cannot come into contact with live parts	2	Leaky roof	Falling liquid shall have a harmful effect up to 15° from vertical
3	2.5 mm	Protection against objects >2.5 mm diam. Tools (e.g. screwdrivers) cannot contact live parts	3		Falling liquid shall have harmful effect up to 60° from vertical
4	1 mm	Protection against objects >1 mm	4		Protection against splashing from any direction
5		Totally enclosed. Dust may enter but not in harmful quantities	5		Protection against hose pipe water from any direction. Water may not enter in harmful quantities
6		Dust may not enter. Total protection	6		Protection against conditions on ships' decks. Occasional immersion. Water must not enter
–	–	–	7	1 m	Permanent immersion up to 1 m. Water must not enter
–	–	–	8	x	Permanent immersion to specified depth and for pressure

Common ratings are IP11, IP21, IP22, IP23, IP44, IP54, IP55.

In either case, an overtemperature alarm should be fitted for protection. Heat-sensitive paints or stickers are also useful for monitoring temperature.

There are two ways in which terminal strips can be laid out. In Figure 8.23(a), the terminal strip has been grouped by plant side cabling (with unused I/O being placed together at the end of the terminal strip). In Figure 8.23(b), the grouping is by PLC I/O, which leads to splits in the

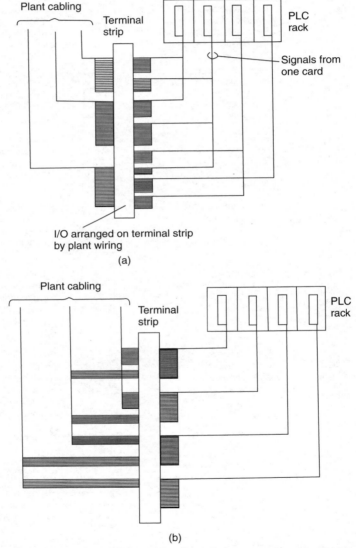

Figure 8.23 *PLC plant I/O connections: (a) cabling by plant I/O; (b) cabling by PLC I/O*

plant cabling. Of these, the author much prefers the second arrangement. To achieve the first successfully, all plant I/O and cabling must be known exactly before construction starts, and any (inevitable) late changes will split the external cables anyway. With Figure 8.23(b), construction can start once the quantity of I/O is known (without knowing its allocation) and the arrangement is clear and self-explanatory. For later modifications, unused I/O is clearly visible. Whichever scheme is used, *all* installed I/O should be brought out to the terminal strip whether it is used or not; 2.5-mm^2 cable added to a card by a shift electrician at 3 a.m. looks distinctly unsightly in comparison to 0.5-mm^2 looms installed by a panelbuilder.

The supply requirements of the outside plant should be provided at the terminal strip. Figure 8.24 shows recommended arrangements for an 8-bit output and 8-bit input card. This is straightforward; the only point requiring comment is the neutral for the input card, which allows easy monitoring with a meter at a plant-mounted junction box.

Terminals allowing local disconnection of a signal are very useful for commissioning and fault finding. The cubicle in Figure 8.25 uses Klippon SAKR disconnect terminals. These allow the cubicle to be powered up for initial testing with the plant totally disconnected, and then areas can be brought in step by step. A similar, but less controlled, approach is to drop all PLC I/O card arms off the card on first power-up and replace them one at a time.

Figure 8.25 also shows the importance of ferruling. This is essential both for installation and maintenance. *All* cable cores on a PLC-based system should be ferruled in a way that relates to the PLC addressing (so if you are working on a solenoid you can see immediately it is bit 05 of card 3 in rack 4). Typical ferrule systems are:

- 12413   input bit 13 of slot 4 in rack 2 on an Allen Bradley PLC5
- A02/12   input bit 12 of word 02 in a GEM-80
- Q63/6   output bit 6 in byte 63 in a Siemens S5

Ferruling is expensive; in the author's experience it costs as much in labour to ferrule a multicore cable as it does to pull and install it. The costs can, however, be recovered at the first major fault. A recent development is computer-generated ferrules, also shown in Figure 8.25.

A useful, and for once inexpensive, aid is to colour-code cores inside a cubicle according to their function. An example is the following:

Supplies (AC and DC)	Red
Returns (Neutral and DC −)	Black
AC Outputs	Orange
AC Inputs	Yellow
DC Outputs	Blue
DC Inputs and analogs	White
Isolated outputs and non-PLC	Violet

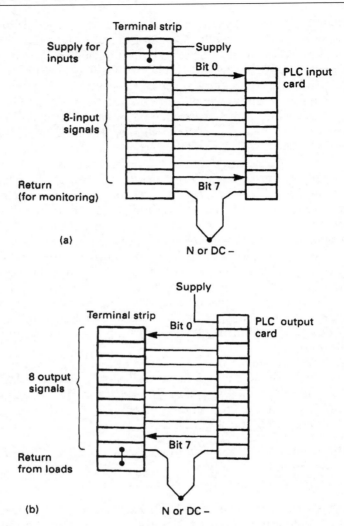

Figure 8.24   *Wiring cards to terminal strips: (a) input card; (b) output card*

The colour coding helps cable location and gives a useful last-minute confirmation that a signal being added during later plant modifications are of the correct signal type for the card (connecting a 110-V AC supplied limit switch to an input card wired in the cubicle with white cores is wrong).

Terminals should have no more than two cores connected, and ideally two cores per terminal should only occur where a linking run is being formed. In a really perfect world, linking bars should be used. Cores

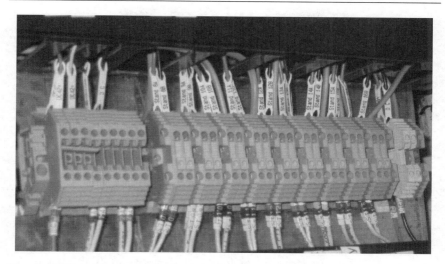

Figure 8.25 *Plant connections inside a cubicle. Note the clear ferruling (computer generated with package from Murrelektronic), the use of SAKR terminals to allow isolation and bootlace ends to prevent stray wires*

should have crimped ends (called bootlace ferrules) as in Figure 8.25 to prevent problems from splayed ends.

8.5 Maintenance and fault finding

8.5.1 Introduction

It is the designer's duty to ensure that in all new plant:

1 There is at least one item which is experimental.
2 There is at least one item which is obsolete.
3 There is at least one item on six months delivery (and this is the one item which has not been placed on stores stock).
4 The drawings do not include site or commissioning modifications.

Perhaps not, there must be a better way.

When a project is completed, the plant becomes the responsibility of the maintenance staff, who always lead a difficult and unappreciated life. They do not really share in the glamour and glory of the new plant, and inevitably get blamed for all the designer's mistakes that do not become apparent until the plant has been in production for a few months.

Production management view maintenance staff as a necessary, and expensive, evil and often express a plant goal of zero fault time. Absolute zero lost time is unachievable, but practically any desired finite level of

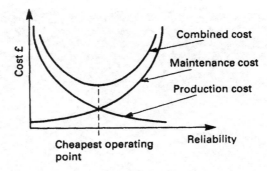

Figure 8.26 *The financial implications of reliability*

reliability can be achieved. Surprisingly, this may not be what is really required.

Low reliability is achieved at low cost, but brings high cost in lost production. As reliability increases, the maintenance costs increase but the production costs fall. Eventually a point is reached where an increase in reliability requires an increase in maintenance costs that exceeds the benefit in reduced production costs, giving curves similar to Figure 8.26. The 'art' of maintenance is to identify, and work at, the point of minimum cost.

This is assisted by designing a plant so it is, to some degree, failure tolerant. Most plants operate in some form of failure mode for a high percentage of the time. Good plant design considers the effect of failures, and provides methods to allow a plant to continue operating economically and safely whilst a fault is located and rectified.

8.5.2 *Statistical representation of reliability*

It is not possible to predict when any one item will fail, so statistical techniques are used to discuss reliability. The reliability of an item, or a complete system, is the probability (0 to 1) that it will perform correctly for a specified period of time. A PLC rack, for example, may have a 0.98 probability of running two years without failure.

Reliability measurements are based on a large number of items. If N items are run in a test period, and at the end of the test N_f have failed, and N_r are still working, the reliability R is defined as

$$R = \frac{N_r}{N} = \frac{N - N_f}{N}$$
(8.1)

and the unreliability, Q, is defined as

$$Q = \frac{N_f}{N} = \frac{N - N_r}{N} \tag{8.2}$$

Obviously $Q + R = 1$.

Reliability is expressed over a period of time (1000 hours, 1 year, 10 years or whatever). An alternative measure is to give an estimate of the expected life expectancy. This is given by the mean time to failure (MTTF) for non-repairable (replaceable) items like lamp bulbs, and mean time between failures (MTBF) for repairable items (or complete systems). Both of these are again the statistical result obtained from tests on a large number of items.

When equipment fails, it is important that it is returned to a working state as soon as possible. The term 'maintainability' describes the ease with which a faulty item of plant can be repaired, and is defined as the probability (0 to 1) that the plant can be returned to an operational state within a specified time.

Mean time to repair (MTTR) is another measure of maintainability, and is defined as the mean time to return a failed piece of equipment to a working state. Like MTTF and MTBF it is a statistical figure based on a large number of observations.

Maintainability is determined both by the designer and the user. Important factors are as follows:

1 The designer should ensure that faults are immediately apparent, and can be quickly localized to a readily changeable item. This requires good documentation, sensible test points and modular construction. We will return to these points later.
2 Vulnerable components should be readily accessible. It is not good for maintainability if the maintenance electrician has to climb a 10-m ladder and remove a cover held in place with 16 screws to reset a tripped breaker.
3 The maintenance staff should be competent, well trained and equipped with suitable tools and test equipment. MTTR is obviously related to how long they take to respond to a fault.
4 Adequate spares should be quickly available. MTTR will be increased by laborious stores withdrawal procedures. MTTR is usually reduced if a policy of unit replacement rather than unit repair is adopted.

Of these, the designer has responsibility for points (1) and (2) with the user being responsible for points (3) and (4).

Plant availability is the percentage of the time that equipment is functional, i.e.

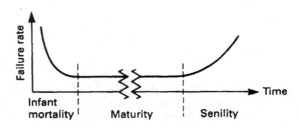

Figure 8.27 *The bathtub curve*

$$\text{Availability} = \frac{\text{Functional time}}{\text{Total time}}$$

$$= \frac{\text{Uptime}}{\text{Uptime} + \text{Downtime}}$$

$$= \frac{\text{MTBF}}{\text{MTBF} + \text{MTTR} + \text{MT}} \tag{8.3}$$

where MT is scheduled maintenance time.

If N components are in operation, and if N_f components fail over time t, the failure rate λ (also called the hazard rate) is defined as

$$\lambda = \frac{1}{N} \times \frac{N_f}{T} \tag{8.4}$$

(Strictly, N_f and T should be defined as incremental failures ΔN_f over time ΔT as ΔT tends towards zero.)

The failure rate for most systems follows the 'bathtub curve' of Figure 8.27. This has three distinct regions. The first, called 'Burn in' or 'Infant mortality', lasts a short time (usually weeks), and has a high failure rate as faulty components, bad soldering, loose connections, etc. become apparent. At the systems level the designer's mistakes and software bugs will also be revealed.

During the centre 'Maturity' region a very low constant failure rate will be observed. In a well-designed system maturity normally lasts for years. The final period, called 'Senility', has a rising failure rate caused by structural old age; oxidizing connectors, electrolytic capacitors drying out, plugs losing the spring in their contacts, breaks in printed circuit board tracks caused by temperature cycling and so on. At this point replacement is normally advisable.

In the 'Maturity' stage, it can be shown that

$$\lambda = \frac{1}{\text{MTBF}} \text{ or } \lambda = \frac{1}{\text{MTTF}} \tag{8.5}$$

depending on whether the item is repairable or replaceable, and the probability that an item will run without failure for time t (i.e. its reliability for time t) is:

$$R = e^{-\lambda t} \tag{8.6}$$

If, for example, a system has a MTBF of 17500 hours (about two years), the probability that it will run for 8750 hours (about one year) is

$$R = \exp(-8750/17\,500)$$
$$\simeq 0.6$$

8.5.3 Maintenance philosophies

Even with the best-planned preventative maintenance procedures, faults will inevitably occur. There is a fundamental difference between a problem on a complex PLC system and, say, a mechanical fault. In the latter case the fault is usually obvious, even to non-technical people, and the cause can be quickly identified. Usually mechanical problems take a long time to repair.

PLC-related problems tend to be more subtle, as far more components are involved. If some actuator does not move, it could be a bug in the PLC program, the PLC itself, an output card fault, the output supply, the actuator or some related part of the sequencing; a limit switch permitting movement having failed, for example. Diagnosis can thus take some time, and whilst it is possible to find a fault eventually by random component changing, a logical fault-finding procedure will shorten the time taken to locate the fault. Once the cause is found, the repair is usually quick and straightforward. Admittedly this broad view is difficult to maintain at 3 a.m. with the shift manager asking the three inevitable questions 'Do you know what's wrong?', 'Do you think you can fix it?' and 'How much longer is it going to take?'

The reliability of modern equipment creates problems for the maintenance staff. With MTBFs measured in years it is likely that a technician will only encounter a piece of equipment for the first time when the first fault occurs (and the maintenance manuals and drawings have been lost or are gathering dust in the chief engineer's bookcase). More reliable equipment also means that a technician can cover, and hence needs to know about, a much larger area of plant. Training is therefore essential, and we will shortly return to this subject.

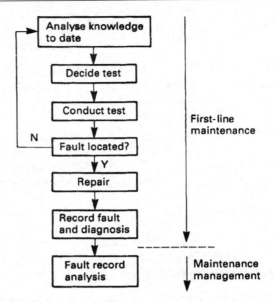

Figure 8.28   *The fault-finding procedure*

Fault finding can be split into first-line maintenance (repairs carried out on site, usually module or unit replacement) and second-line maintenance (repairs carried out to component level in a workshop). In either case it is a logical process which homes in on the fault as shown in Figure 8.28. Symptoms are studied and from these possible causes are identified. Tests are conducted to confirm or discount the possible causes. These tests give more information which allows the possible causes to be narrowed down until the fault is found.

One of the arts of fault finding is balancing the probabilities of the various possible causes of a fault against the time, effort and equipment needed to perform the tests required to confirm or refute them. Figure 8.29 shows the probability of failure of the different parts of a typical PLC system, which, not surprisingly, shows that 95% of 'PLC' faults actually occur on the *plant* items such as actuators and limit switches.

Good equipment design should provide diagnostic aids so that the most probable causes of faults can be checked out quickly without the need of specialized test equipment.

The final stages of Figure 8.28 are concerned with maintenance management, and serve to analyse plant behaviour. Any shift-based technician will only see a quarter of the faults, but a common fault-recording system should reveal recurring failures or a need for training in specific areas.

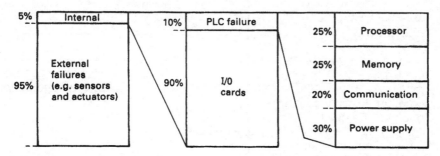

Figure 8.29 *Failure distribution in a typical PLC system. Despite only 5% of the faults being related to the PLC itself, every one goes on record as a 'PLC fault', of course*

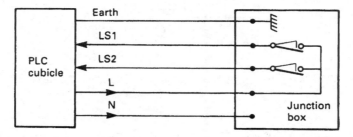

Figure 8.30 *Assisting fault finding by running redundant neutral to junction box*

8.5.4 Designing for faults

All equipment will fail, and the designer of a PLC system should build in methods to allow common faults to be diagnosed quickly. Simple ideas like running the neutral out to a junction box for test purposes as in Figure 8.30 can save precious minutes of time at the first fault. Other simple and cheap ideas are the use of isolating terminals (such as the Klippon SAKR shown earlier in Figure 8.25) and the provision of monitoring lamps on critical signals (particularly local to hydraulic and pneumatic solenoids).

Consider a simple motor starter; at the simplest level this requires two inputs (start and stop buttons) and one output (the contactor) but in the event of a fault shift electricians will have to rely on their own judgement and ideas. With five additional inputs, and three outputs for lamps, much more information can be given and the MTTR reduced. Table 8.6 will cover all common motor faults.

To this should be added an ammeter to allow the motor current to be monitored (and compared against the normal current which all careful

Table 8.6   Common motor faults

Inputs	Outputs
Start PB	Running lamp
Stop PB	Stopped lamp
Contactor auxiliary contact	Fault lamp
Contactor trip healthy	
Local control supply healthy (i.e. MCC is on)	
Emergency stop healthy	
Local isolator healthy	

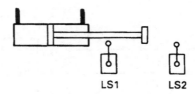

LS1   LS2

Figure 8.31   *Monitoring plant operation with additional devices*

engineers record *before* the first fault). With the above list, the PLC can localize the fault and identify the possible cause through the programming terminal. With VDU screens, full operator messages can be given ('Pump 1 cannot start because the local isolator is open' or 'Conveyor 1 stopped, PLC is energizing the contactor but the auxiliary contact has not made').

Great judgement is needed with alarms, and an alarm should always mean something. Care should be taken with ideas like that in Figure 8.31 which are often used to check actions, typically an alarm condition being 'If extend is called and LS2 does not make within 2.5 s signal Extend Fault'. These ideas can be very useful, but the alarm detection devices (LS1 and LS2 in Figure 8.31) need to be significantly more reliable than the device they are monitoring. If not, false alarms will result and the user's confidence will be lost. There is little worse in maintenance than seeing a plant running with half a dozen alarm messages on the screen and the operator saying 'Oh, ignore them, they're always coming up'. If they are ignored, make them reliable or take them out.

The PLCs themselves provide useful diagnostics of their own, and the plant's performance. Figure 8.32 shows the processor diagnostic page for a PLC-5.

```
ONLINE:Run   Edits:No  Force:No  Proj:VAULTPLC    RUNG 2:0/114        rt   Sta:44
DH+ Station:44      Local Rack:  2-Slot   Power Loss Prot: Protection Disabled
EEPROM Xfer at Powerup:No     RAM Backup:Disabled      Memory:Unprotected
                                                              Scanner Mode
Arithmetic Flags     S:0 Z:1 V:0 C:0
Processor Status     00000000 00000010   PLC-5 in RUN Mode
Minor Fault          00000000 00000000
Major Fault          00000000 00000000
Fault Code........:  0
Where Faulted-File:  0    Rung....:  0    SFC Restart/Continue............:  0
Fault Routine-File:  0    Watchdog:  500  Startup Protection after pwr loss:  0
Sel Timed INT-File:  0    Setpoint:  0
Program Scan [msec] Last:  0        Max:  20

Date/Time            0000-00-00  02:55:00   File Index      47
Active Node List
    0        10        20        30        40        50        60        70
    00000000 00000000 10000000 00000000 00001000 00000000 00000000 00000000
VME Status File: N/A       I/O Status File:  10    Adapter Image File:  0

Sym:                 Des:Key switch in Remote Position
S:1/7 =

  F1       F2       F3       F4       F5       F6       F7       F8       F9      F10
                 Iostat            Clrfalt   Des     Next     Prev   neWaddr   Help
```

Figure 8.32 *PLC-5 diagnostic page as shown on the programming terminal. Fault bits appear in minor/major fault words and are accompanied by text descriptions to the right of the word F1–F10 are soft keys on the programming terminal; Iostat, for example, shows the state of the I/O racks*

8.5.5 Documentation

PLC systems tend to be both complex and reliable, two features that work against the maintenance staff, who do not have the chance to build up experience of common faults. The maintenance technicians will therefore have to rely on the documentation to help locate the fault.

Figure 8.33 is a common drawing, familiar to most engineers, of a car wiring diagram. The drawing has been produced for constructional purposes and is of little use for fault finding. Redrawing it as Figure 8.34, with the drawing laid out by function not location, and a logical flow of signals from left to right, produces documentation which can be used for fault finding.

Figures 8.33 and 8.34 illustrate a common failing. There are two distinct types of drawing. The first is produced by the designer to construct and interconnect the plant. These drawings are essential for the initial construction, but are of little subsequent use unless there is a major disaster (like a fire). These drawings tend to be of a locational nature or panel orientated as in Figure 8.35, which is an extract from the drawings of a typical PLC system. Often such drawings are all that is available, making fault finding a difficult task.

Fault finding requires drawings which group information by a function whilst retaining enough locational information to allow signals to be traced. Figure 8.36 shows the information in Figure 8.35 redrawn to assist fault finding.

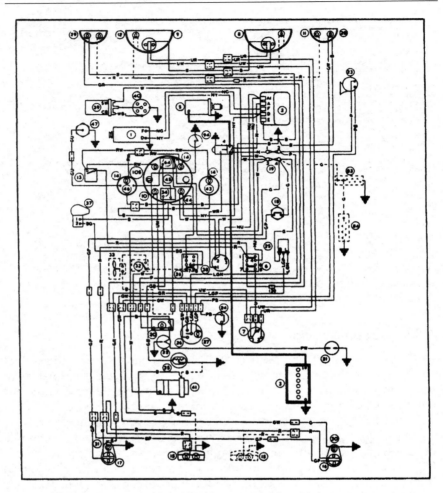

Figure 8.33 *A typical car wiring diagram with which most people are familiar. It emphasizes spatial relationships in that the layout of components follows, to some extent, the physical arrangement in the car. This results in the diagram having a large number of wiring crossovers and parallel runs, and a lack of any 'direction' or functional flow. In consequence it appears cluttered and is difficult to follow*

Unfortunately designers and manufacturers often only provide constructional and locational drawings, making the task of maintenance personnel more difficult than it need be. Ideally both types of drawing are needed.

Drawings should also indicate the sense of signals. Given Figure 8.37(a) (which is based on a real manufacturer's drawing of a hydraulic tank)

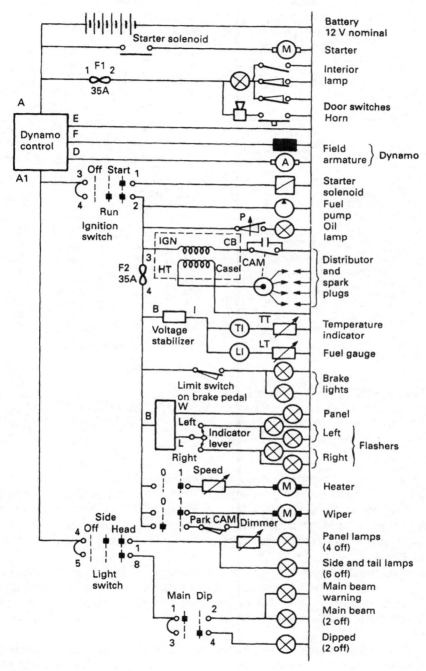

Figure 8.34 *Car wiring diagram redrawn to show functions*

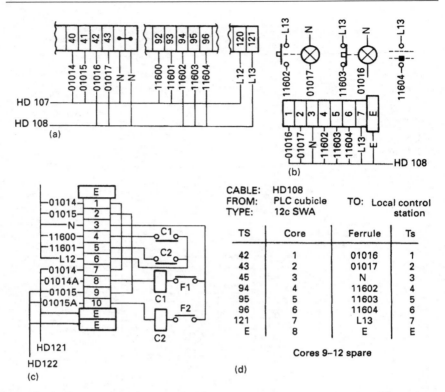

Figure 8.35 *Drawings as normally provided for maintenance. These are constructional drawings. (a) Part of drawing of PLC cubicle. (b) Drawing of local control station. (c) Drawing of starter panel. (d) Cable schedule (one of four needed for fault finding)*

what would you expect to see for normal level (and in the absence of a neutral or DC – in the junction box, how would you check it)? Simple pictograms such as Figure 8.37(b) or pure text like 'Contacts all made in normal operating condition. High level opens for rising level, low level opens for falling level' can save precious minutes of time at the first fault.

PLC programs can normally be documented, with descriptions attached to instructions and rungs/logic blocks. These are vital for easy fault finding. Figures 8.38 and 8.39 are the same part of a program in raw and documented form. The difference for ease of fault finding is obvious.

Most engineering organizations are fairly meticulous about keeping records of drawing revisions and dates of changes (e.g. drawing 702-146 is on issue E revision date 25/2/98). PLC programs are easy to change

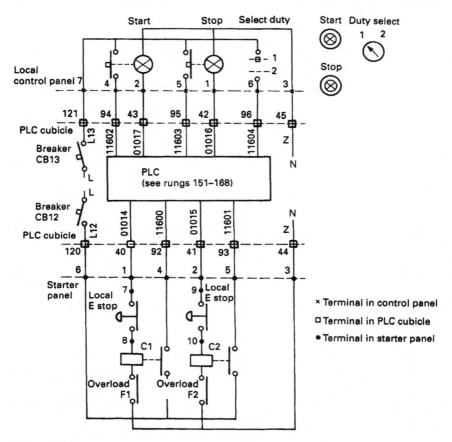

Figure 8.36 *Information from Figure 8.35 redrawn for fault finding*

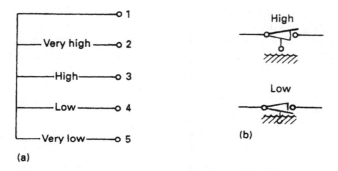

Figure 8.37 *The need for maintenance-friendly drawings.*
(a) Cabling information for level switches; the operation is unknown.
(b) Simple pictograms clarify operation

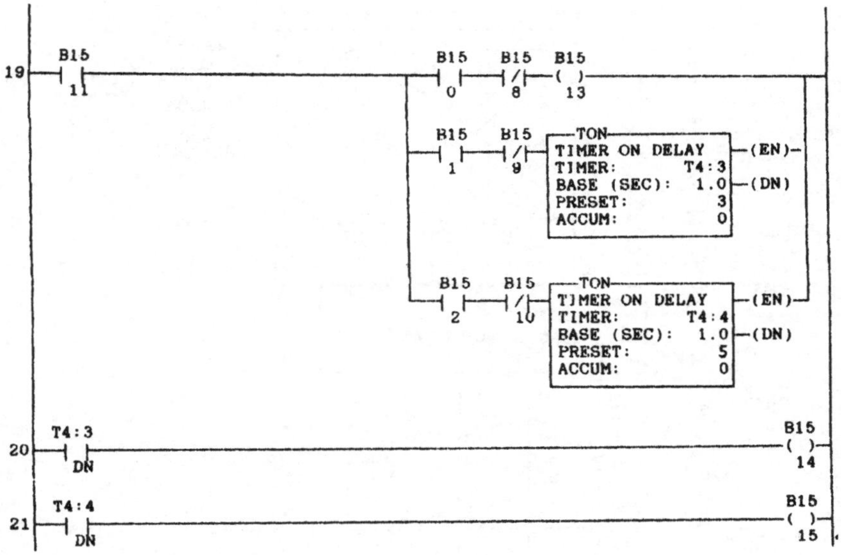

Figure 8.38   *An undocumented PLC printout*

on site, and most companies are very lax about keeping similar control of PLC programs. Figure 8.40 illustrates a common sequence of events. Such clashes can be difficult to resolve, particularly if the Maintenance, Shift and Design Engineers have used the same addresses for different functions.

Figure 8.40 arises out of 'bottom drawer' copies of PLC programs. These should be avoided at all costs. There should be a central store and records plus one (or one plus backup) for reload on site. There should be a recognized procedure for making changes, and a copy of the program taken before the changes are made so that there is a way of undoing the changes if there are unforeseen side effects. PLC programs should be treated as plant drawings and subject to the same type of drawing office control.

8.5.6   Training

Knowledge of a system is obviously required for fault finding. With any complex control system, this knowledge falls into two parts. First is familiarity with the equipment, the PLC, the thyristor drives, the sensors and actuators on the plant. Without this basic knowledge there is little hope of locating a fault. Most plants acknowledge the need for this type of training.

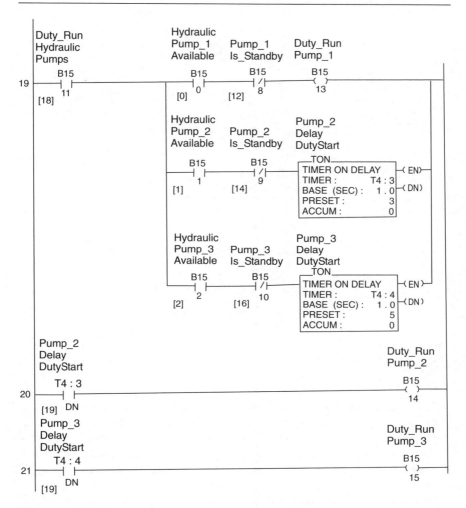

Figure 8.39 *Documented version of Figure 8.38 with descriptions and cross-references [nnn] added*

The second part is usually overlooked. It is necessary to appreciate how these various building blocks link together to produce a complete system. Too often the first-line maintenance engineer gets sent on a PLC course, a thyristor drive course and a hydraulics course, and is told 'OK, you're trained, now look after the Widget Firkilizing Plant'.

Such an approach has real dangers. When a fault occurs, maintenance technicians usually approach it in two stages. Initially, when first called, they are keen and eager to find the fault. If they do not succeed in a short time, they slip into the second stage where they are more concerned about

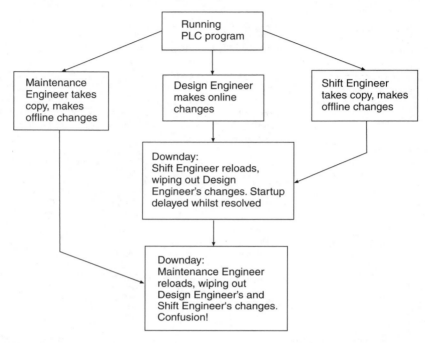

Figure 8.40   *The road to confusion; the result of unofficial program copies*

self-image and not looking a fool than in really finding the fault (and will stay in this mode of operation until the reinforcements arrive). It takes courage to say: 'Sorry I've not got the vaguest idea, send for help'. If technicians working in this mode have just come back from a course on servicing the instrument air compressor they will automatically head for the nearest air compressor and service it.

Consider the following true story. Somewhat simplified in Figure 8.41, an item of plant has X and Y movements of an arm under control of a PLC. The vertical, Y, movement is driven by an electric motor with a non-reversible gearbox, and the horizontal movement by a hydraulic cylinder controlled by proportional valves. The X and Y movements (of around 750 mm) are measured by linear potentiometers. In the night, with the plant not in production, the night shift changed one of the vertical chains but did not check the action of the ultimate limits (linked to strikers on the chain wheels which rotate through about 300 degrees).

As part of its sequence, the PLC drives the arm to a fixed Y position then fully extends the cylinder. At the start of the next day's production, the carriage moved in the Y axis until the (incorrectly aligned) ultimate

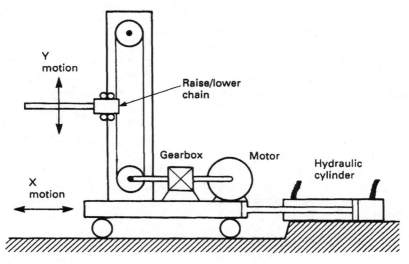

Figure 8.41 *A plant which revealed training needs*

limit operated, then the system froze. The PLC had not achieved the Y position, so it did not operate the X cross-travel.

The operators summoned the maintenance crew, informing them that the X motion was not working (a misleading statement which was taken at face value). Fresh from a course on proportional valves, the crew started stripping down the X motion hydraulics.

The people concerned had been trained in the basics, and were not stupid or bloody minded. The incident raises several important points. When the fault occurred, there had been no training in the machine's operation, so the steps of the sequence were not clear to the maintenance staff. Proportional hydraulic valves had been hammered into them, and this training had produced an attitude of 'With all faults, it must be the proportional valves'. It is, perhaps, significant that at no stage did anyone use the programming terminal to look at what the PLC was trying to do, or even try operating the machine in manual. A related, but separate, issue was the fact that the fitter who changed the chain in the night did not realize that this action could affect the setting of the ultimate limits.

To some extent, this incident arose from job demarcation. The author's plant is pioneering an electro mechanical production (EMP) worker who will cover all aspects of a job and avoid informational difficulties similar to that described above. This requires extensive training and involvement from all employees at all levels.

Other issues that the reader might like to consider are how the system designer can help avoid informational lapses like this. Note that there are space and financial limitations on how far you can go.

The important lesson, though, is that maintenance staff must know the plant, know its operation and be familiar with the documentation. With PLC-based systems, a talk through the PLC program is essential, drawing attention to what is needed for certain actions to take place. Knowledge of the plant is probably more important than knowledge of the internals of the PLC itself, as over 95% of the faults will occur outside the PLC cubicle (see earlier Figure 8.29).

Too often, the first time a maintenance crew sees a new plant properly is on the first fault (when the design team are off on the next job). The time to learn about a plant is when it is being built and when it is being tested and commissioned. All the problems, wrinkles and fixes will be learned then, and the location of all components fixed in the mind. It is an invaluable experience which cannot be repeated.

8.5.7   Fault-finding aids, EDDI and FIMs

The difficulty in the previous section could have been prevented by the addition of two indicator lamps driven by the program as shown in Figure 8.42. This raises the obvious question of how much diagnostic aid the PLC itself can give.

Certainly the PLC is capable of signalling any conditions of which it is aware (and the PLC of Figure 8.41 certainly knew what was going on). The difficulty comes with examples like Figure 8.43 which is a by no means exaggerated example for one hydraulic pump which has 12 conditions that can prevent it from starting. To cover these would require 12 lamps driven by 12 outputs and fed down 12 cores of expensive cable (and 12 indicator rungs compared to two actual plant rungs), all for one pump.

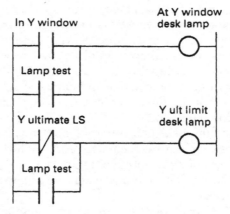

Figure 8.42   *Two rungs which would have helped Figure 8.41. In accordance with good practice, the ultimate limit signal is a second contact on the limit switch which opens when struck*

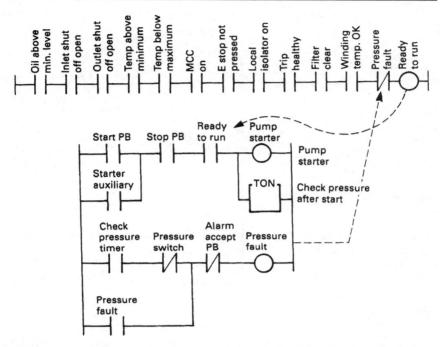

Figure 8.43 *A by no means unusual program for a hydraulic pump*

At the design stage there is a fine balance between annunciating every possible alarm, and annunciating none, with the split being made based on experience of likely faults. One could treat Figure 8.43 with two lamps saying 'Available' and 'Start Inhibited', leaving the faultfinding crew to find the detailed reason for 'Start Inhibited' either by the PLC programming terminal (which shows directly the cause) or via a check list (provided as part of the plant documentation) which can be followed on the indicators on the PLC input cards. For example:

If Pump start is inhibited, check the following inputs are healthy:

Tank Oil Low Level	A3.2
Inlet Valve Open	A3.3
Outlet Valve Open	A5.1
	etc.

There are more solutions when graphics-based workstations are used, and a graphic representation as in Figure 8.44 can be used to show the state of every signal the PLC uses in a relevant place. Significantly such displays take little or no PLC program with, say, the level indicator linked to the state of the level switch and displayed in green or red.

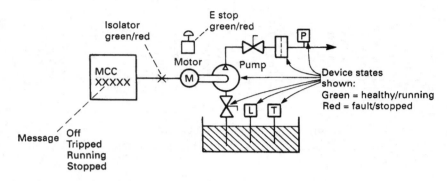

Figure 8.44   *Graphical representation of Figure 8.43*

The graphical displays do, however, take time to build (the most valuable commodity of all).

Most PLCs allow the user to 'force' input and output signals to a required state with a programming terminal. A faulty limit switch, for example, can be forced on to allow a movement, or a motor started by forcing its output on despite the absence of permit signals.

This can be a very powerful aid in commissioning and testing, but can easily become a standard way of working. A plant running with forced PLC signals can behave in a mystifying way if the initiator is not fully aware of what is being done and the possible consequences. It is very easy to trigger some unexpected sequence of events by forcing a limit switch or output signal to overcome some minor fault. Care and thought are needed if forces are to be used safely and sensibly, and forces should only be left for a short period of time.

A plant should not rely on forces for normal working (if it does, the plant is being operated in a risky state, or the original signals were not needed and should be removed).

Sequences (state diagrams, see Section 2.9.2) can cause major fault-finding problems. With workstations, again, very useful displays such as Figure 8.45 can be provided, showing what the machine is doing and what transitional signals it is waiting for.

The Ford Motor Company (a prolific user of PLCs) has defined a standard way of programming sequences. It is known as Error Detection and Diagnostic Indication (EDDI), and is a concept that covers not just maintenance fault-finding display, but the whole way in which the program is laid out, written and documented. It aims to have the same maintenance team interface regardless of which type of PLC is used.

In its simplest form, an EDDI system is built around a state-diagram-based sequence (see Section 2.9.2) with a maintenance display similar to Figure 8.46. The sequence is built up as matrix with the matrix row

<u>Sequence Monitor</u>

Going to:- Charging Position

Step No:- 47

Action:- Lowering Supports

Waiting for:- Supports Down LS

Figure 8.45 *VDU screen monitoring a process. Data after the colons change according to the machine action*

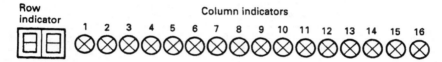

Figure 8.46 *EDDI matrix display and lamp markers*

being the step number, and the matrix column the conditions that have to be met to form a transition to the next step. The seven-segment display shows the current step number (or row) and the indicators the conditions which are yet unfulfilled. If the system freezes due to a faulty input the displays allow the fault to be quickly diagnosed.

There is much more to the EDDI philosophy, though. The standardization of how this is achieved, how the program is laid out and written, what documentation is needed and how this is provided means that staff can readily move between plants with little retraining. Standardization of methods is important for easy fault finding.

In the late 1960s the Royal Navy was concerned about the increasingly complex nature of ship-borne equipment and the related maintenance problems. A team at HMS Collingwood devised an approach called FIMS for Functionally Identified Maintenance System. This is diagnostic documentation supplemented to the main constructional or functional drawings.

It is based on functional modules or blocks whose inputs and outputs can be tested. These blocks are arranged in a hierarchy as summarized in Figure 8.47. Each block represents one drawing on which the location and state of test points can be found. Figure 8.48 shows part of a FIMs scheme for a thyristor drive. The technician follows the hierarchy of

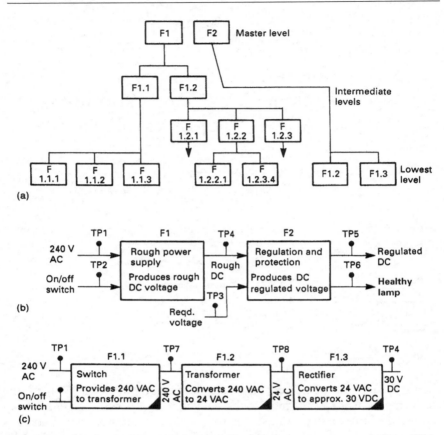

Figure 8.47 *FIMs; functionally identified maintenance system: (a) a FIMs hierarchy; (b) top-level view of a power supply; (c) bottom-level view of F1*

Figure 8.47 down until a replaceable module or card is reached (denoted by a black marker in the corner of the block) or a simple circuit. A complex system is thus broken down into blocks on which fault finding can be carried out without prior experience of the plant.

FIMs is not cheap, but in the author's experience it works well on complex plant. Its very nature imposes modular design at the early stages (always a good idea) and ensures that maintenance is thought out at the early stages. It can be difficult to apply to an already-built non-modular plant.

The HMS Collingwood team also devised the idea of the dependency chart, shown in Figure 8.49. This shows the relationship between actions, functions and event. The charts are used for fault finding by tracing

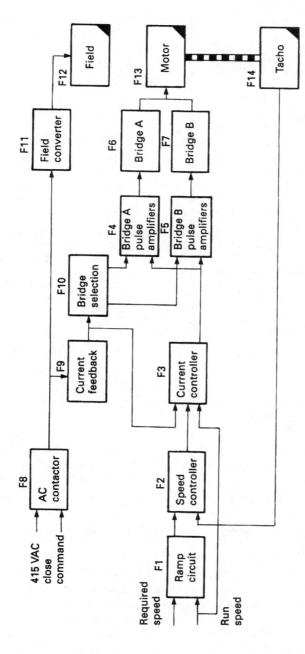

Figure 8.48 Top-level FIMs chart for a thyristor drive. A test point and test conditions will be specified for each arrow

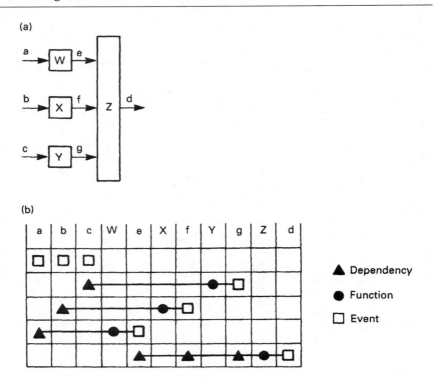

Figure 8.49   *The dependency chart: (a) a simple system; (b) chart for the system in (a)*

a signal back through the chart. Action d, for example, requires function Z, which in turn needs signals e, f and g.

A common fault finding aid is the flow chart of which Figure 8.50 is a common example. These are also known as symptom analysis charts, or algorithmically based diagnostic charts. Their deficiency is that they tend to only cover the simple obvious faults (that would have been found anyway) and ignore the troublesome subtle faults.

8.6   Electromagnetic compatibility (EMC) and CE marking

Electronic equipment is vulnerable to electrical interference. A badly suppressed motor car or poorly installed amateur radio transmitter, for example, will have noticeable effects on a domestic television receiver. Similar effects occur in industry, with power electronics systems or radio communication systems affecting low power or digital electronic circuits. The results can be catastrophic; particularly with computer-based control systems.

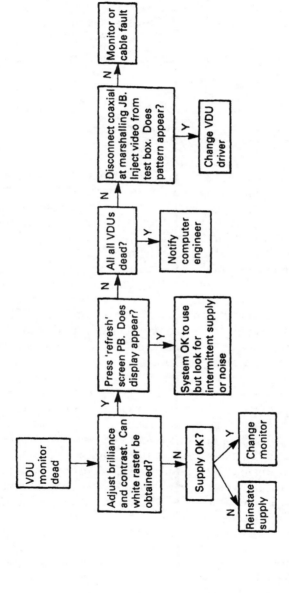

Figure 8.50 *A fault-finding chart*

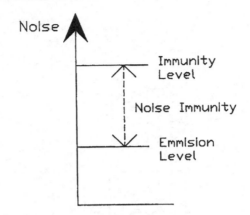

Figure 8.51   *EMC compatibility levels*

As part of the drive towards harmonization of European standards to encourage trade across European frontiers, Directive 89/336/EEC (amended by 92/31/EEG) provides standards designed to ensure that equipment from different manufacturers may be used together without problems from electrical interference between them. To achieve this, the two aspects of Figure 8.51 must be defined.

All electrical equipment causes emissions (often called 'noise'). These emissions can be high frequency (RFI, for radio frequency interference) or at harmonics of mains supply frequency. The latter usually occurs from power electronic systems such as thyristor drives. The noise can radiate through the air, or be conveyed along power or earth cables. Cable-borne noise can cause problems at several kilometres from the source. Equipment meeting the directive must keep its emissions below a defined level.

Immunity defines how susceptible equipment is to external noise from other equipment and the environment. Table 8.7 summarizes the possible effects. Both radiated and line-borne noise must be considered.

Problems occur when one plant item's emission level exceeds a local plant item's immunity level. In the domestic case of an amateur radio transmitter causing television interference, the problem can be solved by reducing the emissions from the transmitter, or increasing the immunity of the television. The practical difficulty in industry is deciding which route to take, and who is responsible.

The directive aims to solve this problem by defining acceptable emission and immunity levels. With the defined emissions level below the defined immunity level of Figure 8.51 a safety margin is formed which ensures that problems should not arise. If there are noise effects the offending items should be easily identifiable.

Table 8.7 Immunity tests

Phenomenon	Simulating	Point applied	Form	Possible outcome
Electrostatic discharge (ESD)	Discharge from a person, common on dry days	Controls and enclosure	High voltage; several kV fast edge, low energy	CMOS circuit failure Microprocessor crash
RF field	Local radio transmitter	Enclosure	Sinusoidal waveform in range 30 MHz–1 GHz	Incorrect operation of analog circuits
Transient burst	Sparking from contacts, brushes, etc.	Power and signal lines	Fast pulses, fast rise time typically 2 kV	CMOS circuit failure Microprocessor crash
Power frequency Magnetic field	Field from power cables, motors, transformers, etc.	Enclosure	50 Hz magnetic field	Distortion of CRT screens Hum on audio circuits
Conducted RF	Radio transmitter (lines acting as aerials)	Power and signal lines	Sinusoidal waveform in range 0.5–230 MHz	Incorrect operation of analog circuits
Mains surge	Lightning strike	Power lines	High voltage; several kV fast edge, high energy	Semiconductor failures (power and control)
Supply interruption	Dips from tap changers, or sudden added load	Power lines	Short duration power loss or reduced voltage	Protection trips, sequencing failures

There are three further immunity aspects. In dry weather, we often experience a short sharp electric shock on leaving a motor car. This is caused by the discharge of very high (several kV) electrostatic voltages attained by the car body, and insulated from earth by the rubber tyres. Similar effects can occur from clothing and carpets. The discharge to earth (called ESD for electrostatic discharge) can severely damage electronic equipment, particularly integrated circuits based on CMOS technology.

When an electrical current passing through an inductive load is interrupted, the inductance (which tries to maintain the current) causes a large transient voltage to be produced. It is this effect which is largely responsible for the clicks and pops heard on telephone lines. In industry this noise is often caused by direct on line (DOL) starting and stopping of motors or the operation of hydraulic and pneumatic solenoids. The effects can be reduced to some extent by suppression of loads (a typical method being shown in Figure 1.19) but it can never be totally eliminated.

Finally there are large voltage spikes conveyed along power or earth lines. These can result from lightning strikes onto power lines or from tap changers, switching of power factor correction equipment or just general load disturbances on the supply network. Direct on line (DOL) starting of large induction motors is a common source of voltage dips.

In each of the above cases the directive defines required levels of immunity.

The regulations concerning EMC are often confused with the related, but broader, topic of CE marking. One of the major barriers to trade is the different standards applied in different countries. This causes great difficulties for manufacturers who have to conform with many different standards making it impossible to make a standardized product. Many European standards (BSI, DIN, VDE, etc.) are being harmonized into a common European set of standards. (XX-ENnnnnn where XX is the national prefix, e.g. BS in the UK. The first one or two digits give an indication of its origin, e.g. 2 is based on ISO, 40 on CENELEC, 50 on CISPR, and 60 on IEC Provisional standards, out for comment, start prEN, and agreed, but not yet adopted, standards start ENV.)

Any product sold within the European Union must conform with any relevant standards. The CE mark is a manufacturer's declaration that the product meets all the requirements of the relevant standards and directives. It is not a quality mark.

Although there are a range of standards and directives (CE marking is found on children's toys, for example), industrial equipment will normally require compliance with at least one of the following directives:

- *LV Directive 73/23/EEC*   This states that electrical equipment shall not endanger the safety of persons, domestic animals or damage property.
- *EMC Directive 89/336/EEC*   This is discussed above.

- *Machinery Directive 89/392/EEC* A machine (an assembly of linked parts or components joined together for a specific application) must satisfy the relevant essential health and safety requirements.

There are many people and organizations involved in the manufacture and use of an industrial machine or complete plant; equipment manufacturers of components such as drives, PLCs or motors, panel builders, machine manufacturers, the installer and the end-users themselves. Each must take responsibility for their actions. For example, the manufacturer of a VF drive unit must provide CE marking for the LV directive, but not for the EMC directive as this is the responsibility of the panel builder or machine manufacturer who must follow correct installation methods for the drive, cabling and motor. Although it is possible, in theory, for a test house to perform relevant tests, it is usual for those responsible to self-certify and apply the CE mark. A technical construction file should be kept to justify the certification. Because it is a self-certification scheme it is likely that enforcement will be complaint driven.

8.7 Other programmable devices

Most modern devices such as control systems, sensors, drives and actuators are programmable. There are many advantages to this approach. For example where maybe ten different pressure transducers previously had to be held on stores stock, only one now needs to be held which can be set for the correct range at the time of installation.

Similarly a motor drive (AC VF or DC) which used to have maybe half a dozen trim potentiometers for acceleration, deceleration, maximum speed and current limit can now have several *hundred* user adjustable parameters including factors such as skip frequencies (to avoid resonant speeds) and various voltage to speed curves for fans, pumps, loads with high starting torque, etc.

Programmable devices bring many benefits but can also bring many problems if care is not taken with the documentation and support. If a flow transducer had to be changed twenty years ago the correct spare would be taken from the stores, the isolation valves closed and the transducer changed, a simple matter of two pipes and two terminals for the electrical signal.

With a programmable device there will be many additional steps. The range (engineering units, maximum and minimum) must be set along with output signal format (voltage, current, serial) filtering, linearization, highway address plus many more parameters. When using programmable devices, therefore the following points should always be considered.

Some method of setting or loading the many device parameters will be needed, either via dedicated front panel controls or via an external programming terminal (often a portable computer). If this terminal has been lost, stolen (take particular care of notebook PCs), damaged or has a flat battery the repair staff have very real problems. Ensure that programming terminals are secure, well cared for and in good working order ready for use at all times. If possible have a spare programming terminal so you are not reliant on one device. Make sure that people know where the programming terminal is, and don't leave it locked in a cupboard where only one person has the key.

The operation of the terminal may be obvious when it has been used several times during commissioning, but three years down the line when experience has vanished and the manual cannot be found it may take many hours to perform a simple device set-up. Try to produce an easily accessible '*Idiots Guide*'. The more powerful the programmable device, the more of a problem this can become.

If an external programming terminal is required, there will be a lead to link the terminal to the device. This will usually use some form of serial communication, often with simple D type connectors. If, however, the lead cannot be found at 3:00 a.m. and the manual showing the connections has been lost, the programming terminal and the programmable device are worse than useless. Therefore:

Keep leads where they can be easily found. Have a home for each lead to which it can be returned after use.

Attach identifying labels so the correct lead can be found. Don't purely go on simple descriptions such as '*9 pin female to 25 pin male*', there is usually a lot more detail required. For example many simple 9 pin D type to 9 pin D type leads have pins 2 and 3 crossed, many don't. Some link CTS/RTS, others use full hardwire handshaking. Leads may look identical to the eye but may not be interchangeable.

Ensure that drawings of the leads are available so that a new lead can be made if the original is lost or damaged. This problem is much more acute if the plugs on the lead are non-standard. As mentioned above common problems are the connections of pins 2 and 3 in a D type connector and what internal links (e.g. CTS–RTS) are required inside the plugs at each end.

Take particular care of fibre optic leads; replacement is near impossible if these are lost or damaged in the middle of the night or at weekends.

Consider having at least two sets of leads. One for everyday use and one stored as a secure backup if the original has been damaged or cannot be found.

Produce documentation showing the parameters which differ from the default. These, in conjunction with the '*Idiot's Guide*', should be written to allow anyone with reasonable knowledge to bring the system back to working order.

Most companies are meticulous about keeping track of drawing modifications (e.g. drawing 702-456 is on revision E dated 23rd November 2001). It is very easy to change the operation of a programmable device and there is usually no record kept of these changes. The earlier Figure 8.40 shows a typical sequence of events leading to a state of confusion in the early hours of the morning. Ensure all modifications are recorded in a log book and ensure some form of version control is enforced to ensure that up to date software and parameters are reloaded into a device if it has to be changed. Don't keep these masters in the bottom drawer of someone's desk, use a secure location (such as a fire safe) and keep adequate back-ups.

Think of the worst situation. The author has had the experience of having all the PCs on a SCADA network stolen (from locked control rooms) during a Christmas shutdown. Nominally each PC held the backups for the others, but with all PCs gone the first line back-ups had also gone! Fortunately this situation was saved by CDROMs in a fire safe. Have a plan as to how you can recover from your worst nightmare.

9 Sample ladder logic

9.1 Introduction

This chapter contains ladder logic demonstrating real life applications and showing how common problems can be solved.

It would be confusing to jump around different PLCs in this chapter so all the examples have been written for the PLC5. They should, though, be easily adaptable to all PLCs with minimal change.

The PLC5 data table areas used in these examples are:

I:rs/bb   Real input from bit bb on card in slot s of rack r (e.g. I:35/07)
O:rs/bb   Real output to bit bb on card in slot s of rack r (e.g. O:41/12)

Inputs and outputs can also be used as a 16 bit word in the form I:rs which is the 16 bit word from slot s in rack r (e.g. I:26) This is useful for obtaining data from decade switches for example.

Bf/b   Bit storage. Bit b in Bit file f (e.g. B13/21)
Nf:n   Integer word n in Integer file f (e.g. N27:5) An integer word can hold a number in the range −32,768 to 32,767
Ff:n   Floating point number n in Float file f (e.g. F8:23). A floating point number has approximately seven significant figures and can be in the range 1.1754933E-38 to 3.4028237E+38 (IEEE single precision 32 bit numbers)

The PLC5 supports other data table types (e.g. ASCII text) but these are not used in these example.

The PLC5 uses the common ladder logic symbols:

-] [-   is true when the signal is present
-]\[-   is true when the signal is absent
-( )-   is an output
-(L)-   is a latched output, once set it remains made and can only be
         cleared by
-(U)-   which unlatches (clears) the address

These can be used with real inputs and outputs or storage bits.

Timers have the form Tf:n where f is the file and n the timer number (e.g. T4:7). All the timers in these examples use a 0.01 s timebase, so a preset of 125 is 1.25 s. Timers can be a delay on (TON) or a delay off (TOF).

A MOV (for Move) instruction moves data from one location to another. In these examples it is usually used to turn a floating point number into an integer number before sending to an analog output card.

Basic arithmetic instructions are ADD, SUB, MUL, DIV. The numbers in the instruction can be any mix of floats and integers. Constants (fixed numbers) can also be used. If a MUL is used with two integer numbers to give an integer result, care must be taken to ensure an overspill does not occur. The DIV (divide) rounds to the nearest number if an integer result is used (e.g. 46 DIV 10 gives the result 5, 44 DIV 10 gives 4). This is discussed further in Section 9.12.

The PLC5 also has a Compute CPT instruction which allows a mathematical equation to be evaluated. This has been used in a few places to stop the ladder code becoming excessively long to the point of disguising the points being made. Any CPT instruction can be expanded into several individual ADD, SUB, MUL or DIV instructions.

Comparison instructions used are:

GRT   Greater than
GEQ   Greater than or equal to
EQU   Equal to
LEQ   Less than or equal to
LES   Less than

The code used in the examples is deliberately 'verbose' so the steps can be clearly seen. This is not necessarily bad practice as good code should be easy to understand. Experienced PLC5 programmers will also note that I have constructed functions which are built into the PLC5 instruction set (e.g. the -[ONS]- oneshot) but the aim is to show how functions can be achieved on less powerful PLCs.

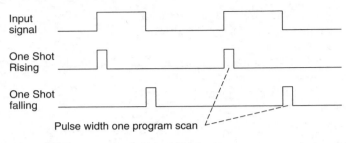

Figure 9.1 *Rising and falling OneShots*

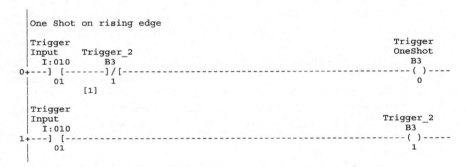

Figure 9.2 *Generation of a OneShot (one PLC scan) pulse on a rising edge*

9.2 One Shot

A One Shot generates a single scan pulse on a rising or falling edge as shown on Figure 9.1. Some PLCs, such as the PLC5, have this function as part of their basic instruction set. For those that don't, however, a single scan pulse is easy to generate with just two rungs.

The rungs in Figure 9.2 show the basic idea. *Trigger* is the input signal. *Trigger_2* in rung 1 is a copy of *Trigger* but this does not become true until rung 0 has been obeyed. *Trigger OneShot* is therefore true for the first scan after *Trigger* becomes true but is false thereafter.

The rungs in Figure 9.3 show typical examples of the use of a one shot. In Figure 9.3(a) the *Raise PB* adds five to the setpoint each time it is pressed. In Figure 9.3(b) output *Lubrication Solenoid* is operated for exactly four seconds each time the *Lubricate* push-button is pressed. In both cases holding the button down has no further effect.

In Figure 9.3(b) a branch has been used to generate the one shot with just one rung.

Changing the normally open contacts of *Trigger* to normally closed contacts as rungs 2 and 3 produces a one scan pulse on the falling edge of *Trigger* as shown on Figure 9.4.

9.3 Toggle action

A toggle action is used on many ball-point pens, the first press of the button extends the head, the next push retracts it. A single button can be used to start and stop a motor, press once to start, press again to stop, sometimes called 'Push On/Push Off'. The action can be summarized by Figure 9.5.

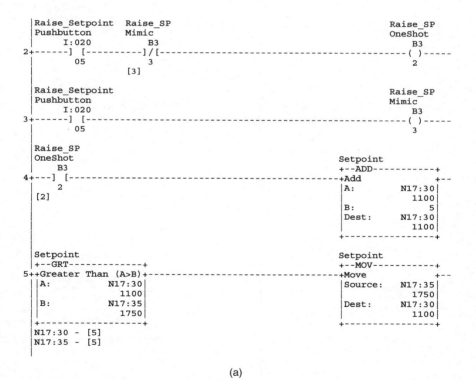

```
|Raise_Setpoint  Raise_SP                                                        Raise_SP
|Pushbutton      Mimic                                                           OneShot
|    I:020          B3                                                              B3
2+------] [----------]/[------------------------------------------------------( )-----
|       05            3                                                            2
|                    [3]
|
|Raise_Setpoint                                                                  Raise_SP
|Pushbutton                                                                      Mimic
|    I:020                                                                          B3
3+------] [-----------------------------------------------------------------( )-----
|       05                                                                         3
|
|Raise_SP                                                             Setpoint
|OneShot                                                              +--ADD----------+
|   B3                                                                +Add            +--
4+---] [--------------------------------------------------------------+A:      N17:30|
|    2                                                                |        1100|
|   [2]                                                               |B:          5|
|                                                                     |Dest:   N17:30|
|                                                                     |        1100|
|                                                                     +--------------+
|
|Setpoint                                                             Setpoint
|+--GRT------------+                                                  +--MOV----------+
5++Greater Than (A>B)+-------------------------------------------------+Move           +--
||A:      N17:30|                                                     |Source:  N17:35|
||        1100|                                                       |         1750|
||B:      N17:35|                                                     |Dest:    N17:30|
||        1750|                                                       |         1100|
|+--------------+                                                     +--------------+
|N17:30 - [5]
|N17:35 - [5]
|
```

(a)

Figure 9.3 *Typical uses of OneShot pulses: (a) changing a number by a fixed amount*

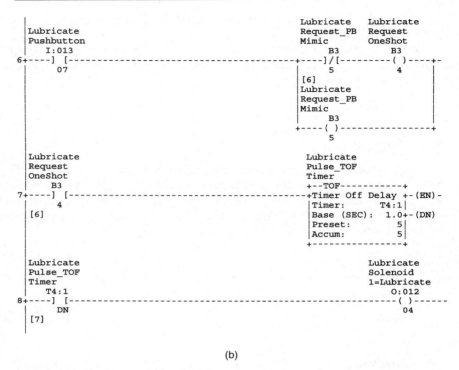

```
                                        Lubricate    Lubricate
       Lubricate                        Request_PB   Request
       Pushbutton                       Mimic        OneShot
          I:013                           B3           B3
    6+----] [--------------------------------------+----]/[---------( )----+-
          07                                       |      5          4     |
                                                   |    [6]                |
                                                   |    Lubricate          |
                                                   |    Request_PB          |
                                                   |    Mimic              |
                                                   |      B3               |
                                                   +----( )----------------+
                                                          5

       Lubricate                        Lubricate
       Request                          Pulse_TOF
       OneShot                          Timer
          B3                            +--TOF-----------+
    7+----] [-----------------------------------------+Timer Off Delay +-(EN)-
          4                             |Timer:        T4:1|
         [6]                            |Base (SEC):  1.0+-(DN)
                                        |Preset:         5|
                                        |Accum:          5|
                                        +----------------+

       Lubricate                        Lubricate
       Pulse_TOF                        Solenoid
       Timer                            1=Lubricate
         T4:1                             O:012
    8+----] [----------------------------------------------------------( )------
          DN                                                            04
         [7]
```

(b)

Figure 9.3 *(cont.) (b) operating a plant device for a fixed duration of time every time an event (in this case a button press) occurs*

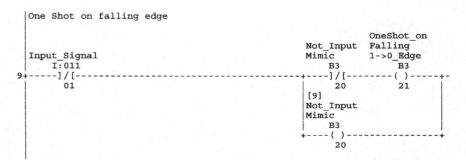

```
    |One Shot on falling edge
    |
    |                                                    OneShot_on
    |                                       Not_Input    Falling
    | Input_Signal                          Mimic        1->0_Edge
    |    I:011                                 B3           B3
   9+-----]/[-------------------------------------+----]/[--------( )-----+-
          01                                      |      20         21    |
                                                  |    [9]                |
                                                  |    Not_Input          |
                                                  |    Mimic              |
                                                  |      B3               |
                                                  +----( )----------------+
                                                         20
```

Figure 9.4 *Generation of a OneShot pulse on a falling edge*

Figure 9.5 *Toggle action*

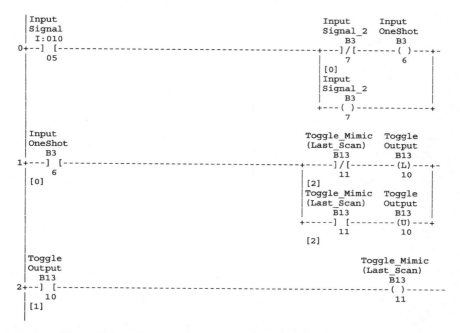

Figure 9.6 *Production of a Toggle (push-on, push-off) action.
The OneShot signal is essential, without it (i.e. replacing B3/6 by I:10/05
in rung 1) would make the output cycle on/off once per program scan
when the input signal is present*

The toggle action is achieved with the three rungs in Figure 9.6. The
first rung simply produces a One Shot on the rising edge of the input
signal as described in Section 9.2. This rung can be omitted if a One
Shot function is available.

The toggle itself is generated in rungs 1 and 2. There is a *Toggle* bit
B13/10 and a *Toggle Mimic* B13/11. Note that, because the program scan
goes from top to bottom, *Toggle Mimic* is updated <u>after</u> *Toggle*. *Toggle
Mimic* therefore shows the state of *Toggle* on the previous scan.

If *Toggle Mimic* is false, i.e. *Toggle* was false last scan, the upper branch
makes *Toggle* true. Similarly if *Toggle Mimic* is true, the lower branch makes

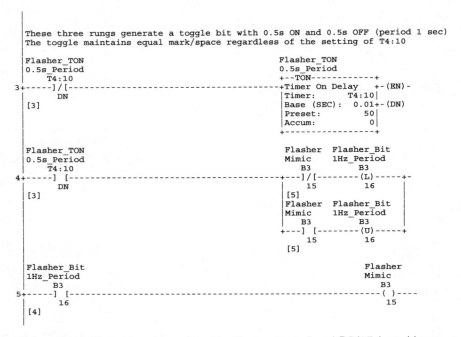

```
|
| These three rungs generate a toggle bit with 0.5s ON and 0.5s OFF (period 1 sec)
| The toggle maintains equal mark/space regardless of the setting of T4:10
|
| Flasher_TON                                     Flasher_TON
| 0.5s_Period                                     0.5s_Period
|    T4:10                                         +--TON------------+
3+-----]/[-----------------------------------------+Timer On Delay   +-(EN)-
|     DN                                          |Timer:      T4:10|
|    [3]                                          |Base (SEC):  0.01+-(DN)
|                                                 |Preset:        50|
|                                                 |Accum:          0|
|                                                 +-----------------+
|
| Flasher_TON                                     Flasher   Flasher_Bit
| 0.5s_Period                                     Mimic     1Hz_Period
|    T4:10                                           B3          B3
4+-----] [-----------------------------------------+---]/[--------(L)-----+-
|     DN                                          |   15          16     |
|    [3]                                          |  [5]                 |
|                                                 |Flasher   Flasher_Bit |
|                                                 |Mimic     1Hz_Period  |
|                                                 |  B3          B3      |
|                                                 +---] [--------(U)-----+
|                                                     15          16
|                                                    [5]
|
| Flasher_Bit                                                     Flasher
| 1Hz_Period                                                      Mimic
|    B3                                                             B3
5+-----] [-----------------------------------------------------------( )----
|    16                                                             15
|   [4]
```

Figure 9.7   *Equal mark/space ratio.* The output signal B3/15 (used here as a flasher bit for driving alarm lamps) has equal mark/space times. This equality is unaffected by the period of the Timer T4:10

Toggle false. *Toggle* (and *Toggle Mimic*) thus change state for each rising edge of the Input signal.

The toggle circuit can also be used to divide an input frequency by two as shown on Figure 9.7. The output pulse train will have an equal mark/space ratio.

9.4   Alarm annunciator

Most control systems will have an alarm annunciator of some type via which problems on the plant can be drawn to the operator's attention. The general principles of alarm annunciators are discussed in Section 6.4, this section turns those ideas into PLC ladder logic.

The basic idea can be summarized on Figure 9.8. When an alarm occurs an indicator on a panel (or a screen) will flash showing a new alarm has occurred and requires acceptance. When the operator accepts the alarm, the indicator goes solid (if the alarm still exists) or goes out (if the alarm condition was transitory).

The ladder logic uses three rungs per alarm as shown on Figure 9.9. The alarm signal arrives via 1:00/14. Like most alarm signals this is

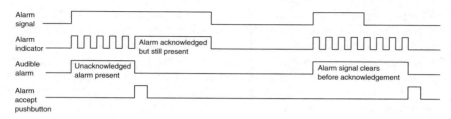

Figure 9.8 *Typical alarm annunciator operation*

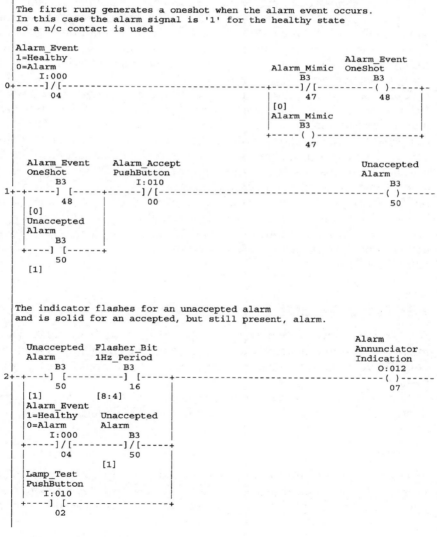

```
  The first rung generates a oneshot when the alarm event occurs.
  In this case the alarm signal is '1' for the healthy state
  so a n/c contact is used

  Alarm_Event
  1=Healthy                                          Alarm_Event
  0=Alarm                           Alarm_Mimic      OneShot
     I:000                             B3               B3
0+-----]/[----------------------------------+-----]/[----------( )-----+-
       04                                    |     47              48   |
                                             |   [0]                    |
                                             |   Alarm_Mimic            |
                                             |     B3                   |
                                             +----- ( )-----------------+
                                                    47


     Alarm_Event    Alarm_Accept
     OneShot        PushButton                                 Unaccepted
        B3            I:010                                     Alarm
                                                                 B3
1+-+-----] [-----+------]/[---------------------------------------( )------
  |      48      |       00                                        50
  |   [0]        |
  |   Unaccepted |
  |   Alarm      |
  |      B3      |
  +----] [------+
        50
      [1]


  The indicator flashes for an unaccepted alarm
  and is solid for an accepted, but still present, alarm.

                                                                Alarm
     Unaccepted    Flasher_Bit                                  Annunciator
     Alarm         1Hz_Period                                   Indication
        B3            B3                                         O:012
2+-+----ˊ] [----------] [-----+-------------------------------------( )------
  |     50            16       |                                     07
  |   [1]           [8:4]      |
  |   Alarm_Event              |
  |   1=Healthy     Unaccepted |
  |   0=Alarm       Alarm      |
  |     I:000         B3       |
  +-----]/[----------]/[-----+
  |     04            50       |
  |                  [1]       |
  |   Lamp_Test                |
  |   PushButton               |
  |     I:010                  |
  +----] [--------------------+
        02
```

Figure 9.9 *Simple alarm annunciator routine*

arranged fail-safe, i.e. it is '1' for a healthy condition and '0' for an alarm. The first rung generates a oneshot when the alarm occurs as described in Section 9.2.

Rung 1 uses the Alarm oneshot to generate the Unacknowledged alarm bit B3/50 which holds itself in until the alarm accept button I:00/00 is pressed. Unacknowledged Alarm bits from all the alarms on an annunciator can be 'ORd' together to sound an audible alarm or flash a beacon when an alarm occurs.

The final rung drives the alarm lamp or indication on a screen. If the unacknowledged alarm bit B3/50 is set the lamp will flash (the flasher bit B3/16 is generated in Figure 9.7). If the alarm is still present and the alarm has been accepted the second rung of the branch keeps the alarm lamp solidly lit. Note the bottom branch is a lamp test button I:10/02 to allow the annnunciator indication to be checked. Often the Alarm Accept PB and the Lamp Test button have the same address to force a lamp test each time an alarm occurs.

Alarms are often based on analog values, e.g. '*Temperature in vessel 4 has risen too high*' or '*Cooling water flow is low*'. The alarm event bits are invariably generated by comparison rungs, but if a simple Pass/Fail test is used nuisance alarms will be generated as the signal wanders in and out of the alarm state as shown on Figure 9.10(a). A better solution is to have a hysteresis as shown on Figure 9.10(b) Here the signal has to go some distance from the alarm condition before the alarm bit resets. Figure 9.11 shows a single rung way of generating an alarm bit with hysteresis. N7:50 is the plant signal showing cooling water flow from an analog input card. The alarm bit is set if N7:50 goes below 1000 litres/min but will not clear until the value rises above 1200 litres/min.

9.5   First order filter

Analog signals invariably have noise superimposed on them, either from interference from other sources or because the signal itself is inherently noisy. Level signals, for example, are prone to noise from waves on the surface.

The simplest way to reduce the effects of noise is a simple first order filter. This can be represented as a mathematical equation:

$$y + T\frac{\mathrm{d}y}{\mathrm{d}t} = f(t) \tag{9.1}$$

This would be difficult to implement directly in a PLC, but a simpler version uses a sampled signal as shown in Figure 9.12. The (noisy) input signal is sampled at regular intervals Δt (typically 0.1 to 5 s

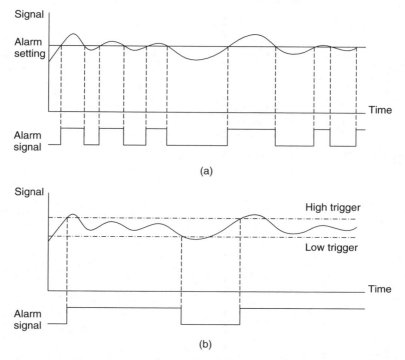

Figure 9.10 *Generation of alarm from analog signals: (a) alarm with one fixed setting; (b) alarm with hysteresis*

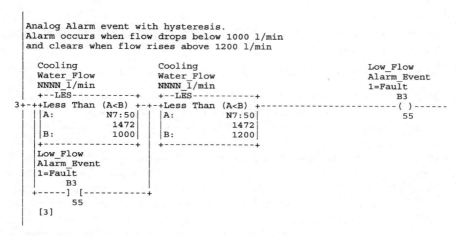

Figure 9.11 *Analog alarm generation with hysteresis*

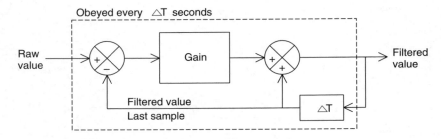

Figure 9.12   *First order filter block diagram*

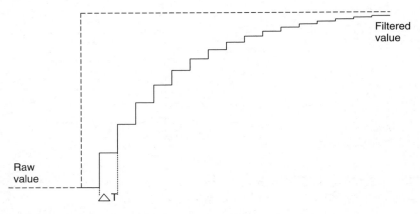

Figure 9.13   *First order filter operation with gain = 0.2*

dependant on the application). The filtered signal is evaluated at each sample by the equation

$$\text{Filtered\_value} = \text{Filtered\_value} + \text{Gain} \times (\text{Raw\_value} - \text{Filtered\_value}) \quad (9.2)$$

This gives a response as Figure 9.13 which approximates to a simple first order filter. As a rough approximation, the time constant is approximately $\Delta t / Gain$ (e.g. for a step change in the raw input signal, a filter with a sample time of 0.5 s and a gain of 0.1 takes ten samples to achieve 66% of the final value, giving a time constant of approximately 5 s). Care must be taken with the choice of ΔT and *Gain*. Samples should be taken sufficiently often to keep up with changes in the input signal (typically Δt will be 0.1 to 5 s) and *Gain* chosen in the range 0.01 to 0.1. Having Δt too small will increase the computational loading on the PLC. Having a small value of Gain means the sample rate is too fast.

Equation (9.2) can be programmed by the five rungs of Figure 9.14. The raw signal from an analog input card appears in N7:5 and the filtered signal is given in N7:6.

Rung 0 is a free running timer used to obey the subsequent rungs for one scan every Preset seconds. It has been set to trigger every 0.5 s. In a real application one timer would probably feed several filters. Rung 1 calculates the difference between the raw and filtered value. Rung 2 multiplies the difference by the *Gain* to calculate what must be added to, or subtracted from, the filtered value. The correction is applied in Rung 3 and Rung 4 simply converts from floating point to integer representation. Note that rungs 2 and 3 must be performed in floating point numbers as the change in the filtered value may be very small.

If a compute (CPT) instruction is available Rungs 1 to 3 inclusive can be performed with the single rung of Figure 9.15.

Double application of Figure 9.14 (or Figure 9.15) acts as an over-damped second order filter and can be used to apply a more severe filter.

9.6   Level control

The level of liquid in a tank must often be controlled. This is usually achieved by varying the speed of a pump by a variable speed drive, or opening/closing an inlet or drain valve. Usually the tank acts as a surge tank, and accurate control of level is not required. In reality all that is needed is some control to ensure the tank does not overflow or run dry.

To maintain strict level control a PID controller must be used, but tuning an absolute level control can be difficult. Level is the integral of flow and the two integral terms (one from the PID controller and one from the flow to level) makes the loop inherently oscillatory. In addition, all liquid surfaces have waves which make the level signal noisy.

A much simpler system directly relates the speed of the pump to the height of the liquid in the tank, the higher the level the faster the pump will run. At the maximum allowable height the pump will be going at full speed and provided the pump has been correctly sized there is no danger of overflowing or the pump running dry.

Figure 9.16(a) shows a typical example where a variable speed pump drains liquid from a sump. The liquid level is measured with an ultra-sonic transducer. Let L_{max} be the highest level we can accept. At this level the pump must be going at full speed V_{max} of, say 50.0 Hz. Let L_{min} be the lowest level we can accept. At this level the pump must be going at near zero speed, say V_{min}. In practice, the pump efficiency falls off as the square of the speed, so a typical value for V_{min} would be 20.0 Hz.

We thus have a level range $R_L = (L_{max} - L_{min})$
and a speed range $R_V = (V_{max} - V_{min})$

```
 |T4:2 sets how often the filter updates (delta T)
 |The preset and the gain F8:17 set the time constant
 |
 |Filter_Update                                        Filter_Update
 |TON_Timer                                            TON_Timer
 |     T4:2                                             +--TON------------+
0+------]/[---------------------------------------------+Timer On Delay    +-(EN)-
 |      DN                                              |Timer:       T4:2|
 |     [0]                                              |Base (SEC):  0.01+-(DN)
 |                                                      |Preset:        50|
 |                                                      |Accum:         38|
 |                                                      +-----------------+
 |
 |
 |N7:5 is the unfiltered input signal,
 |F8:15 is the filtered signal from the last sample
 |
 |Filter_Update                                        Filter
 |TON_Timer                                            Difference
 |     T4:2                                             +--SUB-----------+
1+------] [---------------------------------------------+Sub              +--
 |      DN                                              |A:         N7:5|
 |     [0]                                              |            100|
 |                                                      |B:       F8:15|
 |                                                      |       99.99994|
 |                                                      |Dest:    F8:16|
 |                                                      | 6.103516E-005|
 |                                                      +---------------+
 |
 |
 |F8:17 is the filter gain.
 |This, together with the preset of T4:2 sets the time constant
 |The error times the gain gives the correction to the filtered value
 |
 |Filter_Update                                        Filter
 |TON_Timer                                            Correction
 |     T4:2                                             +--MUL-----------+
2+------] [---------------------------------------------+Mul              +--
 |      DN                                              |A:       F8:16|
 |     [0]                                              | 6.103516E-005|
 |                                                      |B:       F8:17|
 |                                                      |          0.06|
 |                                                      |Dest:    F8:18|
 |                                                      | 3.662109E-006|
 |                                                      +---------------+
 |
 |
```

(a)

Figure 9.14   *Simple first order filter*

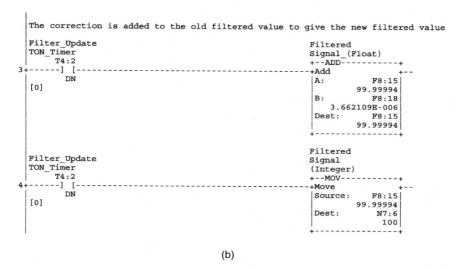

```
The correction is added to the old filtered value to give the new filtered value

   Filter_Update                                          Filtered
   TON_Timer                                              Signal_(Float)
      T4:2                                                +--ADD----------+
3+------] [----------------------------------------------+Add            +--
         DN                                               |A:       F8:15|
   [0]                                                    |     99.99994|
                                                          |B:       F8:18|
                                                          |   3.662109E-006|
                                                          |Dest:    F8:15|
                                                          |     99.99994|
                                                          +---------------+


   Filter_Update                                          Filtered
   TON_Timer                                              Signal
      T4:2                                                (Integer)
                                                          +--MOV----------+
4+------] [----------------------------------------------+Move           +--
         DN                                               |Source:  F8:15|
   [0]                                                    |     99.99994|
                                                          |Dest:    N7:6|
                                                          |         100|
                                                          +---------------+
```

(b)

Figure 9.14 (cont.) Simple first order filter

```
   Unfiltered input signal is in N7:5. As before F8:17 is the Gain
   One more rung is needed if a filtered integer output is required

   Filter_Update                                          Filtered
   TON_Timer                                              Value_(Float)
      T4:2                                                using_CPT
                                                          +--CPT----------------+
5+------] [---------------------------------------------+Compute              +--
         DN                                               |Dest:          F8:19|
   [0]                                                    |           72.28377|
                                                          |Expression:         |
                                                          |  F8:19 + (F8:17 * (N7:5 -|
                                                          |               F8:19))|
                                                          +---------------------+
```

Figure 9.15 One rung filter using CPT (compute) instruction

giving a slope relating them $K = R_V/R_L$

These are shown on Figure 9.16(b).

At any level L, the speed V is then given by

$$V = K(L - L_{min}) + V_{min} \qquad (9.3)$$

The ladder logic in Figure 9.17 achieves this. The level, from an analog input card, is given in engineering units in N17:0. The set-up constants for level L_{max} (3500 mm), L_{min} (2000 mm), and speed V_{max} (500=50.0 Hz) and V_{min} (250=25.0 Hz) are defined in N17:10 to N17:13.

The first three rungs evaluate R_L, R_V and the slope K. Note that a sensibility check is performed in rung 2. This checks that the ranges are

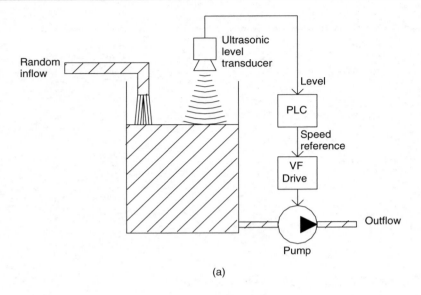

(a)

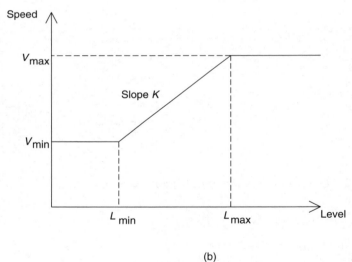

(b)

Figure 9.16   *Surge tank level control: (a) surge tank with random inflow and pumped outflow; (b) system operation*

sensible and non-zero. The latter check is important to avoid a divide by zero error in the DIV instruction at the end of rung 2. The data used in these first three rungs will rarely, if ever, change. Mathematical instructions are computationally time consuming, so B13/0 is used to ensure

The first three rungs calculate the slope and offset of the frequency/level
graph and are only obeyed once after the PLC goes from halt to run
to reduce computational loading on PLC

```
                                                    Frequency
   First_Scan                                       Range
   Completed                                        NN.N_Hz
      B13                                            +--SUB-----------+
0+----]/[-----------------------------------------------+Sub              +--
    0                                                 |A:          N17:12|
   [3]                                                |             500|
                                                      |B:          N17:13|
                                                      |             250|
                                                      |Dest:       N17:16|
                                                      |             250|
                                                      +----------------+

   First_Scan                                       Level_Range
   Completed                                        NNNN_mm
      B13                                            +--SUB-----------+
1+----]/[-----------------------------------------------+Sub              +--
    0                                                 |A:          N17:10|
   [3]                                                |            3500|
                                                      |B:          N17:11|
                                                      |            2000|
                                                      |Dest:       N17:15|
                                                      |            1500|
                                                      +----------------+

                Frequency
   First_Scan   Range               Level_Range         Freq_to
   Completed    NN.N_Hz             NNNN_mm             Level_Slope
      B13       +--GRT-----------+  +--GRT-----------+  +--DIV-----------+
2+----]/[------+Greater Than (A>B)+-+Greater Than (A>B)+-+Div              +--
    0          |A:         N17:16|  |A:         N17:15|  |A:          N17:16|
   [3]         |            250|  |           1500|  |             250|
               |B:             0|  |B:             0|  |B:          N17:15|
               |                 |  |                 |  |            1500|
               +-----------------+  +-----------------+  |Dest:        F18:0|
               N17:16 - [2]         N17:15 - [2]         |        0.1666667|
                                                      +----------------+

                                                    First_Scan
                                                    Completed
                                                       B13
3+-------------------------------------------------------------( )------
                                                                0
```

Figure 9.17 *Surge tank level control*

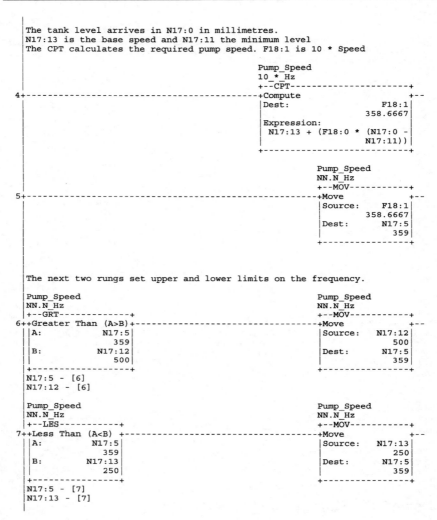

The tank level arrives in N17:0 in millimetres.
N17:13 is the base speed and N17:11 the minimum level
The CPT calculates the required pump speed. F18:1 is 10 * Speed

```
                                            Pump_Speed
                                            10_*_Hz
                                            +--CPT--------------------+
   4+-----------------------------------------+Compute                  +--
                                            |Dest:                 F18:1|
                                            |                   358.6667|
                                            |Expression:                |
                                            | N17:13 + (F18:0 * (N17:0 - |
                                            |                   N17:11)) |
                                            +---------------------------+

                                            Pump_Speed
                                            NN.N_Hz
                                            +--MOV----------+
   5+-----------------------------------------+Move             +--
                                            |Source:   F18:1|
                                            |        358.6667|
                                            |Dest:     N17:5|
                                            |           359|
                                            +---------------+
```

The next two rungs set upper and lower limits on the frequency.

```
 Pump_Speed                                 Pump_Speed
 NN.N_Hz                                     NN.N_Hz
 +--GRT------------+                         +--MOV----------+
6++Greater Than (A>B)+---------------------------------+Move             +--
 |A:          N17:5|                         |Source:   N17:12|
 |             359|                          |            500|
 |B:         N17:12|                         |Dest:     N17:5|
 |             500|                          |           359|
 +----------------+                          +---------------+
 N17:5  - [6]
 N17:12 - [6]

 Pump_Speed                                 Pump_Speed
 NN.N_Hz                                     NN.N_Hz
 +--LES----------+                           +--MOV----------+
7++Less Than (A<B) +---------------------------------+Move             +--
 |A:          N17:5|                         |Source:   N17:13|
 |             359|                          |            250|
 |B:         N17:13|                         |Dest:     N17:5|
 |             250|                          |           359|
 +----------------+                          +---------------+
 N17:5  - [7]
 N17:13 - [7]
```

Figure 9.17   *(cont.) Surge tank level control*

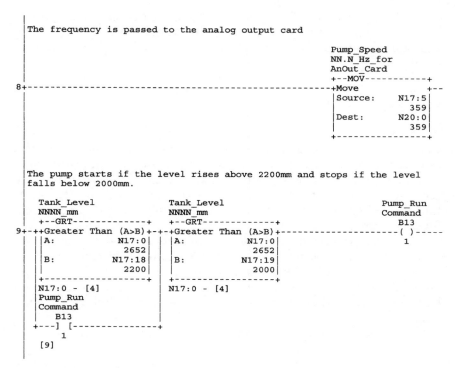

Figure 9.17 (*cont.*) *Surge tank level control*

rungs 0 to 2 are only obeyed once when the PLC goes from Halt to Run. B13/0 will not be energized on the first program scan so rungs 0 to 2 will be obeyed. B13/0 then becomes true and on subsequent scans rungs 0 to 2 will be skipped.

The compute CPT instruction in rung 4 evaluates equation (9.3). Without a CPT a subtraction $(L - L_{min})$ a multiplication by K and an addition of V_{min} would be required. Note that a floating point number must be used for intermediate results and the final result of the CPT.

The resulting floating point number F18:1 (in the range 200 to 500 for 20 to 50 Hz) is converted to an integer number N17:5. This is limited to the range V_{min} to V_{max} by rungs 6 and 7. Rung 8 moves the motor speed to N10:0 for the analog output to the VF drive.

Rung 9 generates the run command for the motor. N17:12 is L_{start}, the level at which the pumping will commence. This is set higher than L_{min}, at 2200 mm giving 200 mm of hysteresis and protection against rapid start/stops on waves. The pump will start once the level rises above 2200 mm and continue until the level falls to L_{min} when it will stop. At very low input flows the level will cycle between L_{min} and L_{start} with the pump starting and stopping. At normal flow rates the pump will run at a speed which makes the pump flow exactly match the input flow.

9.7 Linearization

Analog signals are often non-linear and have to be linearized before they can be used. Typical examples are thermocouples and resistance temperature detectors which have a non-linear response. The linearization is best done with a straight line approximation which can give surprisingly accurate results. The GEM-80 deserves particular praise for its LINCON function which provides a built in straight line $(Ax+B)$ linearization function with limiting.

Figure 9.18 is a typical example of a non-linear relationship between an input x and an output y. The input x could be the millivolts from a thermocouple and y the temperature for example. Four known input points X_1 to X_4 (and their corresponding outputs Y_1 to Y_4) are chosen. Between each of these the response is assumed to follow a straight line.

We start by defining four slopes:

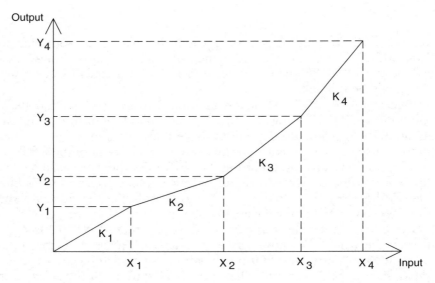

Figure 9.18 *Straight line linearization*

$$K_1 = Y_1/X_1$$
$$K_2 = (Y_2 - Y_1)/(X_2 - X_1)$$
$$K_3 = (Y_3 - Y_2)/(X_3 - X_2)$$
$$K_4 = (Y_4 - Y_3)/(X_4 - X_3)$$

The input signal x is then tested to see which range it lies in, then the output y calculated as below:

if $(x > 0)$ and $(x < X_1)$ then $y = K_1 x$
if $(x >= X_1)$ and $(x < X_2)$ then $y = K_2.(x - X_1) + Y_1$
if $(x >= X_2)$ and $(x < X_3)$ then $y = K_3.(x - X_2) + Y_2$
if $(x >= X_3)$ and $(x < X_4)$ then $y = K_4.(x - X_3) + Y_3$

As an example the rungs in Figure 9.19 convert microvolts to temperature for a type K thermocouple over the range 20 to 400 °C. The signal, in microvolts, from a type K thermocouple over this range is:

Temperature °C (y)	Thermocouple μv (x)
0	0
100	4096
200	8138
300	12209
400	16397

The thermocouple signal comes from the analog input card in N37:0 scaled in microvolts. Cold junction compensation has been applied before N37:0. A temperature of 200 °C would thus give 8138 in N37:0. The four reference microvolts are stored in N37:11 to N37:14 (X_1 to X_4) and the four reference temperatures are stored in N37:21 to N37:24 (Y_1 to Y_4).

The first four rungs calculate the slopes K_1 to K_4 which are stored in F38:1 to F38:4. B33/0 is de-energized for the first scan after the PLC goes from halt to run, so these rungs are only obeyed once and do not load the PLC thereafter. Rung 4 energizes B33/0.

Rung 5 is a validity check on the thermocouple signal. Thermocouples are prone to failing open circuit (which gives a very high temperature reading) or short circuit (which gives zero reading). The output from this rung, B33/1, is used to enable the following rungs and could also be used to trigger an alarm in the event of a bad signal.

Rungs 6 to 9 calculate the temperature for each of the four ranges. Only one of these is active at any one time, the correct rung being selected by the comparison instructions at the beginning of each rung. In each rung the calculated temperature is placed in the floating point number F38:10.

Rung 10 moves a default value into F38:10 if rung 5 determined the input signal was not valid. With temperature measurement the default value is usually high for safety reasons, here a temperature of 500 °C has been used.

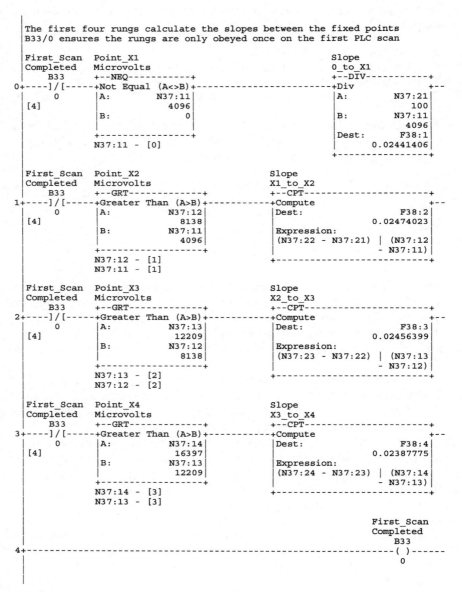

```
The first four rungs calculate the slopes between the fixed points
B33/0 ensures the rungs are only obeyed once on the first PLC scan

   First_Scan  Point_X1                                   Slope
   Completed   Microvolts                                 0_to_X1
|    B33       +--NEQ-----------+                          +--DIV-----------+
0+----]/[-----+Not Equal (A<>B)+--------------------------+Div             +--
|     0        |A:         N37:11|                         |A:         N37:21|
|    [4]       |           4096 |                          |           100 |
|              |B:            0 |                          |B:         N37:11|
|              |                |                          |           4096 |
|              +----------------+                          |Dest:      F38:1|
|              N37:11 - [0]                                 |         0.02441406|
|                                                          +----------------+

   First_Scan  Point_X2                            Slope
   Completed   Microvolts                          X1_to_X2
|    B33       +--GRT-------------+                 +--CPT--------------------+
1+----]/[-----+Greater Than (A>B)+-----------------+Compute                 +--
|     0        |A:         N37:12|                  |Dest:             F38:2|
|    [4]       |           8138 |                   |               0.02474023|
|              |B:         N37:11|                  |Expression:            |
|              |           4096 |                   |(N37:22 - N37:21) | (N37:12|
|              +----------------+                   |               - N37:11)|
|              N37:12 - [1]                         +------------------------+
|              N37:11 - [1]

   First_Scan  Point_X3                            Slope
   Completed   Microvolts                          X2_to_X3
|    B33       +--GRT-------------+                 +--CPT--------------------+
2+----]/[-----+Greater Than (A>B)+-----------------+Compute                 +--
|     0        |A:         N37:13|                  |Dest:             F38:3|
|    [4]       |           12209|                   |               0.02456399|
|              |B:         N37:12|                  |Expression:            |
|              |           8138 |                   |(N37:23 - N37:22) | (N37:13|
|              +----------------+                   |               - N37:12)|
|              N37:13 - [2]                         +------------------------+
|              N37:12 - [2]

   First_Scan  Point_X4                            Slope
   Completed   Microvolts                          X3_to_X4
|    B33       +--GRT-------------+                 +--CPT--------------------+
3+----]/[-----+Greater Than (A>B)+-----------------+Compute                 +--
|     0        |A:         N37:14|                  |Dest:             F38:4|
|    [4]       |           16397|                   |               0.02387775|
|              |B:         N37:13|                  |Expression:            |
|              |           12209|                   |(N37:24 - N37:23) | (N37:14|
|              +----------------+                   |               - N37:13)|
|              N37:14 - [3]                         +------------------------+
|              N37:13 - [3]

                                                   First_Scan
                                                   Completed
                                                   B33
4+-----------------------------------------------------------------( )------
|                                                   0
```

Figure 9.19   *Multi straight line linearization routing*

```
The thermocouple signal arrives in N37:0 as NNNNN microvolts
After a range check the correct range is selected and the temperature calculated

Thermocouple                    Thermocouple                        Input_Signal
Signal_uV                       Signal_uV                           is_Valid
+--GEQ--------------------+     +--LEQ--------------------+             B33
5++Grtr Than or Equal (A>=B)+--+Less Than or Equal (A<=B)+---------( )-------
  |A:                 N37:0|     |A:                 N37:0|              1
  |                   11547|     |                   11547|
  |B:                     0|     |B:                 16397|
+------------------------+       +------------------------+
N37:0 - [6]                     N37:0 - [6]

Input_Signal  Thermocouple                                      Temperature
is_Valid      Signal_uV                                         Degrees_C
|    B33       +--LES-----------+                                +--MUL----------+
6+-----] [------+Less Than (A<B) +---------------------+Mul                 +--
  |    1       |A:         N37:0|                         |A:          N37:0|
  |   [5]      |           11547|                         |            11547|
  |            |B:        N37:11|                         |B:           F38:1|
  |            |           4096|                         |        0.02441406|
  |            +----------------+                         |Dest:      F38:10|
  |            N37:0 - [6]                                |          283.7386|
  |            N37:11 - [7]                               +----------------+

Input_Signal  Thermocouple                  Thermocouple
is_Valid      Signal_uV                     Signal_uV
|    B33       +--GEQ--------------------+   +--LES-----------+            >
7+-----] [------+Grtr Than or Equal (A>=B)+--+Less Than (A<B) +---------->
  |    1       |A:                 N37:0|   |A:         N37:0|            >
  |   [5]      |                   11547|   |           11547|
  |            |B:                N37:11|   |B:        N37:12|
  |            |                   4096|   |            8138|
  |            +------------------------+   +----------------+
  |            N37:0 - [7]                  N37:0 - [7]
  |            N37:11 - [7]                 N37:12 - [8]
  |                                         Temperature
  |                                         Degrees_C
  |                                         <+--CPT--------------------+
  |                                         <+Compute                +--
  |                                         < |Dest:           F38:10|
  |                                           |               283.7386|
  |                                           |Expression:            |
  |                                           | (F38:2 * (N37:0 - N37:11))|
  |                                           |            + N37:21|
  |                                           +------------------------+
```

Figure 9.19 (*cont.*) *Multi straight line linearization routing*

```
|Input_Signal Thermocouple                     Thermocouple
|is_Valid     Signal_uV                         Signal_uV
|    B33      +--GEQ--------------------+ +--LES-----------+           >
8+-----] [------+Grtr Than or Equal (A>=B)+-+Less Than  (A<B) +--------->
|    1        |A:                  N37:0|  |A:           N37:0|         >
|  [5]        |                    11547|  |             11547|
|             |B:                 N37:12|  |B:          N37:13|
|             |                     8138|  |             12209|
|             +-------------------------+  +----------------+
|             N37:0 - [8]                   N37:0 - [8]
|             N37:12 - [8]                  N37:13 - [9]
|                                            Temperature
|                                            Degrees_C
|                                           <+--CPT--------------------+
|                                           <+Compute                +--
|                                           < |Dest:            F38:10|
|                                             |               283.7386|
|                                             |Expression:           |
|                                             |  (F38:3 * (N37:0 - N37:12))|
|                                             |              + N37:22|
|                                             +----------------------+
|
|Input_Signal Thermocouple                     Thermocouple
|is_Valid     Signal_uV                         Signal_uV
|    B33      +--GEQ--------------------+ +--LEQ-----------------+>
9+-----] [------+Grtr Than or Equal (A>=B)+-+Less Than or Equal (A<=B)+>
|    1        |A:                  N37:0|  |A:           N37:0|>
|  [5]        |                    11547|  |             11547|
|             |B:                 N37:13|  |B:          N37:14|
|             |                    12209|  |             16397|
|             +-------------------------+  +----------------------+
|             N37:0 - [9]                   N37:0 - [9]
|             N37:13 - [9]                  N37:14 - [3]
|                                            Temperature
|                                            Degrees_C
|                                           < +--CPT--------------------+
|                                           <-+Compute                +--
|                                           < |Dest:            F38:10|
|                                             |               283.7386|
|                                             |Expression:           |
|                                             |  (F38:4 * (N37:0 - N37:13))|
|                                             |              + N37:23|
|                                             +----------------------+
```

Figure 9.19   (*cont.*) *Multi straight line linearization routing*

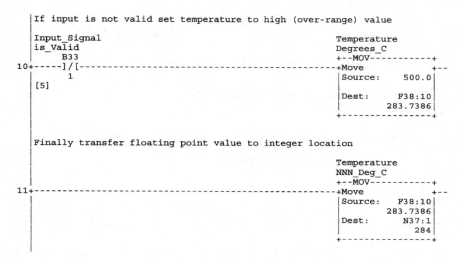

```
   |If input is not valid set temperature to high (over-range) value
   |
   |Input_Signal                                        Temperature
   |is_Valid                                            Degrees_C
   |   B33                                              +--MOV----------+
 10+-----]/[---------------------------------------------+Move          +--
   |    1                                               |Source:   500.0|
   |  [5]                                               |               |
   |                                                    |Dest:     F38:10|
   |                                                    |        283.7386|
   |                                                    +---------------+
   |
   |
   |Finally transfer floating point value to integer location
   |
   |                                                    Temperature
   |                                                    NNN_Deg_C
   |                                                    +--MOV----------+
 11+-----------------------------------------------------+Move          +--
   |                                                    |Source:   F38:10|
   |                                                    |        283.7386|
   |                                                    |Dest:     N37:1|
   |                                                    |            284|
   |                                                    +---------------+
```

Figure 9.19   (*cont.*) *Multi straight line linearization routing*

Finally rung 11 moves the temperature into the integer location N37:1. A floating point number is used for the calculations in rungs 6 to 9 rather than an integer number to avoid rounding errors. The output temperature is given in steps of one degree centigrade and the linearization is accurate to better than one degree.

The program as written could be shortened by doing some off-line work. Rungs 0 to 4 are not strictly necessary as K_1 to K_4 will never change and can be found with a calculator. K_2, for example, is $(200-100)/(8138-4096)=0.024\ 74\ 02$ which can be entered directly into the program. The CPT instruction in rung 7, for example, could have F38:32 replaced by 0.024 74 02 saving storage space and making the program run faster.

9.8   Flow totalization

Flow measurement is very common, and often calculation of the total volume passed in some time is required. This may be for accounting purposes (the total gas used this shift was $40.57\,\mathrm{m}^3$) or production purposes (add 25 litres of product A then 50 litres of product B).

If we have a measured flow rate of F l/min, then $F/60$ litres will pass each second. In general for a flow rate F l/min sampled every Δt seconds, $\Delta t * F/60$ litres will pass in each sample period. We can therefore calculate the total volume with the following pseudo code:

Repeat
    read new flow   {from analog input card}
    wait Δt seconds.
    volume over $dt = \Delta t \cdot$ newflow/60   {assuming new flow is measured
                                                     in l/min}
    total volume = total volume + volume over dt
Until hell freezes over

If the flow is measured in litres per hour line 4 becomes

volume over $dt = \Delta t \cdot$ new flow/3600

as there are 3600 seconds in an hour.

Figure 9.20 shows the procedure. Effectively we are sampling the flow at fixed time intervals and calculating the total volume in each time interval.

The three rungs in Figure 9.21 achieve this. Rung 0 is a free running timer producing a single scan pulse every Δt seconds. Rungs 1 and 2 are thus obeyed every Δt seconds. In the example the preset, and hence Δt is two seconds.

The instantaneous measured flow (in engineering units (litres, gallons, m^3 or whatever) per minute) arrives in N7:0. Rung 1 divides this flow by 30 to give the total volume in the last two seconds in F8:0. The division by 30 is used because the flow rate is in units per minute and we are sampling every two seconds.

Rung 2 adds the volume in the last sample (F8:0) to the overall total volume in F8:1 to give an updated total volume.

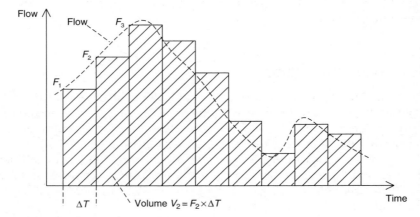

Figure 9.20   *Simple flow totalization*

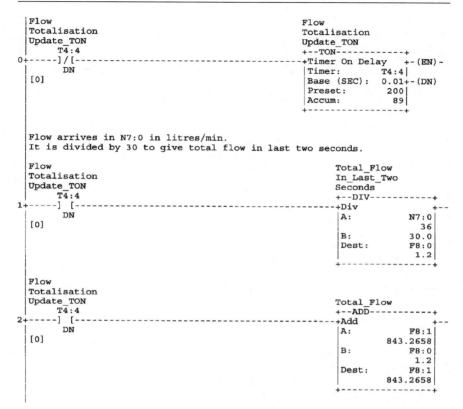

```
 |Flow                                                Flow
 |Totalisation                                        Totalisation
 |Update_TON                                          Update_TON
 |     T4:4                                            +--TON-----------+
0+-----]/[-----------------------------------------+Timer On Delay    +-(EN)-
 |       DN                                          |Timer:       T4:4|
 |      [0]                                           |Base (SEC): 0.01+-(DN)
 |                                                    |Preset:       200|
 |                                                    |Accum:         89|
 |                                                    +----------------+
 |
 |
 |Flow arrives in N7:0 in litres/min.
 |It is divided by 30 to give total flow in last two seconds.
 |
 |Flow                                                Total_Flow
 |Totalisation                                        In_Last_Two
 |Update_TON                                          Seconds
 |     T4:4                                            +--DIV-----------+
1+-----] [-----------------------------------------+Div               +--
 |       DN                                          |A:          N7:0|
 |      [0]                                           |              36|
 |                                                    |B:          30.0|
 |                                                    |Dest:        F8:0|
 |                                                    |             1.2|
 |                                                    +----------------+
 |
 |Flow
 |Totalisation
 |Update_TON                                          Total_Flow
 |     T4:4                                            +--ADD-----------+
2+-----] [-----------------------------------------+Add               +--
 |       DN                                          |A:          F8:1|
 |      [0]                                           |        843.2658|
 |                                                    |B:          F8:0|
 |                                                    |             1.2|
 |                                                    |Dest:        F8:1|
 |                                                    |        843.2658|
 |                                                    +----------------+
 |
```

Figure 9.21   *Simple flow totalization*

This flow totalization is very simple but examination of Figure 9.20 shows we are underestimating the total volume for increasing flow and overestimating the volume for decreasing flow. Over time these errors will more or less cancel, but a more accurate method will use the average flow over the sample time as shown on Figure 9.22. Note that no more samples have to be made as each measurement is both the last reading in the last sample and the first reading in the next sample.

The procedure can be summarized in the pseudo code:

old flow = 0   {done once just to initialize the variable}
repeat
    wait Δt seconds
    read new flow {from analog input card}

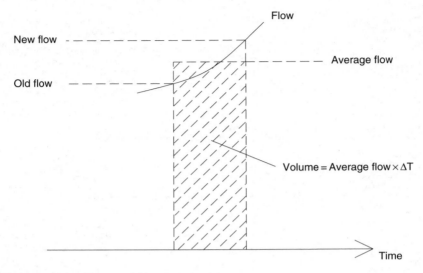

Figure 9.22   *Trapezoid integration*

average flow = (new flow + old flow)/2
volume over dt = Δt · average flow/60
total volume = total volume + volume over dt
old flow = new flow {ready for next loop}
until hell freezes over

This is known as *trapezoid integration* and Figure 9.23 shows how this can be achieved in ladder logic. As before a free running timer is used to generate a single scan pulse every Preset seconds. In this example this has been set at one second. The four rungs are only obeyed for one scan every second.

Rung 1 adds the current flow (in N7:10 from an analog input card) and the flow last scan (in N7:11) ready for the average flow calculation.

Rung 2 divides the summed flows by 120; this is a divide by 60 to get from litres per min to litres passed in a second and a divide by 2 to give the average flow. The result, in F8:5, is the total volume in the last second.

Rung 3 adds the volume in the last second to the total volume in F8:6.

The final rung moves the current flow (N7:10) to the last flow (in N7:11) ready for the next sample time.

Both these approaches work well, but care is needed in the implementation. The first consideration is the sample rate. The accuracy of the

```
T4:6.DN fires once per second
Flow arrives in N7:10 in litres per minute
The flow on the last sample (one second ago) is in N7:11 (see rung 4)

  Trapezoid                                            Last_Flow
  Flow_Totalise                                        Added_to
  TON_Timer                                            This_Flow
      T4:6                                             +--ADD----------+
1+------] [--------------------------------------------+Add           +--
          DN                                           |A:       N7:10|
  [0]                                                  |          563 |
                                                       |B:       N7:11|
                                                       |          563 |
                                                       |Dest:    N7:12|
                                                       |         1126 |
                                                       +--------------+

  Trapezoid                                            Average
  Flow_Totalise                                        Flow_Over
  TON_Timer                                            Last_Second
      T4:6                                             +--DIV----------+
2+------] [--------------------------------------------+Div           +--
          DN                                           |A:       N7:12|
  [0]                                                  |         1126 |
                                                       |B:       120.0|
                                                       |Dest:     F8:5|
                                                       |      9.383333|
                                                       +--------------+

  Trapezoid
  Flow_Totalise                                        Total
  TON_Timer                                            Flow
      T4:6                                             +--ADD----------+
3+------] [--------------------------------------------+Add           +--
          DN                                           |A:        F8:5|
  [0]                                                  |      9.383333|
                                                       |B:        F8:6|
                                                       |     648.7791 |
                                                       |Dest:     F8:6|
                                                       |     648.7791 |
                                                       +--------------+

  Trapezoid
  Flow_Totalise                                        Flow_on
  TON_Timer                                            Last_Sample
      T4:6                                             +--MOV----------+
4+------] [--------------------------------------------+Move          +--
          DN                                           |Source:  N7:10|
  [0]                                                  |          563 |
                                                       |Dest:    N7:11|
                                                       |          563 |
                                                       +--------------+
```

Figure 9.23 *Trapezoid flow totalization*

timers in Figures 9.20 and 9.22 are generally very good, but are influenced by the scan time of the PLC. If the scan time is a typical 20 ms, a nominal one second sample time will vary randomly between 1.00 and 1.02 seconds. This effect can be reduced by having a long sample time. Increasing the sample time, however, makes the system slow to respond to changes in flow. Depending on the physical size of the pipes on the plant, sample times of one to thirty seconds will normally be used.

Some PLCs have system bits which are driven by the PLC, S:23/0 in the PLC5 for example, is a toggling bit with a period of two seconds. These system bits are unaffected by the PLC scan time and their use in place of the free running timers will significantly improve the accuracy.

PLCs which support subroutines often allow program files to be triggered at fixed time intervals. In a PLC5, for example, these are called STI (for Selectable Timed Interrupt) files. These are ideal for flow totalization as the trigger rate is very accurate (repeatability typically better than 1 ms) and is unaffected by the program scan. With a time triggered program file the free running timers and their contacts are not required.

The second consideration is the resolution of floating point numbers. In rung 2 of the first example and rung 3 of the second example we add a small number (the volume passed in the last sample time) to a large number (the total volume since totalization started). A floating point number typically has a resolution of seven digits, so 87 583.52 and 7.405716 are both valid floating point numbers. If they are added, however, eleven digits would be required for the result. This cannot be handled so the result is truncated to 87 590.92. The error here is small but cumulative. Eventually the state is reached where the addition of the small sum cannot be made at all, for example:

12 345 670 + 6.789 gives 12 345 670

and the totalization has totally stopped working.

Let us assume our flow measurement is accurate to about 1%. There is no point in representing it to better than 4 digits. This leaves 3 digits (of the seven) for the total volume if we are not to lose accuracy as the lower digits are dropped. Let us assume a typical volume in the sample period will be N.NNN. The maximum value we can allow the total volume to rise to is NNNN.NNN before we lose digits. This can be achieved by the rung in Figure 9.24 which is again triggered once per sample time.

If the total volume (in F8:6) goes above (or equal to) 1000 (four digits), 1000 is subtracted from F8:6 and one added to F8:7 which holds the total volume divided by 1000. Note that there is no error in this operation. F8:7 holds the flow in thousands of units and F8:6 the balance volume 0 to 999.999. For example

	Vol in dt	F8:6	F8:7	
Scan N		998.327	56	(total 56 998.327)
	7.214			
Scan N+1		5.541	57	(total 57 005.541)

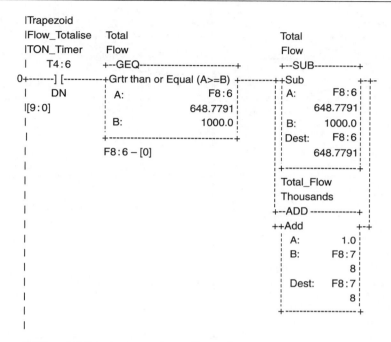

Figure 9.24 *Adding large and small numbers*

Technically the conversion from flow rate to total volume is integration. Identical methods can be used to calculate distance from speed, and speed from acceleration. Double integration can go from acceleration to velocity to distance; a technique used to track rockets.

9.9 Scaling

This example shows how to convert an input analog signal into engineering units for use in the PLC program. Suppose we have a 4–20 mA signal from a temperature transducer with a range 800 to 1500 °C. This is connected by a 250 ohm burden resistor into an analog input card with a 0–10 V range. The analog card gives a 12 bit reading from 0 to 4095. We can thus represent the system by Figure 9.25(a).

The analog temperature signal at the card terminals will go from 1 to 5 V giving a signal range of 410 to 2048 for the program. We wish this to be in engineering units to aid fault finding. To do this we need to follow the graph on Figure 9.25(b).

The range in engineering units, E_{span} is $(1500 - 800) = 700$.
The range from the card I_{span} is $(2048 - 410) = 1638$.

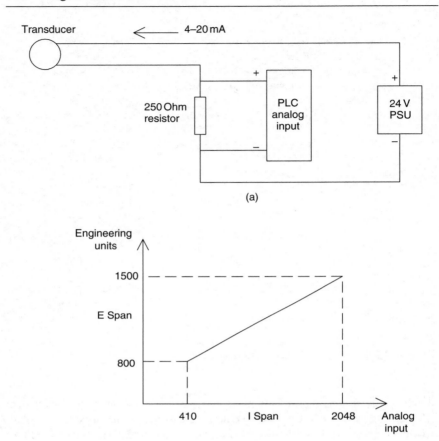

Figure 9.25 *Scaling of an analog input signal to engineering units:
(a) a typical 4–20 mA loop; (b) conversion to engineering units*

If the input signal from the card is I_{in} then the signal in engineering
units is given by

$$E_{out} = E_{min} + (I_{in} - I_{min}) * E_{span}/I_{span}$$

where I_{min} is the minimum card signal (410) and E_{min} is the minimum
input signal in engineering units (800).

The slope E_{span}/I_{span} could be calculated by the PLC on the first scan
(as done for the linearization routine in Section 9.7) but here we will

calculate it off line and enter it as a constant with value 700/1638=0.427 35. Similarly I_{min} and E_{min} are entered into the rungs as constants (410 and 800 respectively).

The routine is shown on Figure 9.26. Rung 0 checks the input signal is within the expected range. If not, rung 3 assigns a default value of 1600.

If the value is good, rung 1 calculates $(I_{in}-I_{min}) * E_{span}/I_{span}$ and rung 2 adds E_{min} to give the result.

The basic four rungs can be used for any linear conversion between two variables of the general form Ax+B. The example converts an analog input. Another common application is for analog outputs where

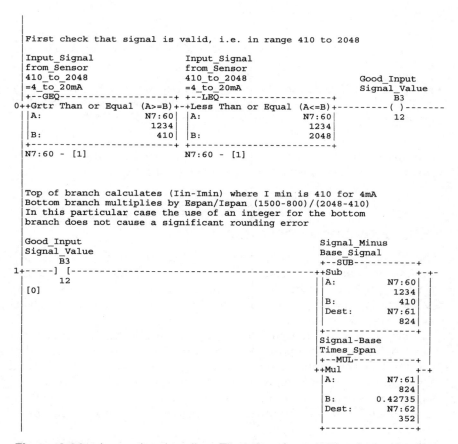

Figure 9.26 *Input signal scaling. These are for a 4–20 mA signal which produces a 1–5 V signal with a 250 ohm burden resistor. The signal is read by a 0–10 V 12 bit analog card with a range of 0–4095*

```
|If input signal is good add in base Engineering Units (800oC) to give result
|                                              Signal_In
| Good_Input                                   Engineering
| Signal_Value                                 Units_oC
|      B3                                       +--ADD----------+
2+-----] [----------------------------------------------+Add              +--
|      12                                       |A:          N7:62|
|    [0]                                         |            352|
|                                                |B:          800.0|
|                                                |Dest:       N7:65|
|                                                |            1152|
|                                                +---------------+
|
|If not a good input signal put safe default value into result
|                                              Signal_In
| Good_Input                                   Engineering
| Signal_Value                                 Units_oC
|      B3                                       +--MOV----------+
3+-----]/[----------------------------------------------+Move             +--
|      12                                       |Source:      1600|
|    [0]                                         |
|                                                |Dest:       N7:65|
|                                                |            1152|
|                                                +---------------+
```

Figure 9.26 *(Cont.) Input signal scaling. These are for a 4–20 mA signal which produces a 1–5 V signal with a 250 ohm burden resistor. The signal is read by a 0–10 V 12 bit analog card with a range of 0–4095*

an output 0–1000 (for, say 0–100.0%) is converted to a range of 410 to 4095 to give 4–20 mA on a 0–20 mA analog output card.

9.10 Gray code conversion

The absolute position of a device is often required in sequencing applications, and this is usually provided by an optical encoder. These consist of an optical grating moving in front of photocells. Figure 9.27 shows a very simple example encoded in binary with four bits giving sixteen possible positions.

Binary encoding, though, has a potential problem. As the encoder goes from position 7 to position 8 the binary count changes from 0111 to 1000. It is very unlikely that all bits will change simultaneously, and PLC input cards inherently turn On and Off with different time delays. Transiently, therefore, we could see 0111>1111>1000 or 0111>0000>1000 or any other combination of four bits. Similar problems can occur on any change.

This problem can be overcome by using a coding where only one bit changes on each transition from one number to the next. Such codes are known as *Unit Distance Codes*. By far the commonest unit distance code is

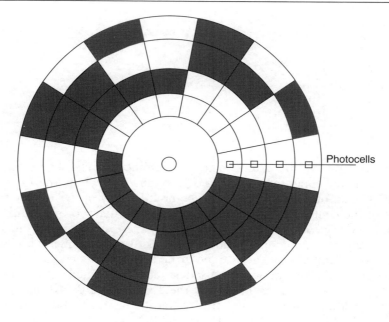

Figure 9.27 *Four bit absolute position encoder*

the Gray code which is built up by reflecting the bit pattern and changing the top bit as below:

Decimal	------- Gray ------		
0	0	00	000
1	1	01	001
2		11	011
3		10	010
4			110
5			111
6			101
7			100

Giving a four bit Gray code

Decimal	Gray	
0	0000	
1	0001	
2	0011	
3	0010	<
4	0110	<
5	0111	<*
6	0101	<*

7	0100	<*
8	1100	<*
9	1101	<*
10	1111	<*
11	1110	<
12	1010	<
13	1011	
14	1001	
15	1000	

The principle can be extended to any number of bits. There are two things to note about the code. First it is symmetrical about the 7/8 transition. The second is that codes can be built for any even number either side of the centre. A ten position Gray code is marked by < and a six position Gray code by *. Note that each of these spills over (from 12 to 3 for example for the ten position code) with only a single bit change. Any even number sequence can be turned into a Gray code.

Before a Gray code can be used it must be turned into a normal binary number. This is based on the Exclusive OR (XOR) gate of Figure 9.28(a) which has the truth table of Figure 9.28(b). In ladder logic and XOR gate is built as Figure 9.28(c).

The conversion of a four bit Gray code to Binary is shown on Figure 9.29. This converts directly to the four rungs of Figure 9.30. The principle can be extended to any number of bits. The most significant Gray code bit is copied to the most significant binary bit. Each subsequent binary bit B_N is the XOR of Gray code bit G_N and B_{N+1}.

The conversion will give the full range of the bits; for four bits the result will be 0 to 15 inclusive. If a restricted range is used an offset must be subtracted. With ten positions and four bits the count will go from 3 to 12 then overspill back to three. Here three must be subtracted from the binary number to give ten positions from 0 to 9.

A common Gray encoder gives 360 positions (0 to 359) per revolution and consequently an output in degrees. This requires nine bits, so the full range is 0 to 511. The 360 positions are centred on the 255/256 tran-

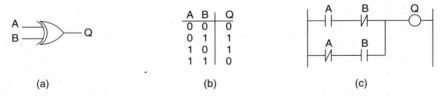

Figure 9.28 *The exclusive OR (XOR) gate: (a) symbol; (b) truth table; (c) implementation in ladder logic*

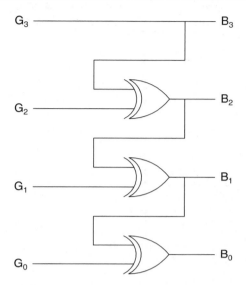

Figure 9.29   *Gray code to binary conversion using XOR gates*

sition and go from 76 (which is zero) to 435 (which is 359). Here the nine bit Gray code must be converted to binary as described above then 76 subtracted from the result to give a position which goes from 0 to 359.

In general for an even number of positions N and G the maximum number of positions of the Gray code (which will always be a multiple of two: 4, 8, 16, 32, 64, 128, 256, 512, 1024, etc.) the offset which must be subtracted is given by:

$$\text{Offset} = (G - N)/2$$

For our 360 degree encoder $\text{Offset} = (512 - 360)/2 = 76$.

There are a few general facts which should be established before using any encoder; Gray, binary or incremental. The vast majority of PLC input cards sink current. Some encoders have an NPN output which also sinks current. If an NPN encoder is connected to a standard input card the signals cannot be read. Pull up resistors, one per bit, may be used, or an input card which sources current.

In these circumstances the polarity of the signal as seen by the PLC must be determined with some care. Some sourcing input cards denote a low voltage (i.e. current being drawn) as a '1' state. These work well with NPN output encoders which often have the low state as the '1'. Some cards don't, however, and these will invert the signal as seen by the PLC. Before the signal is used it must be inverted again by software. This can be done laboriously bit by bit with a n/c contact linked to a coil

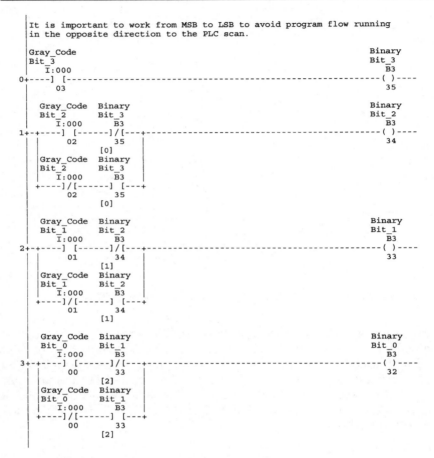

It is important to work from MSB to LSB to avoid program flow running
in the opposite direction to the PLC scan.

```
Gray_Code                                                         Binary
Bit_3                                                             Bit_3
  I:000                                                            B3
0+----] [-----------------------------------------------------------( )----
       03                                                          35

Gray_Code  Binary                                                 Binary
Bit_2      Bit_3                                                  Bit_2
  I:000      B3                                                     B3
1+-+----] [------]/[---+--------------------------------------------( )----
  |    02        35    |                                           34
  |             [0]    |
  |Gray_Code  Binary   |
  |Bit_2      Bit_3    |
  |  I:000      B3     |
  +----]/[------] [---+
       02        35
               [0]

Gray_Code  Binary                                                 Binary
Bit_1      Bit_2                                                  Bit_1
  I:000      B3                                                     B3
2+-+----] [------]/[---+--------------------------------------------( )----
  |    01        34    |                                           33
  |             [1]    |
  |Gray_Code  Binary   |
  |Bit_1      Bit_2    |
  |  I:000      B3     |
  +----]/[------] [---+
       01        34
               [1]

Gray_Code  Binary                                                 Binary
Bit_0      Bit_1                                                  Bit_0
  I:000      B3                                                     B3
3+-+----] [------]/[---+--------------------------------------------( )----
  |    00        33    |                                           32
  |             [2]    |
  |Gray_Code  Binary   |
  |Bit_0      Bit_1    |
  |  I:000      B3     |
  +----]/[------] [---+
       00        33
               [2]
```

Figure 9.30 *Gray code to Binary code conversion*

for each bit as Figure 9.31(a) or, if an XOR function is available, by a
single XOR which operates on the entire word from the encoder. In the
XOR mask each bit used by the encoder must be set to a '1' as shown
on Figure 9.31(b) for a sixteen bit encoder.

9.11 BCD to Binary conversion

Numeric data is often entered by rotary or thumb-wheel switches with a
0–9 range. Each switch gives four binary bits. When these are arranged
as hundreds, tens and units the result is a BCD coded number. Most
PLCs work in two's complement binary, so conversion from BCD to

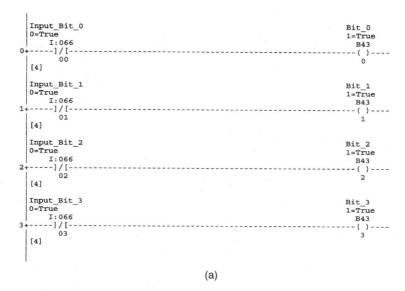

```
   |Input_Bit_0                                                            Bit_0
   |0=True                                                                 1=True
   |   I:066                                                                B43
  0+-----]/[----------------------------------------------------------------( )----
   |     00                                                                  0
   | [4]
   |
   |Input_Bit_1                                                            Bit_1
   |0=True                                                                 1=True
   |   I:066                                                                B43
  1+-----]/[----------------------------------------------------------------( )----
   |     01                                                                  1
   | [4]
   |
   |Input_Bit_2                                                            Bit_2
   |0=True                                                                 1=True
   |   I:066                                                                B43
  2+-----]/[----------------------------------------------------------------( )----
   |     02                                                                  2
   | [4]
   |
   |Input_Bit_3                                                            Bit_3
   |0=True                                                                 1=True
   |   I:066                                                                B43
  3+-----]/[----------------------------------------------------------------( )----
   |     03                                                                  3
   | [4]
   |
```

(a)

Figure 9.31 *Using devices with negative true inputs. (a) Laborious bit by bit inversion of negative true input signal*

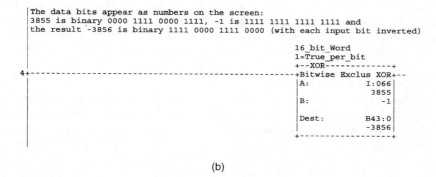

```
   |The data bits appear as numbers on the screen:
   |3855 is binary 0000 1111 0000 1111, -1 is 1111 1111 1111 1111 and
   |the result -3856 is binary 1111 0000 1111 0000 (with each input bit inverted)
   |
   |                                                        16_bit_Word
   |                                                        1=True_per_bit
   |                                                        +--XOR-----------+
  4+--------------------------------------------------------+Bitwise Exclus XOR+--
   |                                                        |A:          I:066|
   |                                                        |            3855 |
   |                                                        |B:            -1 |
   |                                                        |                 |
   |                                                        |Dest:     B43:0  |
   |                                                        |           -3856 |
   |                                                        +-----------------+
```

(b)

Figure 9.31 *(cont.) (b) One rung inversion of 16 bits using an XOR instruction*

Binary is often required. Some PLCs (notably the PLC5 and SLC500 families) have a BCD to Binary function (FRD, FRom Decimal, for the PLC5). For those PLCs without this function, the conversion can be performed as below.

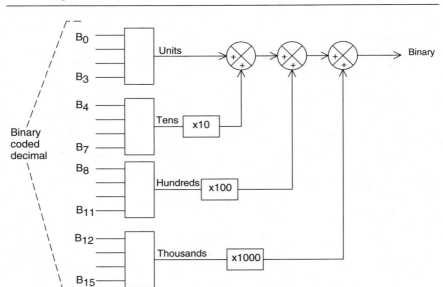

Figure 9.32   *The principle of BCD to binary conversion*

The principle is shown on Figure 9.32. The input signal (range 0–999) occurs on 12 bits from input card 6 in slot 2, i.e. addresses I:26/00, I:26/01, etc. Each group of four bits is converted to a binary number with range 0 to 9 (binary 0000 to 1001). The number for the hundreds (in N17:42) is multiplied by 100, the number for the tens (in N17:41) is multiplied by 10 and both then added to the number for the units in N17:40 to give the binary result in N17:46.

The rungs to achieve this are shown on Figure 9.33. There are 12 rungs which build the numbers in N7:40, 41 and 42. To save space only eight of these are shown. Note that individual bits in an integer number can be accessed; N17:40/0 is the (least significant) bit in the 16 bit integer word N17:40. Store location N17:45 is used as an intermediate store for the part result (N17:40+N17:43). With a CPT instruction all four arithmetic rungs could be done in one rung but the twelve bit rungs are still required.

If a PLC supports subroutines with parameter passing, a BCD to Binary conversion only has to be written once then called when required.

9.12   Binary to BCD conversion

Data is often displayed on seven segment displays using four bits per digit. This requires a binary to BCD conversion routine in the PLC program. Some PLCs (such as the PLC5 family) include this as standard (TOD for TO Decimal). The routine described below can be used where there is no binary to BCD function.

```
| The BCD Number comes from switches via Input Card Rack 2 Slot 6
| I:26/00 to I:26/03 are from units switch
| Units digit is formed in N17:40
|
| BCD_Bit_0                                                  Binary_Units
| Units_Switch                                               Bit_0
|    I:026                                                    N17:40
0+-----] [----------------------------------------------------( )------
|     00                                                        0
|
| BCD_Bit_1                                                  Binary_Units
| Units_Switch                                               Bit_1
|    I:026                                                    N17:40
1+-----] [----------------------------------------------------( )-------
|     01                                                        1
|
| BCD_Bit_2                                                  Binary_Units
| Units_Switch                                               Bit_2
|    I:026                                                    N17:40
2+-----] [----------------------------------------------------( )-------
|     02                                                        2
|
| BCD_Bit_3                                                  Binary_Units
| Units_Switch                                               Bit_3
|    I:026                                                    N17:40
3+-----] [----------------------------------------------------( )-------
|     03                                                        3
|
|
| I:26/04 to I:26/07 are from Tens switch
| Tens digit is formed in N17:41
|
| BCD_Bit_4                                                  Binary_Tens
| Tens_Switch                                                Bit_0
|    I:026                                                    N17:41
4+-----] [----------------------------------------------------( )------
|     04                                                        0
|
| BCD_Bit_4                                                  Binary_Tens
| Tens_Switch                                                Bit_1
|    I:026                                                    N17:41
5+-----] [----------------------------------------------------( )------
|     04                                                        1
|
| BCD_Bit_6                                                  Binary_Tens
| Tens_Switch                                                Bit_2
|    I:026                                                    N17:41
6+-----] [----------------------------------------------------( )------
|     06                                                        2
|
| BCD_Bit_7                                                  Binary_Tens
| Tens_Switch                                                Bit_3
|    I:026                                                    N17:41
7+-----] [----------------------------------------------------( )------
|     07                                                        3
|
```

Figure 9.33 *BCD to binary conversion*

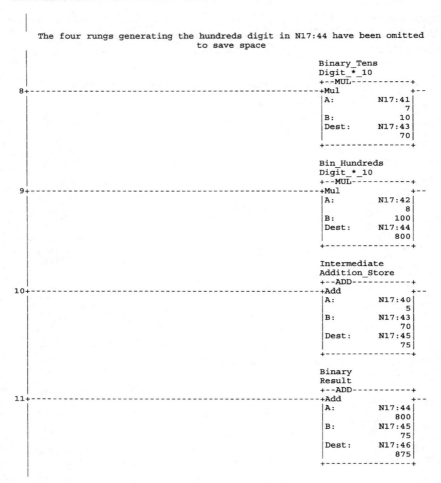

Figure 9.33 *(cont.) BCD to binary conversion*

The principle is, in principle, simple. The number is divided by 10 repeatedly *with no rounding* and the remainder noted each time. The remainders are then the digits for the BCD output. For example:

Binary Number	459
Divide by 10	45 remainder 9
Divide by 10	4 remainder 5
Divide by 10	0 remainder 4

BCD digits are hundreds 4, tens 5 and units 9.

The difficulties come because of the way PLCs handle integer division. Some round down on integer division, e.g. 459/10 gives the result 45. This makes this conversion easy to perform. Some PLCs (of which the Rockwell family is one) round to the nearest number. The sum 459/10 thus gives the result 46 and the sum 454/10 gives the result 45. Both methods have advantages in different applications, but rounding to the nearest number makes Binary to BCD conversion more difficult.

The four rungs in Figure 9.34 extract the first (units) digit for a binary to BCD number for a PLC with rounding to the nearest number integer division. Rung 0 divides the input signal N7:20 by ten giving a result in N7:30. This result is multiplied by ten giving the result in N7:40. If rounding up has occurred, N7:40 will be greater than the input value in N7:20. This is checked in rung 2, and if it has occurred one is subtracted from N7:30. This now contains N7:20 divided by 10 and definitely rounded down. It is multiplied by 10 again giving a result in N7:41. Note that N7:41 and N7:40 are not the same. Subtracting N7:41 from N7:20 gives the first BCD digit in N7:35 which can be transferred bit by bit to the display output in Rack 2 slot 5.

The principle of the first four rungs can be used anywhere where a true modulo/remainder division is required.

The printout in Figure 9.34 is from a running PLC and, at first sight, shows an apparent error at rung 1. At the time that rung 1 has just been obeyed N7:30 will contain 46 (because of the rounding) and hence N7:40 contains 460. After rung 2 the decrement has been performed leaving 45. Because N7:30 contains 46 for just two rungs, and 45 for the (much longer) rest of the program and the program terminal takes 'snapshots' at regular time intervals it is much more likely to see 45 than 46. By eye, the value in N7:30 will flicker between 45 and 46, showing 45 for the majority of the time. Be aware of oddities like this, they can be very confusing when first encountered.

9.13 A hydraulic system

The final example is a real example controlling two hydraulic pumps. The (not very well) defined specification for this would run something along the lines of:

'*The Widget machine has two hydraulic pumps. Only one of these is required for normal operation. The operator should be able to select which is the duty pump. If the duty pump fails for any reason the standby pump should start. In addition to Duty/Standby operation maintenance staff should be able to check a pump is functional without interrupting normal operation*'.

The binary number arrives in N7:20.
The first five rungs (0–4) give the first BCD digit in N7:35 and N7:20/10
rounded down ready for finding the next BCD digit

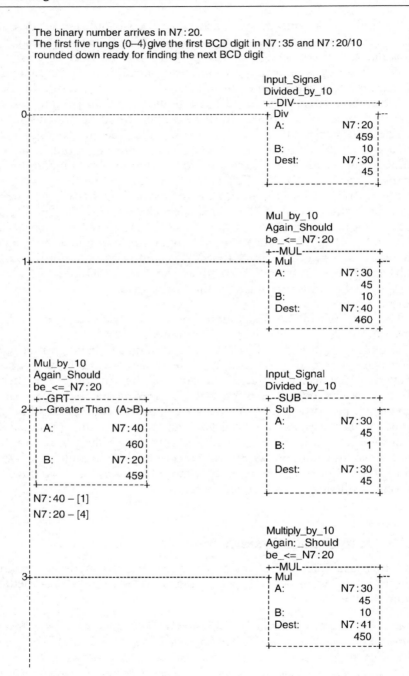

Figure 9.34   *Binary to BCD conversion*

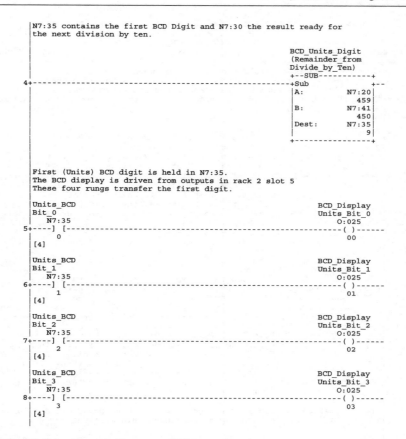

```
     N7:35 contains the first BCD Digit and N7:30 the result ready for
     the next division by ten.

                                                      BCD_Units_Digit
                                                      (Remainder_from
                                                      Divide_by_Ten)
                                                      +--SUB-----------+
   4+-----------------------------------------------+Sub              +--
    |                                                |A:       N7:20|
    |                                                |            459|
    |                                                |B:       N7:41|
    |                                                |            450|
    |                                                |Dest:    N7:35|
    |                                                |              9|
    |                                                +---------------+


     First (Units) BCD digit is held in N7:35.
     The BCD display is driven from outputs in rack 2 slot 5
     These four rungs transfer the first digit.

     Units_BCD                                        BCD_Display
     Bit_0                                            Units_Bit_0
    | N7:35                                           O:025
   5+----] [---------------------------------------------------( )------
    |    0                                              00
    |   [4]

     Units_BCD                                        BCD_Display
     Bit_1                                            Units_Bit_1
    | N7:35                                           O:025
   6+----] [---------------------------------------------------( )------
    |    1                                              01
    |   [4]

     Units_BCD                                        BCD_Display
     Bit_2                                            Units_Bit_2
    | N7:35                                           O:025
   7+----] [---------------------------------------------------( )------
    |    2                                              02
    |   [4]

     Units_BCD                                        BCD_Display
     Bit_3                                            Units_Bit_3
    | N7:35                                           O:025
   8+----] [---------------------------------------------------( )------
    |    3                                              03
    |   [4]
```

Figure 9.34 *(cont.) Binary to BCD conversion*

The final program, some 22 rungs long, is shown on Figure 9.35.
   Remember a PLC program should:

Function correctly
Be understandable
Be easy to modify

Note the documentation tags in the program. Every address is tagged and comments are applied to most rungs. Good documentation is essential to understanding a program. Also note the cross reference tags [N] below contacts. These show where the signal related to the contact originates. For example in rung 12 the contact '*Auto_Run Pump_1 Command*' (B3/23) originates at rung 8 allowing quick backward chasing of signals

```
          PLC-5 LADDER LOGISTICS Report header (c) ICOM Inc. 1987-1993
                          PLC-5 Ladder Listing
                       Duty/Standby Hydraulic Pump
File #2   Proj:HYDPUMP                 Page:001                    09:33 07/20/02
--------------------------------------------------------------------------------
| This PLC program controls two hydraulic pumps.
| The pumps can be run in Auto mode with Duty/Standby auto changeover,
| or locally from alongside the pumps for maintenance work.
| The first two rungs in the program check if the two pumps are available.
| For normal diagnostic work only these two rungs need to be examined.
|
|
|                               Pump_1      Pump_1     Pump_1     Pump_1
|   Pump_1       Pump_1         E_Stop      Isolator   Shut_Off   Filter
|   MCC_Is_On    Oveload_OK     Healthy     Healthy    Valve_Open Healthy
|   I:023        I:023          I:035       I:035      I:035      I:035       >
0+----] [--------] [--------] [--------] [------] [--------] [--------] [------->
|     01          02            10          11         12         13         >
|                               Pump_1      Pump_1     Oil_Level  Return
|                               Failed_to   Pressure   Healthy    Filter    Pump_1
|                               Start       Fault      In_Tank    Healthy   Available
|                       <        B3          B3        I:035      I:035       B3
|                       <----]/[-------]/[--------] [-------] [-------( )-----
|                       <        31          41         05         06         1
|                              [14]        [18]
|
|
|                               Pump_2      Pump_2     Pump_2     Pump_2
|   Pump_2       Pump_2         Isolator    E_Stop     Shut_Off   Filter
|   MCC_Is_On    Oveload_OK     Healthy     Healthy    Valve_Open Healthy
|   I:023        I:023          I:035       I:035      I:035      I:035       >
1+----] [--------] [--------] [--------] [-------] [-------] [--------] [------->
|     03          04            14          15         16         17         >
|                               Pump_2      Pump_2     Oil_Level  Return
|                               Failed_to   Pressure   Healthy    Filter    Pump_2
|                               Start       Fault      In_Tank    Healthy   Available
|                       <        B3          B3        I:035      I:035       B3
|                       <----]/[-------]/[--------] [-------] [-------( )-----
|                       <        32          42         05         06         2
|                              [15]        [19]
|
| It is usual to create flags saying if both/one/no pumps are available
| These can be used for display screens or by higher level computer.
|
|                                                              At_Least
|   Pump_1                                                     One_Pump
|   Available                                                  Available
|   B3                                                         B3
2+-+----] [----+------------------------------------------------( )-----
| |    1        |                                               3
| | [0]         |
| | Pump_2      |
| | Available   |
| |   B3        |
| +----] [----+
|      2
|    [1]
--------------------------------------------------------------------------------
          PLC-5 LADDER LOGISTICS Report header (c) ICOM Inc. 1987-1993
                          PLC-5 Ladder Listing
                       Duty/Standby Hydraulic Pump
File #2   Proj:HYDPUMP                 Page:001                    09:33 07/20/02
```

Figure 9.35 *Complete program for two hydraulic pumps with duty/standby operation. Note the use of annotation and comments, very important for fault diagnosis. The program is laid out so only the first two rungs need to be examined for common faults*

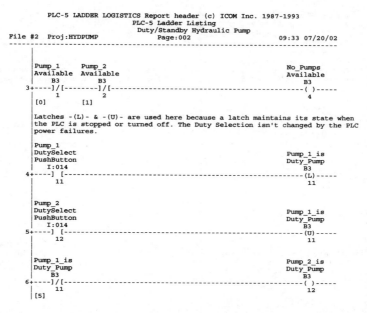

```
       PLC-5 LADDER LOGISTICS Report header (c) ICOM Inc. 1987-1993
                     PLC-5 Ladder Listing
                  Duty/Standby Hydraulic Pump
File #2  Proj:HYDPUMP         Page:002                 09:33 07/20/02
---------------------------------------------------------------------
  |
  | Pump_1     Pump_2                                   No_Pumps
  | Available  Available                                Available
  |    B3         B3                                       B3
 3+----]/[--------]/[------------------------------------( )-----
  |    1           2                                       4
  |  [0]         [1]
  |
  | Latches -(L)- & -(U)- are used here because a latch maintains its state when
  | the PLC is stopped or turned off. The Duty Selection isn't changed by the PLC
  | power failures.
  |
  | Pump_1
  | DutySelect
  | PushButton                                          Pump_1_is
  |  I:014                                              Duty_Pump
  |   11                                                   B3
 4+----] [------------------------------------------------(L)-----
  |   11                                                   11
  |
  | Pump_2
  | DutySelect
  | PushButton                                          Pump_1_is
  |  I:014                                              Duty_Pump
  |   12                                                   B3
 5+----] [------------------------------------------------(U)-----
  |   12                                                   11
  |
  | Pump_1_is                                           Pump_2_is
  | Duty_Pump                                           Duty_Pump
  |    B3                                                  B3
 6+----]/[------------------------------------------------( )-----
  |   11                                                   12
  |  [5]
```

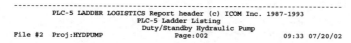

```
---------------------------------------------------------------------
       PLC-5 LADDER LOGISTICS Report header (c) ICOM Inc. 1987-1993
                     PLC-5 Ladder Listing
                  Duty/Standby Hydraulic Pump
File #2  Proj:HYDPUMP         Page:002                 09:33 07/20/02
```

Figure 9.35 *(cont.) Complete program for two hydraulic pumps with duty/standby operation*

```
          PLC-5 LADDER LOGISTICS Report header (c) ICOM Inc. 1987-1993
                         PLC-5 Ladder Listing
                       Duty/Standby Hydraulic Pump
File #2  Proj:HYDPUMP                Page:003                    09:33 07/20/02
---------------------------------------------------------------------------
      |Duty Run tries to start the duty pump. If it fails, (i.e. the duty pump is,
      |or becomes, not available), an attempt is made to start the standby pump.
      |Note that Duty Run is not permitted unless at least one pump is available.
      |
      |
      | Duty_Run      Duty_Run     At_Least
      | Start         Stop         One_Pump                                  Duty_Run
      | PushButton    PushButton   Available                                 Command
      |   I:014         I:014        B3                                        B3
     7+-+----] [-----+-----] [---------] [--------------------------------( )-----
      | |     13    |       14                                              20
      | |           |                   [2]
      | |Duty_Run   |
      | |Command    |
      | |  B3       |
      | +---] [------+
      |      20
      |     [7]
      |
      |
      |                                                                    Auto_Run
      | Duty_Run    Pump_1_is  Pump_1                                      Pump_1
      | Command     Duty_Pump  Available                                   Command
      |   B3          B3         B3                                          B3
     8+---] [-----+----] [--------] [----+----------------------------( )-----
      |    20     |    11        1       |                              23
      |   [7]     | [5]        [0]       |
      |           |Pump_2_is  Pump_2     |
      |           |Duty_Pump  Available  |
      |           +----] [--------]/[----+
      |                12         2
      |               [6]        [1]
      |
      |
      |                                                                    Auto_Run
      | Duty_Run    Pump_2_is  Pump_2                                      Pump_2
      | Command     Duty_Pump  Available                                   Command
      |   B3          B3         B3                                          B3
     9+---] [-----+----] [--------] [----+----------------------------( )-----
      |    20     |    12        2       |                              24
      |   [7]     | [6]        [1]       |
      |           |Pump_1_is  Pump_1     |
      |           |Duty_Pump  Available  |
      |           +----] [--------]/[----+
      |                11         1
      |               [5]        [0]
      |
```

```
---------------------------------------------------------------------------
          PLC-5 LADDER LOGISTICS Report header (c) ICOM Inc. 1987-1993
                         PLC-5 Ladder Listing
                       Duty/Standby Hydraulic Pump
File #2  Proj:HYDPUMP                Page:003                    09:33 07/20/02
```

Figure 9.35   *(cont.) Complete program for two hydraulic pumps with duty/standby operation*

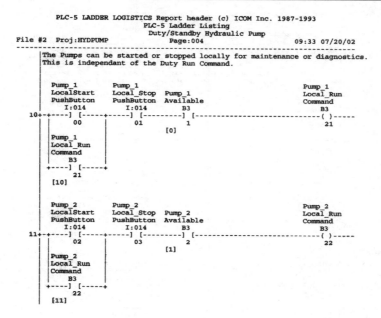

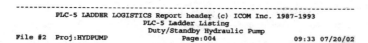

Figure 9.35 *(cont.) Complete program for two hydraulic pumps with duty/standby operation*

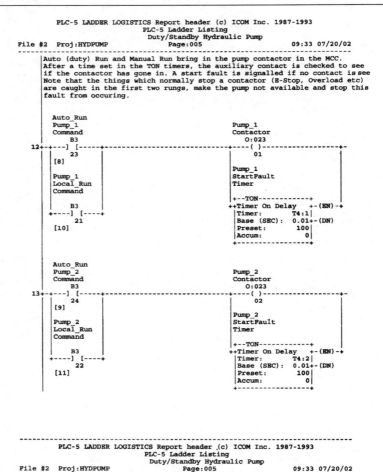

```
           PLC-5 LADDER LOGISTICS Report header (c) ICOM Inc. 1987-1993
                         PLC-5 Ladder Listing
                       Duty/Standby Hydraulic Pump
   File #2  Proj:HYDPUMP                Page:005                 09:33 07/20/02
   ----------------------------------------------------------------------------
           |Auto (duty) Run and Manual Run bring in the pump contactor in the MCC.
           |After a time set in the TON timers, the auxiliary contact is checked to see
           |if the contactor has gone in. A start fault is signalled if no contact is see
           |Note that the things which normally stop a contactor (E-Stop, Overload etc)
           |are caught in the first two rungs, make the pump not available and stop this
           |fault from occuring.


              Auto_Run
              Pump_1                               Pump_1
              Command                              Contactor
                B3                                 O:023
          12+-+---] [----+-------------------------+----( )-----------------+-
            |     23     |                         |     01                 |
            | [8]        |                         |                        |
            |            |                         |  Pump_1                |
            | Pump_1     |                         |  StartFault            |
            | Local_Run  |                         |  Timer                 |
            | Command    |                         |                        |
            |   B3       |                         | +--TON-----------+     |
            +----] [----+                          ++Timer On Delay  +-(EN)-+
                 21                                 |Timer:      T4:1|
                [10]                                |Base (SEC):  0.01+-(DN)
                                                    |Preset:       100|
                                                    |Accum:          0|
                                                    +-----------------+


              Auto_Run
              Pump_2                               Pump_2
              Command                              Contactor
                B3                                 O:023
          13+-+---] [----+-------------------------+----( )-----------------+-
            |     24     |                         |     02                 |
            | [9]        |                         |                        |
            |            |                         |  Pump_2                |
            | Pump_2     |                         |  StartFault            |
            | Local_Run  |                         |  Timer                 |
            | Command    |                         |                        |
            |   B3       |                         | +--TON-----------+     |
            +----] [----+                          ++Timer On Delay  +-(EN)-+
                 22                                 |Timer:      T4:2|
                [11]                                |Base (SEC):  0.01+-(DN)
                                                    |Preset:       100|
                                                    |Accum:          0|
                                                    +-----------------+




   ----------------------------------------------------------------------------
           PLC-5 LADDER LOGISTICS Report header (c) ICOM Inc. 1987-1993
                         PLC-5 Ladder Listing
                       Duty/Standby Hydraulic Pump
   File #2  Proj:HYDPUMP                Page:005                 09:33 07/20/02
```

Figure 9.35 *(cont.) Complete program for two hydraulic pumps with duty/standby operation*

```
          PLC-5 LADDER LOGISTICS Report header (c) ICOM Inc. 1987-1993
                          PLC-5 Ladder Listing
                       Duty/Standby Hydraulic Pump
File #2  Proj:HYDPUMP                 Page:006                   09:33 07/20/02
-------------------------------------------------------------------------------
    |Failed to Start Flags (see earlier comments on timers).
    |Note these make the pump not available in the first two rungs
    |
    |
    |   Pump_1        Pump_1         Alarm                        Pump_1
    |   StartFault    Contactor      Accept                       Failed_to
    |   Timer         AuxContact     PushButton                   Start
    |    T4:1          I:024          I:015                        B3
  14+-+----] [---------]/[-----+-----]/[-------------------------( )-----
    | |        DN         01    |      00                          31
    | |[12]                     |
    | |Pump_1                   |
    | |Failed_to                |
    | |Start                    |
    | |   B3                    |
    | +----] [-----------------+
    |       31
    |    [14]
    |
    |
    |   Pump_2        Pump_2         Alarm                        Pump_2
    |   StartFault    Contactor      Accept                       Failed_to
    |   Timer         AuxContact     PushButton                   Start
    |    T4:2          I:024          I:015                        B3
  15+-+----] [---------]/[-----+-----]/[-------------------------( )-----
    | |        DN         02    |      00                          32
    | |[13]                     |
    | |Pump_2                   |
    | |Failed_to                |
    | |Start                    |
    | |   B3                    |
    | +----] [-----------------+
    |       32
    |    [15]
    |
    |The next block of program checks that the pumps are delivering pressure.
    |The Timers T11/T12 allow time for the system to reach working pressure.
    |
    |Pump_1                                      Pump_1
    |Contactor                                   Pressure
    |AuxContact                                  FaultTimer
    | I:024                                      +--TON-----------+
  16+----] [----------------------------------------+Timer On Delay  +-(EN)-
    |    01                                      |Timer:      T4:11|
    |                                            |Base (SEC):  1.0+-(DN)
    |                                            |Preset:        5|
    |                                            |Accum:         0|
    |                                            +---------------+
    |
```

```
-------------------------------------------------------------------------------
          PLC-5 LADDER LOGISTICS Report header (c) ICOM Inc. 1987-1993
                          PLC-5 Ladder Listing
                       Duty/Standby Hydraulic Pump
File #2  Proj:HYDPUMP                 Page:006                   09:33 07/20/02
```

Figure 9.35 *(cont.) Complete program for two hydraulic pumps with duty/standby operation*

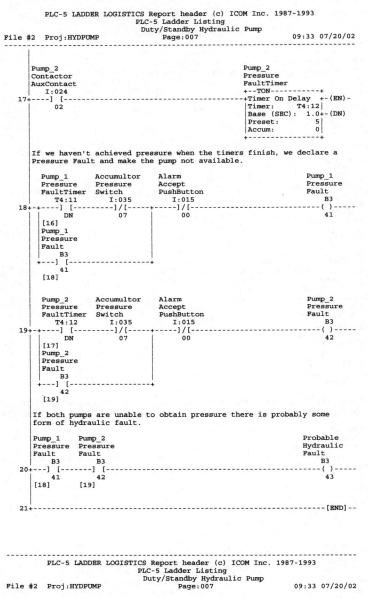

```
                 PLC-5 LADDER LOGISTICS Report header (c) ICOM Inc. 1987-1993
                              PLC-5 Ladder Listing
                            Duty/Standby Hydraulic Pump
     File #2  Proj:HYDPUMP                   Page:007              09:33 07/20/02
     --------------------------------------------------------------------------
            |
            |   Pump_2                                        Pump_2
            |   Contactor                                     Pressure
            |   AuxContact                                    FaultTimer
            |    I:024                                        +--TON----------+
         17+----] [----------------------------------------+|Timer On Delay  +-(EN)-
            |     02                                         |Timer:     T4:12|
            |                                                |Base (SEC):  1.0+-(DN)
            |                                                |Preset:       5|
            |                                                |Accum:        0|
            |                                                +---------------+
            |
            |   If we haven't achieved pressure when the timers finish, we declare a
            |   Pressure Fault and make the pump not available.
            |
            |   Pump_1        Accumultor    Alarm                   Pump_1
            |   Pressure      Pressure      Accept                  Pressure
            |   FaultTimer    Switch        PushButton              Fault
            |    T4:11         I:035         I:015                   B3
         18+-+----] [---------]/[-----+-----]/[--------------------------( )-----
            | |    DN           07     |      00                       41
            | |[16]                    |
            | |Pump_1                  |
            | |Pressure                |
            | |Fault                   |
            | |  B3                    |
            | +---] [------------------+
            |    41
            |   [18]
            |
            |   Pump_2        Accumultor    Alarm                   Pump_2
            |   Pressure      Pressure      Accept                  Pressure
            |   FaultTimer    Switch        PushButton              Fault
            |    T4:12         I:035         I:015                   B3
         19+-+----] [---------]/[-----+-----]/[--------------------------( )-----
            | |    DN           07     |      00                       42
            | |[17]                    |
            | |Pump_2                  |
            | |Pressure                |
            | |Fault                   |
            | |  B3                    |
            | +---] [------------------+
            |    42
            |   [19]
            |
            |   If both pumps are unable to obtain pressure there is probably some
            |   form of hydraulic fault.
            |
            |   Pump_1     Pump_2                              Probable
            |   Pressure   Pressure                            Hydraulic
            |   Fault      Fault                               Fault
            |    B3         B3                                  B3
         20+---] [-------] [-------------------------------------------( )-----
            |    41         42                                  43
            |   [18]       [19]
            |
         21+------------------------------------------------------------[END]--
```

```
     --------------------------------------------------------------------------
                 PLC-5 LADDER LOGISTICS Report header (c) ICOM Inc. 1987-1993
                              PLC-5 Ladder Listing
                            Duty/Standby Hydraulic Pump
     File #2  Proj:HYDPUMP                   Page:007              09:33 07/20/02
```

Figure 9.35 (*cont.*) *Complete program for two hydraulic pumps with duty/standby operation*

```
               PLC-5 LADDER LOGISTICS Report header (c) ICOM Inc. 1987-1993
                        Miscellaneous Report Information and Format Key
                                  Duty/Standby Hydraulic Pump
Misc Info Report                        Page:008                      09:33 07/20/02
-------------------------------------------------------------------------------
PLC-5 Program Information
------------------------
Program Name.............: C:\PLC5\PROGS\HYDPUMP
Processor Type...........: PLC-5/25 Series:B Revision:A
  35 Data Table Files use: 152 Words
   3 Program Files use...: 145 Words
Total number of rungs....: 21
PLC-5 Memory Free........: 13445 Words
PLC-5 Memory Size........: 13824 Words
Time done printing Report: 09:33:48
Company Name.............: PLC Book Revision


Compressed Cross Reference Format Key:
--------------------------------------
 #   - Xref Address used in rung #
/#   - Xref Address used in rung # as a XIO contact (-|/|-)
w#   - Xref Address used in rung # as a word.
wf#  - Xref Address used in rung # as a word in a group of words used in a file
        such as being used in the 'FILE A:' parameter in a File-To-File Move Instruction.
{b}  - All following xref information is for bit/subelement 'b' in the Xref Address
[f]  - All following xref information is for program file # 'f'

Examples:

I:011/00 - 2,3,4    <---- Normal Rung #'s indicate usage as OTE,OTL,OTU,XIC
           /7,/8    <---- Slashes indicate usage as XIO
           w25      <---- Indicates bit address is used as a word in this rung, such as a
                          MOV,TON,etc...

I:011/00 used as an XIC,OTE,OTL,or OTU in rungs 2,3 and 4.
         Used as an XIO (-|/|-) in rungs 7 and 8
         Used in rung 25 in a word instruction (such as a MOV)

I:010 - 5,10,15,f30 <---- Indicates address used as a word, such as in a TON,MOV,etc...
        {3},10,15   <---- Indicates a specific bit in the cross referenced word is
                          used in an OTE,OTL,OTU,XIC (or XIO if '/' precedes rung #)
        {5},23,/55  <---- Reference for address I:010 would be read as follows:

I:010 used as a word address in rungs 5,10,15 and 30, and rung 30 was a file reference.
Bit I:010/03 used in rungs 10 and 15.
Bit I:010/05 used in rungs 23 and 55, and rung 55 was an XIO (-|/|-)




-------------------------------------------------------------------------------
               PLC-5 LADDER LOGISTICS Report header (c) ICOM Inc. 1987-1993
                        Miscellaneous Report Information and Format Key
                                  Duty/Standby Hydraulic Pump
Misc Info Report                        Page:008                      09:33 07/20/02
```

Figure 9.35 (*cont.*) *Complete program for two hydraulic pumps with duty/standby operation*

when a fault occurs. All modern programming software has similar facilities.

Rungs 0 and 1 check if a pump is available to run. There are six real plant inputs per pump (MCC On to Filter Healthy) and two inputs common to both pumps (Low Oil Level and Return Filter). For each pump there are two alarm signals (Failed to Start and Pressure Fault).

These two rungs are very important and have been deliberately put at the beginning of the program so a technician investigating a fault at 3:00 a.m. only has to look at these two rungs to have a very good idea of what is going on.

Rungs 2 and 3 generate status signals about the availability of the pumps. Only one of these (B3/3) is actually used but B3/4 was added because it was thought to be a signal which might be useful for future development; alarm generation on a SCADA system for example.

Rungs 4, 5 and 6 select which pump is the duty pump. Latches –(L)– and –(U)– are used here so the duty pump selection is maintained through supply interruptions. Note that the logic ensures pump 1 or pump 2 is the duty pump; if it is not one it is the other. It is not possible to select no pump as duty. Small points like this are often forgotten!

Rung 7 generates the 'Duty Run' command. There are two points to note here. The Stop Button follows good practice by using a normally closed (fail safe) contact so it appears as a normally open contact –] [– in the program. We must also have at least one pump available (B3/3 from rung 2).

Rungs 8 and 9 generate the auto run signals for each pump. Note the branches in each rung. If the duty pump is available the duty pump is asked to run. If the duty pump is not available when an auto run is required, the standby pump will run.

Rungs 10 and 11 are the manual (maintenance) run commands. These do not consider which is the duty or standby pumps, and there is no duty/standby changeover.

The auto and manual run commands are brought together at rungs 12 and 13. Note that the auto and manual run commands are separate and there is no interaction between them. The program for manual or auto operation can be changed without any side effects on the other. Many people would try to compact rungs 7 to 13 inclusive into just two rungs, eminently feasible but not easy to understand or modify. Keep rungs simple, logical and easy to change.

The two timers (T4:1 and T4:2) are part of the first part of the fault checking. The two pump contactors have auxiliary contacts which should make shortly after the contactor is energized. The two timers will time out (i.e. energize the done (DN) bit) one second after the coil is energized. This is more than ample time for the contacts to make.

Rungs 14 and 15 check that the auxiliary contact on the starter has made. If the auxiliary contact has not made when the corresponding timer times out, the fault bits B3/31 or B3/32 will energize and latch in. These in turn will make the corresponding pump not available at rung 0 or 1 and cause a changeover to the standby pump at rung 8 or 9. The fault bits are cleared by the Alarm Accept pushbutton.

The two pumps feed a common line and the hydraulic pressure is checked by a pressure switch I:35/07. This should make (meaning good pressure) within two seconds of a pump starting and stay made thereafter. Timers T4:11 and T4:12 in rungs 16 and 17 check the state of the pressure switch five seconds after a pump starts (and continually thereafter) in rungs 18 and 19. If, say, there is a coupling failure between motor and pump the pressure will fall, the pressure fault bit will be set making the pump not available at rungs 0 or 1, and a changeover to the standby pump will occur.

If both pumps generate a pressure fault there could be a common hydraulic fault such as a serious leak. This is detected at rung 20. A hydraulic fault makes both pumps not available at rungs 0 and 1.

Appendix   Number systems

We are so used to the decimal number system that it is difficult to conceive of any other way of counting. Normal everyday arithmetic is based on multiples of ten; for example, the number 4057 means:

4 thousands	$=4\times10\times10\times10=4000$	
plus 0 hundreds	$=0\times10\times10$	$=\ 000$
plus 5 tens	$=5\times10$	$=\ \ \ 50$
plus 7 units	$=7$	$=\ \ \ \ \ 7$
Total		$=4057$

Each position in a decimal number represents a power of ten. Our day-to-day calculations are performed to a base of ten because we have ten fingers, but counting can be done to any number base. Of particular interest are number bases of eight (called octal), sixteen (called hexadecimal or hex for short) and two (called binary). In the discussion below we will use the suffix o for an octal number, h for a hex number and d or text for a decimal number where there is any possibility of confusion. 124o is thus octal, 306h is hex and 255d or twelve are decimal.

Octal, to base eight, uses the digits 0–7. In octal you count 0, 1, 2, 3, 4, 5, 6, 7, 10, 11, 12, 13, 14 and so on. The octal number 14o means one eight and four units, which is decimal 12d. Similarly 317o means:

$3\times8\times8$	$=192d$
plus $1\times8=8d$	
plus 7	$=7d$
Total	$=207d$

Hex deals with numbers in multiples of sixteen. We thus need some way of writing a single digit to represent decimal numbers ten to fifteen. For these the capital letters A to F are used. In hex you count 0, 1, 2, 3, 4, 5, 6, 7, 8, 9, A, B, C, D, E, F, 10, 11, 12, 13, etc. The hex number 12h is one sixteen plus two units which is decimal 18d. Similarly C52h means:

$12 \times 16 \times 16 = 3072d$ (Ch is 12d)
plus 5×16 = 90d
plus 2 = 2d
Total $= 3164d$

We will return to octal and hex shortly.

Binary, to base two, only needs two symbols, 0 and 1. Each position in a binary number represents a power of two and is called a bit (for binary digit). In binary you count 0, 1, 10, 11, 100, 101, 110, 111, etc. A binary number rapidly grows in length. A binary number such as 101101 is evaluated in exactly the same way as we saw earlier for decimal, octal and hex. Binary 101101 is:

$1 \times 2 \times 2 \times 2 \times 2 \times 2 = 32d$
plus $0 \times 2 \times 2 \times 2 \times 2$ = 0
plus $1 \times 2 \times 2 \times 2$ = 8d
plus $1 \times 2 \times 2$ = 4d
plus 0×2 = 0
plus 1 = 1
Total $= 45d$

Similarly 1101011 is (noting $2 \times 2 = 4d$, $2 \times 2 \times 2 = 8d$ and so on):

$1 \times 64d =$ 64d
plus $1 \times 32d =$ 32d
plus $0 \times 16d =$ 0
plus $1 \times 8d$ = 8d
plus $0 \times 4d$ = 0
plus $1 \times 2d$ = 2d
plus 1 = 1
Total $= 107d$

Conversion from decimal to binary is achieved by successive division by two, and noting the remainders. Reading the remainder from the top (LSB, least significant bit) to bottom (MSB, most significant bit) gives the binary equivalent. For example, 23d:

23
11 rl (LSB)
 5 rl
 2 rl
 1 r0
 0 rl (MSB)

Decimal 23 is thus binary 10111.

Binary numbers are used in computers and PLCs because the two states, 0 and 1, are easy to handle with simple circuits. Commonly, eight binary bits (called a byte) and sixteen binary bits (called a word) are used. A byte can represent a number from 0 to 255, and a sixteen-bit word a number from 0 to 65535.

Octal and hex give a simple way of representing binary numbers. To convert a given binary number to octal, the binary number is written in groups of three bits (from the LSB) and the octal representation written directly underneath. For example, 11010110:

grouped in threes	11	010	110 (LSB)
octal	3	2	6

giving 326o directly in octal.

Hex conversion is similar, but groupings of four are used. Taking the same binary number 11010110:

grouped in fours	1101	0110
hex	D	6

giving D6 in hex.

Octal 326o and hex D6h are both representations of the same binary number 11010110.

PLCs (like all computers) work internally in binary, but this is difficult for human beings to deal with. Octal and hex are therefore used in many places as a halfway house between the internal workings of the machine and our decimal system. Siemens, for example, use octal bytes, and Allen Bradley label I/O addresses in octal.

A single decimal digit can lie between 0 and 9 inclusive. Four binary bits are therefore needed to represent one decimal digit. Decimal displays and keypads frequently ignore bit combinations 1010 (10d) to 1111 (15d) giving binary coded decimal or BCD. In BCD, each *decade* is coded independently into binary. For example:

Decimal	9	4	0	7	6
BCD	1001	0100	0000	0111	0110

BCD is not as efficient as pure binary. Twelve bits can represent 0–4095 in binary, but only 0–999 in BCD. It is, however, much easier to interface to external devices.

Binary arithmetic is similar to decimal arithmetic. Consider the decimal sum:

$$\begin{array}{r} 345 \\ +272 \\ \hline 617 \end{array}$$

This is evaluated in three steps:

(a) 5+2=7, no carry
(b) 4+7=11, one down (as result) plus carry
(c) 3+2+carry=6

At each stage we consider three 'inputs', the two digits to be added and a possible carry from the previous (lesser significant) column. Each column has two outputs; a result and a carry to the next column.

Binary addition is similar, except there are only two possible states for each digit and the carry, allowing us to build a simple truth table with just eight entries (Table A.1).

Table A.1

	Inputs		Outputs	
Digit 1	*Digit 2*	*Carry in*	*Sum*	*Carry out*
0	0	0	0	0
0	0	1	1	0
0	1	0	1	0
0	1	1	0	1
1	0	0	1	0
1	0	1	0	1
1	1	0	0	1
1	1	1	1	1

An example of binary arithmetic is:

```
1 0 1 1 0 1 0
0 1 0 1 0 1 1
```
```
1 0 0 0 0 1 0 1   Sum (result)
1 1 1 1 0 1 0     Carry
```

The implementation of an adder truth table is a simple problem of combinational logic.

Negative numbers are generally represented in a form called two's complement. The most significant digit represents the sign, being 0 for positive numbers and 1 for negative numbers. The value part of a negative number is complemented (1s changed to 0s and vice versa) and 1 added. For example, +12d in 8-bit binary two's complement is 00001100 and −12d is 11110100.

As expected, addition of a positive and negative number of the same value will give a result of zero:

+12d	00001100
−12d	11110100
Lost →	$\overline{1}$ 00000000

The top ninth bit is lost with an 8-bit byte, giving the expected result of zero.

Two's complement therefore allows subtraction to be performed with simple addition. For example, 12d − 3d:

+12d	00001100
−3d	11111101
Lost →	$\overline{1}$ 00001001

The top bit is again lost giving the correct result of +9d. An 8-bit byte in two's complement form can represent −128d to +127d, and a 16-bit word from −32768d to +32767d. Integers inside a PLC are generally held in 16-bit two's complement form.

Index